T0331391

Decision-Making in Energy Systems

Decision-Making in Energy Systems

Vivek D. Bhise

CRC Press
Taylor & Francis Group
Boca Raton London New York

CRC Press is an imprint of the
Taylor & Francis Group, an **informa** business

First edition published 2022
by CRC Press
6000 Broken Sound Parkway NW, Suite 300, Boca Raton, FL 33487-2742

and by CRC Press
2 Park Square, Milton Park, Abingdon, Oxon, OX14 4RN

Library of Congress Cataloging-in-Publication Data
Names: Bhise, Vivek D. (Vivek Dattatray), 1944- author.
Title: Decision-making in energy systems / Vivek D. Bhise.
Description: First edition. | Boca Raton : CRC Press, [2022] | Includes index.
Identifiers: LCCN 2021032249 (print) | LCCN 2021032250 (ebook) | ISBN 9780367620158 (hbk) | ISBN 9780367620189 (pbk) | ISBN 9781003107514 (ebk)
Subjects: LCSH: Power resources. | Electric power systems. | Energy industries–Decision making.
Classification: LCC TJ163.2 .B54 2022 (print) | LCC TJ163.2 (ebook) | DDC 333.79–dc23
LC record available at https://lccn.loc.gov/2021032249
LC ebook record available at https://lccn.loc.gov/2021032250

ISBN: 978-0-367-62015-8 (hbk)
ISBN: 978-0-367-62018-9 (pbk)
ISBN: 978-1-003-10751-4 (ebk)

DOI: 10.1201/9781003107514

Publisher's note: This book has been prepared from camera-ready copy provided by the authors.

Contents

SECTION II Measurements, Analysis, and Decision-Making

SECTION III   Customers, Governments, and Future Changes

SECTION IV   *Current Issues Facing the Energy Industries*

SECTION V   Applications of Methods: Examples and Illustrations

Preface

This book is a result of my several years of developing and teaching a course in our Energy Systems Engineering Master's Degree Program. The course is formally titled "Energy Evaluation, Risk Analysis and Optimization" and is referred as ESE 504 course at the University of Michigan-Dearborn campus. The course gave me a unique opportunity to apply and teach many systems engineering, safety engineering, and industrial engineering concepts and techniques and to discuss many problems in energy systems that involve risk analysis and decision-making.

As I started developing the course, I found that the U.S. Department of Energy through its Energy Information Agency (EIA) and National Renewable Energy Laboratory (NREL) have generated many comprehensive reports, databases, and models on energy sources, energy consumption, electricity generating plants, and energy distribution technologies. Similarly, the U.S. Environmental Protection Agency (EPA) has developed many reports, databases, and models to provide data on pollution levels within different counties of all states, their dose–response relationships, acceptable levels of toxins and pollutants, and cancer risks for many commonly emitted pollutants. These data and materials were useful in obtaining quantitative estimates needed for teaching analytic and decision-making methods and in solving many energy systems problems. The decision-making methods applied in this course include Pugh analysis, risk analysis, cost–benefit analysis, and subjective evaluations based on paired comparison techniques.

Two major objectives in writing this book are as follows:

1. Provide the knowledge and skills required to identify, analyze, assess, and manage costs, benefits, and risks inherent in selecting various energy sources and solving energy-related problems.
2. Introduce tools and techniques for applications to energy, environment, and resource management policy and investment decisions that involve multi-criteria considerations such as (a) reducing capital, operating, and maintenance costs, (b) increasing revenues, (c) reducing energy consumption, and (d) reducing health-, safety-, and environment-related risks.

   Some examples of problems discussed in this book are as follows:
   a. Deciding whether to invest in a solar photovoltaic electric generator for residential use
   b. Selecting type of energy source (e.g., natural gas, wind, solar, and geothermal) for a future electric power plant
   c. Minimizing risk of lung cancer due to emissions produced by energy systems
   d. Establishing automotive fuel economy and emission requirements based on estimations of costs associated with introducing new technologies and benefits from their implementation

There are many types of risks that need to be considered in solving problems in energy systems areas. Some examples of the risks considered in this book are as follows:

1. *Construction risk*: property damage or liability from errors and accidents during the building of new projects
2. *Company risk*: viability of the project developer, e.g., risks related to key personnel leaving the company, financial solidity, and technical ability, to execute new energy generation plans.
3. *Environmental and human effects risk*: environmental damage caused by a power plant including any liability following such damage and its effect on human health.
4. *Financial risk*: insufficient access to financial resources to construct a new power plant, increasing operating costs, and variability in electricity demand over the life cycle of the plant.
5. *Market risk*: cost increases for key input factors such as labor or materials, or changes in rates ($/kWh) and demands (kWh) for generated electricity.
6. *Operational risk*: unscheduled plant closures due to the lack of resources, equipment component failures, or damages due to accidents.
7. *Technology risk*: systems, such as a geothermal power plant or an oil well, generating less output over time than expected.
8. *Political and regulatory risk*: changes in policy that may affect the profitability of the project, e.g., changes in levels of tax credit, changes in policy as related to permitting and interconnection.
9. *Climate and weather risk*: changes in electricity generation due to lack of sun illumination or snow covering solar panels for long periods of time.
10. *Sabotage, terrorism, and theft risk*: all or parts of the solar park subjected to sabotage, terrorism, or theft and thus generating less electricity than planned.

The book is divided into five sections and 17 chapters. "Section I: Core Building Blocks" provides background material to understand energy sources, supply, demand, and energy source allocation options as well as the interrelationships between the energy sectors, environment, and the economy. This material enables understanding different types of costs, benefits, and risks and formulation of economically efficient strategies and development of future energy systems. It also covers basic considerations in implementation of systems engineering and safety engineering to understand the big picture and to integrate technological considerations in developing electricity-generating plants along with issues such as energy supply, demand, environment, and the economy.

"Section II: Measurements, Analysis and Decision Making" provides the knowledge and skills to identify, analyze, and assess costs and manage the risks inherent in selecting various energy sources, projects, and portfolios of projects. Thus, this section of the book covers costs, revenues, incentives, risk assessments, and cost–benefit analyses. It also covers applications of techniques based on subjective judgments (e.g., ratings and paired comparisons under several structured situations) to extract and understand how the experts (in the related energy fields) judge importance of relevant variables and strength of relationships between the variables.

"Section III: Customers, Governments and Future Changes" provides an insight into customers of the energy industries and their needs. The section also covers future government regulations in the energy field and on automotive products in terms of fuel economy and emission requirements.

"Section IV: Current Issues Facing the Energy Industry," presents information on current topics such as smart grid, electric storage technologies, and electric vehicles that will create major engineering challenges in the development and use of future energy systems and also provide options/alternatives to customers and users of the energy systems.

"Section V: Applications of Methods: Examples and Illustrations," provides several analytical studies demonstrating applications of important decision-making tools covered in this book to solve problems in the energy systems area by presenting a detailed cost–benefit analysis and applications of sensitivity analyses and the Monte Carlo simulation technique.

This book is especially designed for graduate students and practicing engineers, researchers, and mangers working in various energy-generating, distributing, and consuming companies to help them understand, evaluate, and decide on solutions to their energy-related problems.

Vivek D. Bhise
Ann Arbor, Michigan

Author

Vivek D. Bhise is currently a LEO Lecturer/Visiting Professor and a Professor in post-retirement of Industrial and Manufacturing Systems Engineering at the University of Michigan-Dearborn. He received his BTech in Mechanical Engineering (1965) from the Indian Institute of Technology, Bombay, India; MS in Industrial Engineering (1966) from the University of California, Berkeley, California; and PhD in Industrial and Systems Engineering (1971) from The Ohio State University, Columbus, Ohio.

From 1973 to 2001, he held several management and research positions at the Ford Motor Company in Dearborn, Michigan. He was the manager of Consumer Ergonomics Strategy and Technology within the Corporate Quality Office, and the manager of the Human Factors Engineering and Ergonomics in Corporate Design where he was responsible for the ergonomics attribute in the design of car and truck products.

Dr. Bhise is the author of *Ergonomics in the Automotive Design Process* (ISBN: 978-1-4398-4210-2. Boca Raton, FL: CRC Press, 2012), *Designing Complex Products with Systems Engineering Processes and Techniques* (ISBN: 978-1-4665-0703-6. Boca Raton, FL: CRC Press, 2014.) and *Automotive Product Development: A Systems Engineering Implementation* (ISBN: 978-1-4987-0681-0. Boca Raton, FL: CRC Press, 2017).

Dr. Bhise has taught graduate courses in vehicle ergonomics, vehicle package engineering, automotive systems engineering, automotive assembly systems, management of product and process design, work methods and industrial ergonomics, human factors engineering, total quality management and Six Sigma, quantitative methods in quality engineering, energy evaluation, risk analysis and optimization, product design and evaluations, safety engineering, computer-aided product design and manufacturing, and statistics and probability theory over the past 40 years (1980–2001 as an adjunct professor, 2001–2009 as a professor, and 2009–present as a visiting professor in post-retirement) at the University of Michigan-Dearborn. He also worked on several research projects in human factors with late Prof. Thomas Rockwell at the Driving Research Laboratory at The Ohio State University (1968–1973).

His publications include over 100 technical papers in the design and evaluation of automotive interiors, parametric modeling of vehicle packaging, vehicle lighting systems, field of view from vehicles, and modeling of human performance in different driver/user tasks.

Dr. Bhise has also served as an expert witness on cases involving automotive ergonomics, quality, and product safety, patent infringement, and highway safety.

He received the Human Factors Society's A. R. Lauer Award for Outstanding Contributions to the Understanding of Driver Behavior in 1987. He has served on several committees of the Society of Automotive Engineers (SAE), the Motor Vehicles Manufacturers Association (MVMA), the Transportation Research Board (TRB) of the National Academies, and the Human Factors and Ergonomics Society (HFES).

Section I

Core Building Blocks

1 Energy Systems, Problems, and Risks

INTRODUCTION

This book is about decision-making involved in solving energy systems problems. The book provides many problem-solving methodologies. Applications of the methodologies require gathering data, analyzing the data, and evaluating various alternatives. Based on the evaluations, the best alternative is usually selected by considering one or more decision principles selected by the decision-maker. The evaluations can be conducted by using subjective and/or objective methods. The subjective methods are based on the judgments or ratings (i.e., data and/or decisions) provided by persons who are assumed to be very knowledgeable about the problem. The objective methods are based on the data collected by using measurement instruments such as flow meters, force gauges, pressure transducers, temperature measurement sensors, and so forth, and estimates of costs and revenues associated with each alternative are used in solving the problem.

PROBLEM-SOLVING

Engineers and managers working in the energy systems area solve many problems and solving the problems requires them to make many decisions. Decision-making involves determining what to do, i.e., what alternative (or solution) to select. The scientific method is an efficient method for problem-solving. Applications of the scientific method typically include the following five steps:

1. *Define the problem* – involves preparing a statement describing the problem as clearly and precisely as possible
2. *Analyze data* – involves studying the problem by analyzing the data to understand the processes included in the problem, their outputs and inputs, and relationships between the inputs and outputs and other variables that can affect the processes and their outputs
3. *Make search for solutions* – involves exploring all possible approaches and methods and changes in system designs to solve the problem
4. *Evaluate and determine the best solution* – involves evaluating the leading solutions or designs of the proposed system (by using available methods, such as physical testing of prototype hardware, exercising simulation models, and conducting cost–benefit analysis) and selecting the best or the most desired solution
5. *Specify and implement the selected solution* – involves incorporating (i.e., designing, engineering, and installing) the selected solution within an existing system and at a location specified in the problem

DOI: 10.1201/9781003107514-2

It should be noted that the above five-step process used in engineering problem-solving can be remembered by its acronym DAMES (Define, Analyze, Make, Evaluate and Specify).

Application Scientific Method (DAMES) in Engineering Problem-Solving: An Example

Let us study a problem in the energy systems area of determining if a photovoltaic solar panel system should be installed in an average-size home in Phoenix, Arizona, to generate electric power.

The first step of the problem-solving method would be to state the problem clearly and precisely. The problem can be stated as "To decide if an installation of a photovoltaic solar panel system in 2021 would be a profitable investment for an owner of an average-size home in Phoenix, Az."

The second step will involve conducting a detailed analysis of costs and benefits associated with selecting, purchasing, and installing an optimal capacity photovoltaic solar panel system, and operating it over the useful life of the system (e.g., 30 years). Other issues that should be considered here are (a) available solar energy from sunlight; (b) cost savings due to avoided electric power from the utility company; (c) revenue generated by selling excess generated power to the utility company; (d) incentives (e.g., rebates or tax credits) offered by the government agencies (local, state, and federal) for installing the renewable system in a residential property; (e) operating, maintenance, and insurance costs; and (f) discount rate to determine present values of costs and benefits incurred over the life of the system.

The third step of the above-described problem-solving process will involve making search for possible solar photovoltaic systems and their configurations (e.g., solar systems from different manufacturers and models with or without tracking system) available in the market and selecting leading candidates for further evaluation.

The fourth step will be to evaluate the leading candidate systems and select the most profitable system (alternative) based on predetermined selection requirements. The selection will be based on evaluation of different characteristics (or attributes of the alternate systems). The evaluation will involve determination and comparisons of costs and benefits of installing various models of different capacities (e.g., 5 or 10 kW) of photovoltaic solar panel systems available in the market and comparing their performance and power output characteristics.

The fifth or the last step then will involve implementing the solution, i.e., installing the selected system. To ensure that the correct decision was made, the homeowner should keep track of costs and benefits every year and compare those with the calculated values obtained during earlier steps. [Note: Chapter 8 provides a detailed cost–benefit analysis of the above problem.]

Objectives of This Book

The objectives of the book are (a) to provide basic understanding, knowledge, and skills to identify, analyze, assess, and manage the risks, costs, and benefits inherent in selecting various energy sources and energy-related projects, and (b) to introduce tools and techniques for problem-solving in areas such as energy, environment, and resource management policy and investment decisions that involve multi-criteria including safety and societal costs and environmental impacts.

RISK TAKING

Development and operation of any system involve many risks. Risks are generally regarded as occurrences of undesirable events (e.g., accidents) because they can cause unexpected losses, harm to people (e.g., pain, sufferings, injuries, diseases, and even deaths), and interruptions and delays in accomplishing business activities. The undesirable events usually incur additional costs over normal recurring costs (e.g., operating and maintenance costs) due to damages, losses, delays, and reduction in business reputation.

Organizations that use, operate, and provide energy services also face many risks. For example, since natural gas and coal have been used as the primary fuels for electric power generation, the emissions of toxic substances such as CO_2, nitrous oxides, SO_2, and particulate matters generated by the power plants cause many risks due to environmental damages and health effects (e.g., respiratory diseases and cancer). Use of nuclear power plants, hydroelectric power plants, other renewable energy sources such as solar, wind turbines, geothermal also are not free from many types of risks. For example, injuries resulting from accidents and unexpected revenue loss caused by system shutdowns due to adverse weather conditions (e.g., flooding, power line breakages) can affect operations of various energy systems.

ENERGY SYSTEMS OVERVIEW

Energy systems field covers a variety of systems involved in generating, transmission, distribution, and consumption of energy. The operation of the systems used in this field is based on one or more of the technologies related to energy sources, such as oil, coal, natural gas, wind, and solar. An energy system is a system primarily designed to supply energy services to its users (customers or consumers). The energy services typically involve the following types of energy conversions: (a) burning fuel to create sources of heat, (b) using heat to generate pressurized steam, (c) converting mechanical energy to electrical energy, (d) converting wind/solar energy to electrical energy, (e) using nuclear reaction to produce steam to drive generators to make electrical energy, and so forth. The converted energy is transmitted and distributed to its customers/users. The primary end users (or external customers) of the energy systems are the residential customers, commercial customers, industrial customers, institutional customers, and users of various types of vehicles and many other service providers. The internal customers of the energy companies are typically employees (e.g., engineers, operators of equipment, installers, and maintenance and repair personnel) and also the shareholders of the company.

WHAT IS A SYSTEM?

A system consists of a set of components (or elements) that work together to perform one or more functions. The components of a system generally consist of people, hardware (e.g., parts, tools, machines, computers, and facilities), and software (i.e., codes, instructions, programs, databases) and the environment within which the system operates. The system also requires operating procedures (or methods) and organization policies (e.g., documents with goals, requirements, and rules) to implement its processes to get its work done. The system also works under a specified range of environmental and situational conditions (e.g., temperature and humidity conditions, vibrations, magnetic fields, power/traffic flow patterns). The system must be clearly defined in terms of its purpose, functions, and performance capability (i.e., abilities to perform or produce output at specified level over a specified period of time and under specified operating environment).

Some definitions of a system are provided below.

1. A system is a set of functional elements organized to satisfy specified objectives. The elements include hardware, software, people, facilities, and data.
2. A system is a set of interrelated components working together toward some common objective(s) or purpose(s) (Blanchard and Fabrcky, 2011).
3. A system is a set of different elements so connected or related as to perform a unique function not performable by the elements alone (Rechtin, 1991).
4. A system is a set of objects with relationships between the objects and between their attributes (Hall, 1962).

The set of components have the following properties (Blanchard and Fabrycky, 2011):

a. Each component has an effect on the whole system.
b. Each component depends on other components.
c. The components cannot be divided into independent subsystems.

EXAMPLE OF AN ENERGY SYSTEM: A POWER GRID

Figure 1.1 presents a sketch showing a power grid. A power grid is a system involving interconnected network for delivering electricity from producers to consumers. It includes power plants using different technologies (e.g., coal, natural gas, hydro, wind, and solar), high-voltage power lines (transmission lines), substations with voltage conversion transformers, cables/wires (distribution lines), and consumer locations (e.g., homes, offices, shops, and factories) with power meters, switches, fuse boxes, and electric equipment/loads (e.g., lamps, computers, appliances, heaters, air conditioners, and so forth).

Each of these items (or systems) included in the above description of the power grid, thus, can be considered subsystems of the power grid. The subsystems can be further divided into sub-subsystems. The decomposition of the systems can be continued into lower-level systems until the component level is reached (e.g., a steam turbine casing, or a pipe fitting in a boiler) (see Systems Decomposition section). The lowest level in the system hierarchy is the component level as a component by definition [(c) above] cannot be divided into an independent subsystem.

Other examples of energy systems are (a) natural gas pipelines transporting the natural gas from its extraction sites (e.g., oil/gas wells) to refining plants, storage

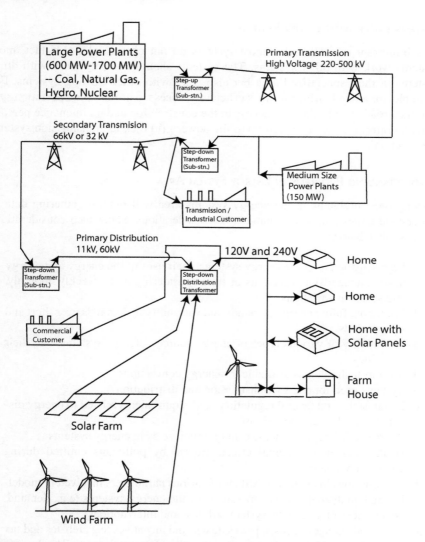

FIGURE 1.1   A power grid supplying electricity from power plants to consumers.

tanks, homes, businesses, and industrial customers; (b) petroleum/oil/gasoline transportation systems (involving pipelines, valves, pumps, trains, trucks, and so forth); (c) electric power generation plants (e.g., coal-fired power plant, natural gas combustion turbine power plant, hydroelectric power plant, and so forth); (d) electric power transmission systems with substations for step-up and distribution systems with step-down transformers; (e) energy storage systems (e.g., batteries, capacitors, fuel tanks, water reservoirs); (f) power demand monitoring and electric power switching and distribution control stations; (g) gas stations; and (h) electric vehicle charging systems and charging stations with other convenience facilities (e.g., food stores).

Systems Work with Other Systems

It is important to realize that most systems do not operate alone. If fact, most systems work with other systems. The systems should not only work with other systems in their specialized area but also work with other existing systems. For example, for a wind turbine to work efficiently, access to ground transportation system (i.e., roads and bridges for access to the construction and maintenance personnel) and interface (i.e., connection) to the power grid (i.e., its transmission system) are essential.

Understanding Problems in Energy System Area

Solving most problems in the energy systems area will require gathering data to understand issues related to a number of considerations. Some such considerations are described below:

1. Users/customers of the energy systems, their uses of the energy and energy needs (demands) as functions of time considering daily, weekly, monthly, and seasonal variations
2. Predicting future energy demands and variability in demands over short and long periods
3. Availability of various types of energy sources, storage systems, and their capacities
4. Various technologies available for energy conversions
5. Systems used for energy transmission and distribution
6. Local, state, and federal regulations and requirements related to energy distribution, sales, and emissions
7. Sources of disruptions, losses, and inefficiencies in energy systems
8. Health and environmental effects caused by pollutants emitted during energy conversions
9. Uncertainties in estimating values of various parameters involved in modeling and analyzing various operations of the energy systems (e.g., demand, occurrences of accidents, system failures, and interruptions)
10. Costs and changes in costs, prices, taxes, and incentives (e.g., rebates and tax credits for use of renewable energy sources) and fluctuations in oil/fuel prices
11. Available data and obtaining the needed data to conduct analyses (e.g., reports and databases created by the U.S. Department of Energy [DOE] and its Energy Information Agency [EIA] and National Renewable Energy Laboratory [NREL])

In addition, the following steps involved in systems engineering implementation need to be considered.

Societal and Customer Benefits from Energy Systems

Systems engineering implementation begins with understanding customer needs. People use energy to operate many systems to enable performance of many functions.

These functions range from providing comfortable environment to enabling performance of many activities within our residential, commercial, and industrial activities (e.g., provide light, heat, operate machines, computers, and entertainment devices; see Chapter 10 for more details). The energy systems have evolved and continuously improved since the industrial revolution. They are more efficient (e.g., provide more electrical energy (kWh) per unit of input [lbs. of fuel]), less costly to construct, operate, and maintain, and provide safer and healthful environments. However, most users, operators, and customers of the energy providers are not free from many types of risks. The risks can occur due to human failures (e.g., human errors, negligence, distractions, failures to take right decisions at right time), equipment failures (e.g., malfunctions or breakdowns), and natural disasters (e.g., floods, hurricanes, and earthquakes) (see Chapter 3 for more details).

SYSTEM REQUIREMENTS DEVELOPMENT

During the development of a system, the overall concept of the whole system is first developed to ensure that it will meet the overall requirements on the whole system (or the entire product such as a power plant or a wind farm). The whole system is then divided (or decomposed) into many lower-level systems to manage their development to ensure that when all the lower-level systems are assembled to form the whole system, the whole product will meet its stated requirements (Bhise, 2014; INCOSE, 2006; NASA, 2016; Ulrich and Eppinger, 2015) (see Chapter 4 for more details).

CONCEPT DEVELOPMENT FOR A NEW SYSTEM

The concept of how a system, such as a power plant, should be configured, how it will work and meet its stated requirements should be concurred with the engineering and senior management of the organization developing the whole system. Once the concept is accepted, the specifications and requirements on all attributes (i.e., characteristics of the whole system such as its dimensions, performance, and operating capabilities, weight, costs, and so forth) should be documented. These documented attribute requirements must be then cascaded (or assigned) to each lower-level system (e.g., boilers, burners, turbines, pumps, pipes, generators, and so forth) of the whole system to ensure that the lower-level systems will function to provide the needed attributes (Bhise, 2014) (see Chapter 4 for more details).

SYSTEMS DECOMPOSITION

Systems can be divided or decomposed into lower-level systems (subsystems). A subsystem can be decomposed into sub-subsystem and so on until the elements (or components) of the lowest-level system cannot or need not be decomposed further at a lower level. The system decomposition helps in managing development of complex systems that have many systems and lower-level systems. Different engineering groups (or teams) can be made responsible for design, development, and testing (e.g., verification tests to ensure that the designed system will meet its stated requirements) of the systems (e.g., higher [larger] or lower-level systems included in the whole system) (see Chapter 4 for more details).

FUNCTION ALLOCATION

Each system exists to serve one or more functions. Thus, function of each system must be clearly defined to ensure that it is designed to carry out its specified functions.

Systems interface with other systems to perform their functions. Thus, interfaces must be designed to ensure that the systems can be designed to interface (i.e., connect to other systems) and function together to meet higher-level functions.

Many other topics and considerations in implementation of systems engineering in solving energy systems problems are covered in Chapter 4.

RISKS IN ENERGY SYSTEMS

There are many risks in designing, building, and operating energy systems. Some examples of risks in energy systems are:

1. Utility company can fail to provide required level of electric power due to power overload on very cold or very hot days.
2. A natural gas pipeline can rupture, cause fire, and fail to provide required amounts of natural gas to homes and industries.
3. A natural disaster (e.g., hurricane) can snap a power line.
4. A construction company can fail to complete a new power plant on the schedule date and thus cause delays and cost overruns.
5. A coal-fired power plant can emit large amounts of pollutants in the environment, which in turn can cause disproportionately higher cases of respiratory diseases.
6. A wind turbine can collapse during construction.
7. A line repair worker can fall from high locations during repair to an overhead power line.
8. The output of an oil well can suddenly drop below the previously estimated level.
9. A geothermal plant can fail to provide required amount of heat (hot water flow and water temperature) over its operating life.
10. A steam turbine in a fossil-fuel-fired plant can develop a large leak.
11. A train carrying oil can derail, spill oil, and cause fire.
12. A solar plant can fail to provide a minimum required level of output.

The risks can be classified in different ways. For example:

1. *Construction risk*: property damage or liability from errors during the building of new projects
2. *Company risk*: viability of the project developer, for example, risks related to key personnel leaving, financial solidity, and technical ability to execute on plans

3. *Environmental and human effects risk*: environmental damage caused by a solar park including any liability following such damage and its effect on human health

4. *Financial risk*: insufficient access to investment and operating capital

5. *Market risk*: changing demands, cost increases for key input factors such as labor or materials, or rate decreases for electricity generated

6. *Operational risk*: unscheduled plant closure due to the lack of resources, equipment damages, or component failures

7. *Technology risk*: degradation of components affecting electricity output over time than expected or technological obsolescence affecting operation/efficiency of a selected system

8. *Political and regulatory risk*: change in policy, political pressure, or regulation may affect the profitability of the project, for example, changes in levels of tax credit. Also, this includes changes in policy as related to permitting and interconnection.

9. *Climate and weather risk*: changes in electricity generation due to lack of sunshine or snow-covering solar panels for long periods of time

10. *Sabotage, terrorism, and theft risk*: all or parts of the solar park will be subject to sabotage, terrorism, or theft and thus generate less electricity than planned.

UNDERSTANDINGS RISKS

Occurrences of unplanned, unexpected, and undesirable situations can interrupt operations, increase costs, and/or reduce revenues. Such hazardous situations are generally identified by:

1. Brainstorming of possible causes of undesirable situations that can increase costs and decrease benefits by using multifunctional team of experts.

2. Use of checklists on possible risk-related situations (e.g., hazards related to fires and explosions, electrical shocks and burns, falls, and so forth)

3. Use of available reliable databases on past occurrences (historic data) of possible risks, e.g., severity, frequency, detection of power line failures.

4. Discovering unknown variables, their values and relationships through design reviews using experienced experts and systems safety tools (e.g., Failure Modes and Effects Analysis. See Chapter 3). The design reviews are very useful especially for new systems with new technologies where accelerated tests still may not be representative and hence may not predict failures well.

5. Studying analogous situations to help understand risks and their related characteristics

6. Possible effects of adverse weather situations, e.g., floods and hurricanes

7. Studying past damages caused by earthquakes (e.g., Fukushima nuclear power plant disaster in Japan)

8. Extensive testing for reliability and durability can help uncover weaknesses in designs, allow fixing design and/or manufacturing defects, and thus reduce risks.

IMPORTANCE OF RISK ANALYSIS AND DECISION-MAKING

Solving problems involves decision-making, i.e., deciding on what alternative to select. Thus, decision-making includes understanding all possible alternatives and possible outcomes and costs and benefits associated with each combination of alternatives and outcomes. Understanding costs involves knowing processes associated with the systems involved in the problem and risks associated in their operations.

This book will take the reader systematically in different areas to build knowledge about various energy systems issues and variables and provide various methods used in problem-solving.

The book covers both subjective and objective methods for problem-solving. The subjective methods rely on the judgments of experts. On the other hand, the objective methods are based on measured values of parameters obtained by physical measurements and costs and benefits values measured in monetary units. A cost–benefit analysis is a very useful tool for decision-makers.

The advantages (or strengths) of cost–benefit analysis are:

a. It provides quantitative data on costs and benefits associated with the risk.
b. It allows conducting sensitivity analyses to determine effects of changes in values of input parameters (see Chapter 17).
c. It allows determination of best (optimistic), worse (pessimistic), and average (or nominal) cost-to-benefit ratios.
d. It allows selection of best alternative among different selected alternatives.
e. It forces decision-makers to reduce or account/plan for future risky situations.

The disadvantages (or weaknesses) of the costs–benefits analysis are:

a. Limited number of variables are generally used in conducting risk analyses. In real world, there are generally many more variables that can affect the outcomes (and thus costs and benefits) may not be estimated with high accuracy.
b. Many variables are unknown and/or uncontrollable. Especially in projects that involve implementation of new technologies and unfamiliarity of developers. Lack of any historic data would make the system development work more challenging and difficult.
c. Many difficulties are encountered in accurately estimating costs and probabilities of future outcomes.

CONCLUDING REMARKS

System designers and decision-makers must consider all issues (e.g., requirements, functionality, operation, timings, costs, risks, failures, and so forth) that a new system will face from its conceptualization to its disposal (see Chapter 4). Risky situations must be considered and planned for like any "rainy day" fund. This book is about understanding issues, problems, and risks in designing and operating complex energy systems. A comprehensive cost–benefit analysis must be conducted during early phases of every major energy systems project to help decision-makers in understanding magnitude of the economic and safety picture (see Chapters 8 and 16).

REFERENCES

Bhise, V.D. 2014. *Designing Complex Products with Systems Engineering Processes and Techniques.* ISBN: 978-1-4665-0703-6. Boca Raton, FL: CRC Press, Taylor and Francis Group.

Blanchard, B.S. and W.J. Fabrycky. 2011. *Systems Engineering and Analysis*, 5th Edition. Upper Saddle River, NJ: Prentice Hall PTR.

Hall, A.D. 1962. *A Methodology for Systems Engineering.* New York, NY: D. Van Nostrand Company, Inc.

International Council of Systems Engineering (INCOSE). 2006. *Systems Engineering Handbook.* Website: http://disi.unal.edu.co/dacursci/sistemasycomputacion/docs/SystemsEng/SEHandbookv3_2006.pdf (Accessed: June 6, 2016).

National Aeronautics and Space Administration (NASA). 2007. *Systems Engineering Handbook.* Report SP-2007-6105, Rev 1. Website: http://ntrs.nasa.gov/archive/nasa/casi.ntrs.nasa.gov/20080008301.pdf (Accessed: June 6, 2016).

Rechtin, E. 1991. *Systems Architecting, Creating and Building Complex Systems.* Englewood Cliffs, NJ: Prentice Hall.

Ulrich, K.T. and S.D. Eppinger. 2015. *Product Design and Development*, 6th Edition. New York: McGraw-Hill.

2 The Energy Picture

INTRODUCTION

The number of energy sources available in the United States to provide energy to homes, offices, buildings, factories, roads and operate equipment such as lighting, heating, and air-conditioning equipment, various machines, computers, and different types of vehicles have been increasing over time. During the early days (around 1775–1880), wood was the primary source of energy. From 1850 onward, coal was used to run engines that could power locomotives, automobiles, factories, and power plants to generate electricity. From around 1900, petroleum and various petroleum products were available to run automobiles, airplanes, ships, and electricity generating plants. Natural gas was being used as fuel to generate heat and electricity required for homes, offices, and factories. After 1925, many dams were built with hydroelectric power plants to generate electricity. And many nuclear power plants were built during 1960–2005 to generate electric power. Now since about 1980, due to emphasis on cleaner energy and gradual change in climate, more renewable energy plants (e.g., wind turbines, solar photovoltaic (PV) and concentrating solar plants) are being constructed to generate electricity. Many of the older coal-fired power plants are being retired and replaced with either natural gas fired or renewable energy power plants. Thus, our energy source picture is changing in terms of its composition and complexity in terms of availability of different types of power sources. The demand for energy is also increasing due to increasing uses and number of newer devices for purposes, such as communications, information processing, lighting, heating and air-conditioning, transportation, and processing equipment in various facilities and industries.

U.S. TOTAL ENERGY PRODUCTION AND CONSUMPTION

According to U.S. EIA (2020a), the total U.S. energy production and consumption were 8.12 and 8.31 quadrillion Btu, respectively (see Table 2.1. Note: quadrillion is 10^{15}). The fossil fuel, which accounts for 79.3% of the total production of energy and 79.9% of the total consumption, dominates the U.S. energy picture.

The sources of energy and percent of energy provided for consumption in the United States in 2020 were as follows:

1. Petroleum 36.7%
2. Natural Gas 32.1%
3. Renewable Energy 11.5%
4. Coal 11.3%
5. Nuclear Electric Power. 8.5%

DOI: 10.1201/9781003107514-3

TABLE 2.1
Total U.S. Energy Review

U.S. Energy Production/Imports/Exports/ Change and Consumption	Quadrillion Btu
Total fossil fuels production	6.43
Nuclear electric power production	0.69
Total renewable energy production	0.99
Total primary energy production	3.12
Primary energy imports	1.72
Primary energy exports	2.17
Primary energy net imports	−0.44
Primary energy stock change and other	0.64
Total fossil fuels consumption	6.64
Nuclear electric power consumption	0.69
Total renewable energy consumption	0.97
Total primary energy consumption	8.31

The uses and percent consumption of the energy available from all the above sources were as follows:

1. Transportation 37%
2. Industrial 35%
3. Residential 16%
4. Commercial 12%

Out of the above sources of energy, the following percentages were used for electric power generation:

1. 1% of Petroleum
2. 35% of Natural Gas
3. 56% of Renewable Energy
4. 9% of Coal
5. 100% of Nuclear Electric Power

Customers of Electric Energy

Electric energy is used in homes, offices, buildings, and factories to operate many electrical devices, such as lamps, heaters, air-conditioners, computers, appliances, and machines. The customers of the electric power thus can be classified as residential, commercial, and industrial. Table 2.2 presents electricity generating sources, customers, their locations, and how they consume the energy by using different equipment.

Figure 2.1 presents four pie charts based on the data provided by EIA (2020f). The larger middle pie chart provides the distribution of electricity used by three types of customers – residential, commercial, and industrial. It shows that on the average in the

TABLE 2.2
Electricity Generating Energy Sources, Their Customers, and Uses

	Electricity Generating Energy Source						
	Oil and Gas Energy Sources		Mined Sources		Renewable Energy Sources		
	Drilled Wells – Fossil Fuels						
	Petroleum Products – Kerosene, Gasoline, Diesel, Heavy Fuel Oil, Propane, LPG	Gases – Natural Gas, LNG	Coal	Nuclear Fission (Nuclear)	Water-based (Hydro and Tidal)	Plant-based (Biomass- Biofuels, Wood, Biomass Waste)	Intermittent (Wind Turbines, Solar – Photovoltaic and Concentrating Thermal) — Drilled (Geothermal)

Energy Consuming Uses and Equipment

Location	Customers and Uses	
Home systems	Residential customer	Thermal (heating and cooling), lighting, cooking communications and appliances
Work/Business systems	Commercial users in offices, and services, customers in malls, restaurants, entertainment and sports centers	Thermal, lighting, cooking communications and appliances and equipment
	Industrial and construction users in plants, farms, and factories	Thermal, lighting, communications, operating machines, and equipment

(Continued)

TABLE 2.2 (*Continued*)
Electricity Generating Energy Sources, Their Customers, and Uses

Electricity Generating Energy Source

	Oil and Gas Energy Sources		Mined Sources		Water-based	Plant-based	Renewable Energy Sources	
	Drilled Wells – Fossil Fuels			**Nuclear**			**Intermittent**	**Drilled**
	Petroleum Products – Kerosene, Gasoline, Diesel, Heavy Fuel Oil, Propane, LPG	**Gases – Natural Gas, LNG**	**Coal**	**Nuclear Fission**	**Hydro and Tidal**	**Biomass-Biofuels, Wood, Biomass Waste**	**Wind Turbines** **Solar – Photovoltaic and Concentrating Thermal**	**Geothermal**

Energy Consuming Uses and Equipment

Location	Customers and Uses	
Transportation systems	Drivers and passengers in automotive products – cars, trucks, vans, and busses	Operating gasoline, diesel flex fuel, and electric vehicles
	Railway transportation – passengers and goods trains	Operating, diesel, flex fuel, and electric vehicles
	Air transportation – passengers and cargo	Operating airplanes using jet fuel and gasoline and other supporting using gasoline/diesel, flex fuel, and electricity
	Sea transportation – passengers and freight ships	Operating gasoline, diesel, flex fuel, and electric boats and ships

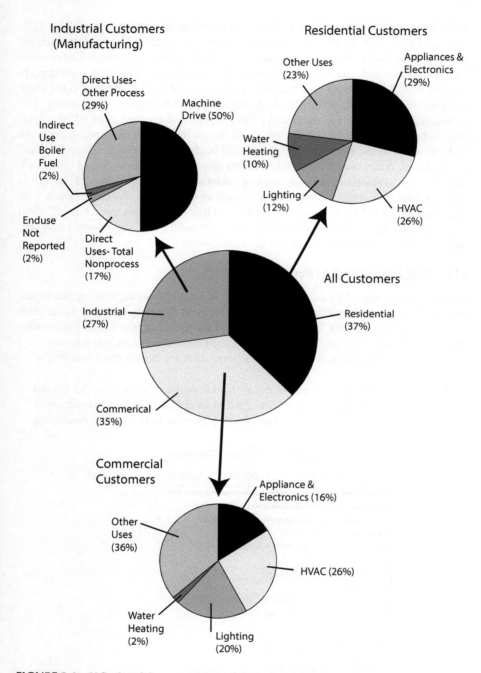

FIGURE 2.1 U.S. electricity consumers and their electricity uses.

United States, the residential, commercial, and industrial customers use about 37%, 35%, and 27%, respectively, of the electrical energy consumed. The smaller three pie charts provide distribution of energy use for the three types of customers. The residential customers use electrical energy running (a) appliances and electronics devices, (b) heating ventilating and air-conditioning systems (HVAC), (c) lighting, and (d) water heating, for about 29%, 26%, 12%, and 10% of their consumption, respectively. The commercial customers use about 26%, 20%, and 16% of the energy in HVAC, lighting, and running appliances and electronics, respectively. And, depending upon the industrial sector, the electrical energy use will be distributed among categories, such as machines, direct uses for processes, direct uses for non-processes, and other indirect uses. In addition to the three types of customers shown in Figure 2.1, energy is also consumed by institutional customers such as educational institutes, health care facilities, public safety, and worship facilities. Their energy uses are relatively small as compared with the three types of customers. Therefore, the institutional customers are typically accounted through other uses categories.

ELECTRICITY GENERATING SOURCES

Electricity is generated by using several different sources from different technologies. Figure 2.2 presents electricity generation in the United States by different power plant technologies (EIA, 2019a). The figure shows that the three largest electricity generation technologies used in the United States in 2019 were natural gas, coal, and nuclear.

The energy sources used for electricity generation are briefly described below:

Natural gas: The natural gas is a significant source of energy in the United States, accounting for nearly one-third of total U.S. energy consumption in 2019. Natural gas is used to fuel power plants, to heat interiors (including

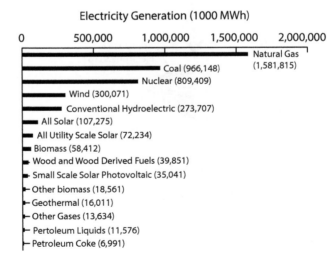

FIGURE 2.2 Electricity generated in 2019 by power plant technologies in the United States (EIA, 2019a).

providing hot water) for homes and buildings, and as feedstock for industrial facilities and other uses. Natural gas consists mostly of methane and is drawn from gas wells or in conjunction with crude oil production. It is delivered through the pipeline system. It contains hydrocarbons such as ethane and propane as well as other gases such as nitrogen, helium, carbon dioxide, sulfur compounds, and water vapor. A sulfur-based odorant is normally added to natural gas to facilitate leak detection. Natural gas is lighter than air and thus will normally dissipate in the case of a leak, giving it a significant safety advantage over gasoline or LPG (Liquefied Petroleum Gas). The natural gas power production process begins with the extraction of natural gas, continues with its treatment and transport to the power plants, and ends with its combustion in boilers to generate steam to operate steam turbines or in combustion chambers of gas turbines to run electricity generators.

Liquified natural gas: Liquefied natural gas (LNG) is natural gas (predominantly methane, CH_4) that has been converted to liquid form for ease of storage and transport. It takes up about 1/600th the volume of natural gas in its gaseous state. Thus, LNG is a preferred and economical mode to transport the natural gas (e.g., using LNG tankers). It is odorless, colorless, nontoxic, and noncorrosive. LNG is natural gas stored as a super-cooled (i.e., cryogenic) liquid. The temperature required to condense natural gas depends on its precise composition, but it is typically between −120°C and −170°C (−184°F and −274°F). This process reduces its volume by a factor of more than 600 (like reducing the natural gas filling a beach ball into a liquid filling a ping-pong ball). The advantage of LNG is that it offers an energy density comparable to gasoline and diesel fuels. The disadvantage, however, is the high cost of cryogenic storage on vehicles and the lack of major infrastructure for LNG dispensing stations, production plants, and transportation facilities.

Coal: Coal was the second-largest energy source for U.S. electricity generation in 2019 – about 23%. Nearly all coal-fired power plants use steam turbines (EIA, 2019a). A few coal-fired power plants convert coal to a gas for use in a gas turbine to generate electricity. Coal is classified into different types, largely based on the amount of carbon it contains and how much heat energy it produces when combusted. Most of the coal produced in the United States is bituminous or subbituminous. Bituminous coal, which accounted for an estimated 48% of total U.S. coal production in 2019, ranges from 45% to 86% carbon and is primarily produced in West Virginia, Illinois, Pennsylvania, Kentucky, and Indiana. Subbituminous coal, which accounted for an estimated 44% of production in 2019, ranges from 35% to 45% carbon, has a lower heating value than bituminous coal, and is mostly produced in Wyoming. Smaller amounts of lower-energy lignite (25%–35% carbon) and higher-energy anthracite (86%–97% carbon) coal are also produced in the United States. Most U.S. production of lignite comes from North Dakota and Texas; all U.S. production of anthracite comes from Pennsylvania.

Biomass: Biomass is renewable organic material that comes from plants and animals. Biomass contains stored chemical energy from the sun. Plants produce biomass through photosynthesis. Biomass can be burned directly for heat or converted to renewable liquid and gaseous fuels through various processes.

Biomass sources for energy include:

a. *Wood and wood processing wastes* – firewood, wood pellets, and wood chips, lumber and furniture mill sawdust and waste, and black liquor from pulp and paper mills
b. *Agricultural crops and waste materials* – corn, soybeans, sugar cane, switchgrass, woody plants, and algae, and crop and food processing residues
c. *Biogenic materials in municipal solid waste* – paper, cotton, and wool products, and food, yard, and wood wastes
d. Animal manure and human sewage

Biomass was the largest source of total annual U.S. energy consumption until the mid-1800s. Biomass continues to be an important fuel in many countries, especially for cooking and heating in developing countries. The use of biomass fuels for transportation and for electricity generation is increasing in many developed countries as a means of avoiding carbon dioxide emissions from fossil fuel use. In 2019, biomass provided nearly 5 quadrillion British thermal units (Btu) and about 5% of total primary energy use in the United States (EIA, 2020h).

Petroleum: Petroleum is also called crude oil, and it is a fossil fuel. The word petroleum comes from the Latin *petra*, meaning "rock," and *oleum*, meaning "oil." It is a thick, flammable, yellow-to-black mixture of gaseous, liquid, and solid hydrocarbons that occurs naturally beneath the earth's surface. It can be separated into fractions including natural gas, gasoline, naphtha, kerosene, fuel and lubricating oils, paraffin wax, and asphalt and is used as raw material for a wide variety of derivative products. The oil industry classifies "crude" by the location of its origin and by its relative weight or viscosity ("light", "intermediate" or "heavy"). The relative content of sulfur in natural oil deposits also results in referring to oil as "sweet," which means it contains relatively little sulfur, or as "sour," which means it contains substantial amounts of sulfur. Petroleum was the source of less than 1% of U.S. electricity generation in 2019. Residual fuel oil and petroleum coke are used in steam turbines. Distillate or diesel, fuel oil is used in diesel-engine generators. Residual fuel oil and distillates can also be burned in gas turbines. In 2019, petroleum products accounted for about 91% of the total U.S. transportation sector (i.e., automotive products, locomotives, airplanes, and ships) energy use (EIA, 2019b).

Nuclear: Nuclear energy originates from the splitting of uranium atoms using a process called fission. This generates heat, and it is used to produce steam, which is used by a turbine generator to generate electricity. Uranium is the fuel most widely used by nuclear plants for nuclear fission. Uranium is considered a nonrenewable energy source, even though it is a common metal found in rocks worldwide. Nuclear power plants use a certain kind of uranium, referred to as U-235, for fuel because its atoms can be easily split apart. Although uranium is about 100 times more common than silver, U-235 is relatively rare. Because nuclear power plants do not burn fuel, they do not produce greenhouse gas emissions. Nuclear energy was the source of about 20% of U.S. electricity generation in 2019.

Renewable energy sources: Renewable energy, often referred to as clean energy, comes from natural sources or processes that are constantly replenished. Many renewable energy sources (e.g., wind, solar, biomass, geothermal, and hydroelectric) are used to generate electricity and were the source of about 17% of total U.S. electricity generation in 2019. Biomass is organic material that comes from plants and animals and includes crops, waste wood, and trees. When biomass is burned, the chemical energy is released as heat and can generate electricity with a steam turbine. Biomass is often mistakenly described as a clean fuel and a greener alternative to coal and other fossil fuels for producing electricity. However, many forms of biomass – especially from forests – produce higher carbon emissions than fossil fuels.

Additional information on power generating technologies using the above-described energy sources is provided in a later section of this chapter.

TRANSMISSION AND DISTRIBUTION OF ENERGY

The electricity generated by every source needs to be transmitted and distributed to various users. Electricity can be transmitted over longer distances using electric power lines. On the other hand, the natural gas and liquid fuel products are distributed over longer distances using pipelines, rail cars, tanker trucks, and ships. These energy transportation and distribution systems have provided flexibility and more opportunities to build power plants near locations where energy sources are available. For example, geothermal power plants are located where underground hot water is available, solar plants are located where higher sun illumination and large land areas are available to mount many large reflectors. And wind turbines are located on-shore and off-shore locations where less variable and higher wind velocities are prevalent.

ELECTRICITY DEMAND AND SOURCES FOR MEETING THE DEMAND

The electricity demands in different parts of the United States vary by time of the day and seasons depending upon the needs of the customers. Figure 2.3 presents a graph illustrating overall daily electricity demand (on June 1, 2020) in the United States (EIA, 2020b). The utility companies need to decide on how to obtain the required amount of electric energy from different power plants at any given time point in the day and distribute the energy to their customers through their complex network of transmission and distribution lines.

The energy companies thus are faced with more complex tasks of deciding the composition of available power sources to operate and meet the varying power demand in normal and abnormal (e.g., weather emergency) situations. Figure 2.4 shows an example of the how the outputs of power plants with different technologies

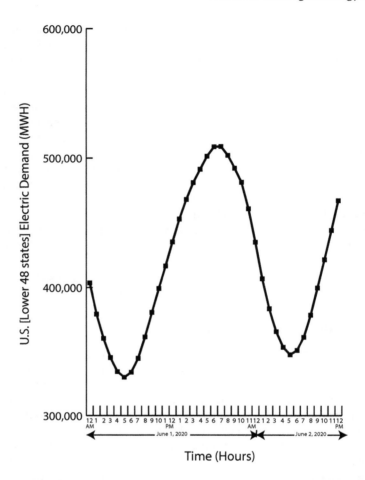

FIGURE 2.3   U.S. electricity hourly demand (in lower 48 states). (From: EIA, 2020b.)

generated and provided power on 1 day (May 25, 2020) to meet the varying demand levels in the United States (EIA, 2020b). In future, greater use of smart grid technologies (e.g., computers and communication systems) will help takeover many of the power source selection and power distribution tasks and improve flexibility, efficiency of the grid and reduce power outages (see Chapter 13 for more details).

HOUSEHOLD CONSUMPTION

The household energy survey database created by the EIA provides information on household energy consumption in 16 states in the United States. For each of the 16 states, the database provides (a) all energy consumed [in MBtu/year from 1979 to 2015] by an average household in each state (excluding energy used for transportation) and its cost in dollars, (b) electrical energy consumed [in MBtu/year] by an average household in each state (excluding energy used for transportation) and its

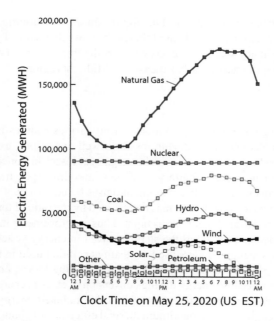

FIGURE 2.4 U.S. hourly electricity generation by source. (From: EIA, 2020b.)

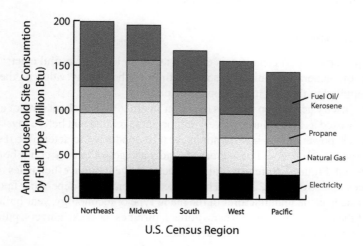

FIGURE 2.5 Annual household site consumption by fuel type (From: EIA, 2015.)

cost in dollars, (c) energy consumption by end use, such as air-conditioning, water heating, appliances, electronics, and lighting and space heating, (d) main heating fuel used (e.g., natural gas, electricity, other, or none), and (e) cooling equipment used (e.g., central air-conditioning, window/wall air-conditioning units or none). Figure 2.5 presents an example of data available from the household annual energy consumption website (EIA, 2015). The figure shows consumption of energy by source type in five regions of the United States. It shows that natural gas and fuel oil/kerosene are two major sources of energy across the five regions. The natural gas is the primary source of energy for Northeast and Midwest regions.

WASTED ENERGY

Considerable percentage of energy available from various sources is wasted during extraction (e.g., leaks and spills during fossil oil well operations, explosions during coal mining), transformation (or conversion, e.g., heat losses in boilers and turbines), storage (losses during storage, battery efficiency, and drainage), transmission (e.g., voltage drops), transportation, distribution, and uses by end users.

The energy efficiency of any system is determined by the ratio of total useful energy output to total energy available for use as input. Thus, improvements in energy efficiencies of various systems used in all the steps involved in energy systems supply chain are important. For example, boilers used in steam generation used in fossil fuel power plants have losses due to flue losses, wall losses, operating losses, and cooling water losses. Lighting devices lose energy during conversion of electric energy into luminous energy due to heat generated and output of invisible radiated energy (e.g., infrared). Modern gasoline engines have a maximum thermal efficiency of about 20%–35% when used to power a car. The U.S. Energy Information Administration (EIA) estimates that electricity transmission and distribution (T&D) losses average about 5% of the electricity that is transmitted and distributed annually in the United States.

GENERATING EFFICIENCY

Electric power plant efficiency, η, is defined as the ratio of the useful electricity output from the generating unit in a specific time to the energy value of the energy source supplied to the unit in the same time period.

For electricity generation systems based on steam turbines, about 65% of all prime energy is wasted as heat. Thus, it is very important to realize that before investing and building new power plants, investigations must be conducted to determine if efficiencies of existing plants could be increased, and wastages of energy could be reduced or eliminated. Figure 2.6 presents data on electricity generation efficiencies of different sources (Electropaedia, 2020). The efficiency values vary depending upon many variables such as the manufacturer of the plant, type of technology, year built, design details of the equipment (e.g., burners, boilers, turbines, heat exchangers, pumps, and so forth), and characteristics of the fuel used.

ENERGY SYSTEMS ISSUES

Some important issues facing the energy system industries are as follows:

1. *Many types of technologies used in generating energy*: The primary sources of energy include nuclear energy, fossil energy (e.g., oil, coal, and natural gas), and renewable sources (e.g., biomass, wind, solar, geothermal, and

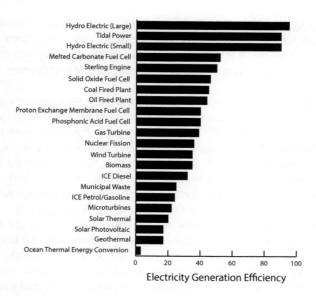

FIGURE 2.6 Efficiencies of power sources.

hydropower). Electric energy is the secondary source of energy used for heating, air-conditioning, lighting, and running various appliances and equipment. Thus, depending upon availability of an energy source, the required equipment needs to be designed or modified to operate by using the source.

2. *Availability of greater number of sources of energy*: The variety of energy sources from small to very large electricity generating plants are now connected to the power grid. Depending upon their ability to quickly adjust (i.e., increase or decrease) their outputs and the electricity selling prices, the power companies need to decide on the amount of electricity to produce or purchase from various plants and sell to customers to improve their efficiency and revenues.

3. *Greater and more variable energy demand*: With the increasing number of customers and their constantly changing energy demands, the variability in the overall market demand for the power has also increased. This makes the tasks of predicting, monitoring, and distribution of the electric energy without brownouts or blackouts more challenging.

4. *Limited energy storage systems and capacities*: With the variations in the electricity demand and variations in power generating capacities of renewable power sources (e.g., wind and solar), the need for efficient and cheaper electricity storage systems is becoming more urgent.

5. *Limited energy transmission and distribution networks*: The power grid in most parts of the United States is getting older, and it needs upgrading and renovation to transform it into a "smart" grid (see Chapter 13).

6. *Uncertainties in determining values of various parameters involved in analyzing and solving energy problems*: The energy markets are increasingly

affected by changes in economic and political situations, government policies, technological innovations, and social changes and climatic effects. These changes make it difficult to predict values of many variables, especially in the longer terms, such as prices of fuel, plant efficiency, operating costs, and incentives.

7. *Power disruptions, losses, and inefficiencies in power generation and distribution*: The fluctuations and quick changes in power demands with older power grid make the task of providing reliable power to many customers more difficult. (Note: Many businesses and homes have now backup generators to supply electricity in case the power from the utility company is interrupted.) Opportunities to eliminate wastages of energy and improving efficiencies of power generation and power consuming equipment are being considered as possible solutions to reduce energy shortages.

8. *Difficulties in predicting health and environmental effects*: The fossil fuel power plants emit larger quantities of greenhouse gases and pollutants as compared with the renewable power sources. The constantly changing availability of mix of power sources, variability in demand, and increasingly mobile population make it difficult to predict the magnitude of health and environmental effects on the population surrounding the power plants.

9. *Security and safety of power sources and power distribution network*: Increasing treats on the security of the power grid from sources, such as terrorist attacks and computer system hackers, make it difficult to maintain high level of reliability over the vast power grid with many distributed electric generators and miles of transmission and distribution cables.

POWER GENERATION TECHNOLOGIES

This section is intended to provide more information on how various power generation technologies work and important issues associated with producing electric power.

NUCLEAR POWER PLANTS

The nuclear power plants generate high amount of electricity by using uranium as fuel and a nuclear fission reaction. Nuclear fission releases heat energy by splitting atoms. This resulting energy is then used to heat water in nuclear reactors and generate steam to run steam turbines that drive electricity generators. The nuclear power plants produce low greenhouse gas emissions, and thus the energy is considered environmentally friendly. When compared with renewable sources of energy such as solar and wind, the power generation from nuclear power plants is also considered more reliable as they can run continuously over long periods of time. Though the investments required to set up nuclear power plants are huge, the costs involved in operating them are low. Nuclear energy sources also have higher density than fossil fuels and release massive amounts of energy. Due to this, nuclear power plants require low quantities of fuel but produce enormous amounts of power, making them particularly efficient once up and running.

Nuclear reactors generating electricity in the United States fall into two main categories: boiling water reactors (BWRs) and pressurized water reactors (PWRs). Both systems boil water to make steam (BWRs within the reactor and PWRs outside the reactor); in both cases, this steam must be cooled after it runs through a turbine to produce electricity.

Like other thermoelectric power plants, nuclear reactors use once-through or recirculating cooling systems. About 40% of nuclear reactors in the United States use recirculating cooling systems; 46%, once-through cooling (Union of Concerned Scientists, 2012). BWRs and PWRs use comparable amounts of water to produce a unit of electricity. Nuclear plants as a whole withdraw and consume more water per unit of electricity produced than coal plants using similar cooling technologies because nuclear plants operate at a lower temperature and lower turbine efficiency and do not lose heat via smokestacks (World Nuclear Association, 2013). Nuclear power plants, thus, require large amount of water to control and cool the reactor and therefore are located near lakefronts or seashores.

HYDROELECTRIC POWER PLANTS

Hydroelectric power plants harness the gravitational force of flowing water. They capture the energy of falling water to generate electricity. A turbine converts the kinetic energy of falling water into mechanical energy. Then a generator converts the mechanical energy from the turbine into electrical energy. The most common type of hydroelectric power plant is an impoundment facility. An impoundment facility, typically a large hydropower system, uses a dam to store river water in a reservoir. Water released from the reservoir flows through a turbine, spinning it, which in turn activates a generator to produce electricity.

Compared with fossil-fuel-powered energy plants, hydroelectric power plants emit fewer greenhouse gases. The construction of hydroelectric power plants and dams requires huge investment. According to the International Hydropower Association's 2018 report (IHA, 2018), an estimated 4,185 terawatt hours (TWh) electricity was generated from hydropower in 2017, avoiding up to 4 billion tonnes of greenhouse gases as well as harmful pollutants. Worldwide hydropower installed capacity rose to 1,267 gigawatts (GW) in 2017, including 153 GW of pumped storage. During the year, 21.9 GW of capacity was added including 3.2 GW of pumped storage. China is the world's largest producer of hydropower and accounted for nearly half of global added installed capacity, at 9.1 GW. It was followed by Brazil (3.4 GW), India (1.9 GW), Portugal (1.1 GW), and Angola (1.0 GW).

FOSSIL-FUEL-FIRED POWER PLANTS

Fossil fuel power stations provide most of the electrical energy used in the world. Some fossil-fired power stations are designed for continuous operation as baseload power plants, while others are used as "peaker" plants, which only provide supplementary power when the demand for electricity increases for a short period of time.

Fossil fuel power stations have equipment to convert the heat energy from the fuel combustion into mechanical energy, which then operates an electrical generator.

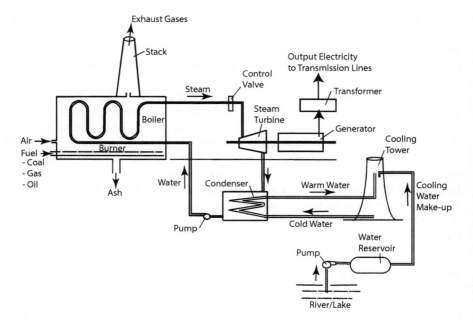

FIGURE 2.7   Fossil-fuel-powered steam turbine electricity generation plant.

The prime mover may be a steam turbine, a gas turbine, or in small plants, a recip-rocating gas engine. Figure 2.7 shows a flow diagram of a fossil-fuel-powered steam turbine-based power plant. The fuel, such as coal, oil, natural gas, or biomass, is burnt to produce heat, which converts water into steam in a boiler. The pressurized steam is sent to run a steam turbine. The steam turbine output shaft is connected to an electricity generator. The steam exiting from the steam turbine is sent to a con-denser where it is converted into water. The water is pumped back into the boiler. The cooling of the steam is accomplished by circulating cold water in the condenser from the cooling tower.

The flue gas from combustion of the fossil fuels contains carbon dioxide and water vapor, as well as pollutants such as nitrogen oxides (NO_x), sulfur oxides (SO_x), and, for coal-fired plants, mercury, traces of other metals, and fly ash. Usually, all the car-bon dioxide and some of the other pollution are discharged to the air. Solid waste ash from coal-fired boilers must also be removed. Therefore, the fossil-fuel-fired plants are being required to meet health-hazards-related government requirements. Some examples of such requirements are as follows:

1. Removal of heavy metals (lead, mercury) and SO_2 to meet EPA's Mercury and Air Toxic standards (MATS).
2. Meet requirements developed to meet Clean Water Act (CWA)
3. Reduce coal combustion residue (coal ash) (future EPA rulemaking action).
4. Meet local and state regulations on CO_2 emissions
5. Requirements in EPA's proposed clean power plan (reduce CO_2 emissions)

DIESEL-FIRED POWER PLANTS

With diesel as fuel, this type of power plant is used for small-scale production of electric power. They are installed in places where there is no easy availability of alternative power sources and are mainly used as a backup for uninterrupted power supply whenever there are outages. Diesel plants require only a small area to be installed as compared with coal-fired power plants. Due to high maintenance costs and diesel prices, the diesel power plants have not gained popularity at the same rate as other types of power generation plants such as steam and hydro.

NATURAL-GAS-FIRED COMBUSTION TURBINE PLANTS

The natural-gas-fired power plants use a combustion turbine (CT) (see Figure 2.8). The CT receives pressurized air from a compression turbine mounted on the same shaft as the CT. Natural gas is injected in the combustion chamber along with the pressurized air of the CT. The generator is attached to the same shaft as that of the CT. The natural gas CT shown in Figure 2.8 is also called a simple cycle turbine.

The natural gas power plants can be classified into the following three types: (a) a simple cycle natural gas combustion turbine (gas-CT), (b) a natural gas combined cycle system (gas-CC; see next section, Figure 2.9), and (c) a natural gas combined cycled system with carbon capture and sequestration (gas-CC-CCS). The carbon capture sequestration is described in a later section of this chapter.

Gas turbines can be used for large-scale power generation ranging from about 200 to over 5,000 MW. EIA's long-term projections show that most of the electricity generating capacity additions installed in the United States through 2050 will be natural gas combined cycle and solar PV (EIA, 2021). Smaller natural gas power plants such as 150 MW installations are not normally used for base load electricity generation, but for bringing power to remote sites such as oil and gas fields. They do however find use in the major electricity grids in peak shaving applications to provide emergency peak power. Low-power gas turbine generating sets with capacities up to 5 MW can be accommodated in transportation containers to provide mobile emergency electricity supplies, which can delivered by truck to the point of need.

Almost all gas turbine installations use fossil fuels. One further advantage of gas turbines is their fuel flexibility. They can be adapted to use almost any flammable gas

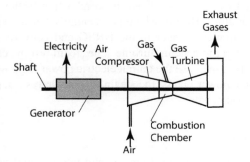

FIGURE 2.8   Natural gas combustion turbine power plant.

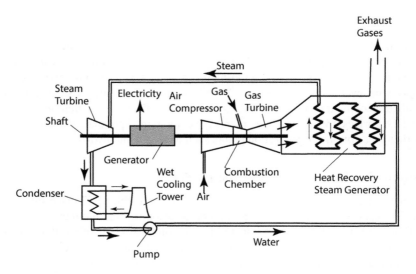

FIGURE 2.9   Natural gas combined cycle power plant.

or light distillate petroleum products such as gasoline (petrol), diesel, and kerosene (paraffin), which happen to be available locally, though natural gas is the most used fuel. Crude and other heavy oils can also be used to fuel gas turbines if they are first heated to reduce their viscosity to a level suitable for burning in the turbine combustion chambers.

COMBINED-CYCLE POWER PLANTS

Using both gas and steam turbines, combined-cycle power plants produce higher amounts of electricity from a single fuel source compared with a traditional single cycle natural gas power plant. In a combined-cycle natural-gas-fueled power plant, the waste heat of simple cycle system (i.e., combustion turbine) in the exhaust gases is used to generate steam and to drive a steam turbine electricity generating set. In such cases, the exhaust temperature may be reduced to as low as 140°C enabling efficiencies of up to 60% to be achieved in combined-cycle systems.

Figure 2.9 presents a flow diagram of a natural-gas-powered combined-cycle power plant. The hot exhaust gases from the combustion turbine are sent to a heat recovery steam generator (HRSG), which is essentially a heat exchanger that raises the temperature of circulated water in the HRSG and converts it into steam. The steam is used as an input to a steam turbine. The steam turbine can be connected to the same output shaft to increase the output electric power as shown in Figure 2.9. In some combined-cycle power plants, the steam turbine drives a separate generator to gain additional electric power.

SOLAR POWER PLANTS

Solar energy plants convert energy from the sun into thermal or electrical energy. Thus, they are cleanest and most abundant renewable energy sources. They generally do not require high maintenance and last for about 20–25 years. The International Energy Agency (IEA) projected in 2014 that by 2050, solar PV and solar thermal would contribute about 16% and 11%, respectively, to the worldwide electricity consumption, and solar would be the world's largest source of electricity (IEA, 2014). However, initial costs involved in setting up solar power plants are high, and the installation of solar power systems requires a lot of land space.

Solar energy can be captured in two forms, either as heat by using thermal systems or as electrical energy by using photovoltaic systems. The two systems are briefly described below.

Thermal systems: Thermal systems capture the sun's heat energy (infrared radiation) in some form of solar collectors (e.g., flat or parabolic mirrors) and use them mostly to provide hot water (e.g., for space heating) or to generate electricity by heating the working fluid in a heat engine, which in turn drives a generator. Solar thermal can also be a system of giant mirrors arranged in such a way to concentrate the sun's rays on a very small area to create significant amount of heat that can be used to create steam to power a turbine and an electricity generator.

Photovoltaic systems: PV systems capture the sun's higher frequency radiation (visible and ultraviolet) in an array of semiconductor photovoltaic cells (solar panels), which convert the radiant energy directly into electricity. There are many different manufacturers of photovoltaic systems, and the systems differ in their characteristics (e.g., single crystal vs. multi-crystalline, number of junctions, film/layer thickness, and coatings) due to use of different materials and manufacturing processes. The efficiency of the solar panels has been steadily increasing over the past four decades. The efficiencies of currently available PV materials range from about 11% to 46%. Over the past decade, the cost of the utility-scale PV solar panels has decreased substantially from about $5 to $1/watt DC (Fu et al., 2017).

WIND POWER PLANTS

In recent years, there has been a rapid growth in the number of wind farms across the world, due technological advancements, increase in turbine size, and decreases in capital costs. As wind is naturally occurring source of energy, there are no limitations to harness its power. Operational costs involved in maintaining wind power plants are low after the erection of wind turbines, and they are generally considered cost-effective. Wind farms can also be built on agricultural lands, without causing any interruption to cultivation activities. However, maintenance of wind turbines may vary, as some need frequent checks, and the wind power projects typically require huge capital expenditure.

Wind power has the advantage that it is normally available 24 hours/day, unlike solar power, which is only available during daylight hours. Unfortunately, the availability of wind energy is less predictable than solar energy – as the sun rises and sets every day. However, based on data collected over many years, some predictions about the frequency of the wind at various speeds, if not the timing, are possible to install wind turbines. Wind turbines are installed both on land (on-shore) and over water (off-shore). Off-shore wind turbines tend to be large with higher outputs (e.g., over 600 feet tall and producing 6 MW) and generally have higher capacity factor because of more sustained and higher wind velocity over water as compared with the on-shore wind turbines.

A wind turbine converts kinetic energy from the wind into electricity. The blades of a wind turbine turn between 13 and 20 revolutions per minute, depending on their technology, at a constant or variable velocity, where the velocity of the rotor varies in relation to the velocity of the wind in order to reach a greater efficiency. The wind turbine is automatically oriented to take maximum advantage of the kinetic energy of the wind, from the data registered by the vane and anemometer that are installed at the top. The nacelle turns around a crown located at the top end of the tower.

The turbine blades typically begin to move with wind speeds of around 3.5 m/s and provide maximum power at about 11 m/s. With very strong winds (25 m/s), the blades are feathered, and the wind turbine slows down in order to prevent excessive voltages and blade breakages. The rotor (unit of three blades set in the hub) is connected to a gear box that increases the angular velocity from about 13 to 1,500 revolutions per minute. The output shaft of the gearbox is connected to the generator, which produces the electricity. The generated electricity runs through cables (mounted inside the tower) from the top to the base of the tower. From there, the electric energy runs through an underground line to a substation, where its voltage is raised in order to connect it to the electrical grid for distribution to customers. All the critical functions of the wind turbine are monitored and supervised from the substation and the control center in order to detect and resolve any operational problems.

GEOTHERMAL POWER PLANTS

Geothermal energy is the heat that comes from the subsurface of the earth. The heat is contained in the rocks and fluids beneath the earth's crust and can be found as far down to the earth's hot molten rock, magna. To produce power from geothermal energy, wells are dug a mile deep into underground reservoirs to access the steam and hot water, which can then be used to drive turbines connected to electricity generators. As of May 2015, 24 countries were home to a combined geothermal power capacity of 12.8 GW, according to a report by Geothermal Energy Association (Matek, 2015). Geothermal power plants are considered environmentally friendly and emit lower levels of harmful gases compared with coal-fired power plants.

There are three types of geothermal power plants: (a) dry steam, (b) flash, and (c) binary. Dry steam is the oldest form of geothermal technology and takes steam out of the ground and uses it to directly drive a turbine. Flash plants take high-pressure hot water from deep inside the earth and convert it to steam to drive generator turbines. When the steam cools, it condenses to water and is injected back into the ground to be used again. Most geothermal power plants are flash steam plants. Binary cycle power plants transfer the heat from geothermal hot water to another liquid. The heat causes the second liquid to turn to steam, which is used to drive a generator turbine.

Suitability of application of geothermal power is significantly dependent upon the quality of the energy source. Although site investigations may give evidence for site suitability, the true characteristics of the site (the energy source) are not known until years after power production begins. The characteristics of suitable applications are (a) high heat (generally at least 320°F – the higher the better), (b) sufficient reservoir (rock) permeability over a large underground volume, (c) sufficient water flow and reserves, (d) low NCG (non-condensable gas) content of brine, (e) low total dissolved solids (TDS) in the brine, and (f) close proximity to market demand and transmission lines.

TIDAL POWER PLANTS

Tidal energy is generated from converting energy from the force of tides into power. The tidal power generation is considered more predictable compared with wind energy and solar power. Tidal stream generators are very similar to wind turbines except they are below the water surface instead of above or on land. The turbine-driven generator converts the kinetic energy in movement of water coming from change in tide into electricity. Tidal power is also relatively greater at low speeds in contrast to wind power. Water has one thousand times higher density than air, and tidal turbines can generate electricity at speeds as low as 1 m/s (2.2 mph). In contrast, most wind turbines begin generating electricity at about 3–4 m/s (7–9 mph).

Tidal power is a known green energy source and does not emit greenhouse gases. It also does not take up that much space. The largest tidal project in the world is the Sihwa Lake Tidal Power Station in South Korea, with an installed capacity of 254 MW. The project, established in 2011, was added to a 12.5 km-long seawall built in 1994 to protect the coast against flooding and to support agricultural irrigation. While the true effects of tidal barrages and turbines on the marine environment have not been fully explored, there has been some research into how barrages manipulate ocean levels and can have similar negative effects as hydroelectric power. The electromagnetic field effects and effects of toxic substances used in water turbines have been mentioned to have possible negative effect on the marine life. Despite this, tidal power is still not exploited widely even as the world's first large-scale tidal power plant (La Rance Tidal Power Plant of 240 MW capacity in Brittany, France) became operational in 1966. However, increased focus on generating power from renewable sources is expected to accelerate the development of new methods to exploit the tidal energy in the future.

OTHER POWER PLANT-RELATED ISSUES

ENERGY PLANT INPUTS AND OUTPUTS

The characteristics of inputs and outputs of a power plant will depend upon the technology used to produce electricity. However, in general terms, the following are lists of inputs and outputs that should be considered during the power plant planning stages.

Inputs of an electric power plant:

1. Fuel (e.g., coal, natural gas, oil)
2. Air (for combustion)

3. Reagents and additives (to facilitate processes in plant operation)
4. Electric power (e.g., for starting the turbine, operating pumps, and during maintenance work)
5. Water (makeup water and return water; source of kinetic energy for hydro-electric and tidal power plants; source of heat for geothermal plant)
6. Steam (return for fossil plants; geothermal plants)

Outputs of an electric power plant:

1. Electric power
2. Flue gas
3. Fuel gas
4. Hot water
5. Hot steam
6. Combustion residue
7. Heat – convection and radiation losses
8. Gas treatment residues and ashes
9. Blowdown, leakage, and drainage

Carbon Capture and Sequestration

A wide range of mitigation strategies have been developed to reduce CO_2 emissions from fossil-fuel-operated power plants. Technological alternatives for reducing CO_2 emissions from power plants to the atmosphere include the following: (a) switching to less carbon-intensive fuels, e.g., natural gas instead of coal, (b) increasing the use of renewable energy sources (e.g., wind or solar) or nuclear energy, each of which emits little to no net CO_2, and (c) capturing and sequestrating CO_2.

The CO_2 capture and sequestration (CCS) is an efficient strategy to limit climate destabilization due to high levels of energy-related CO_2 emissions. CCS is a highly promising approach to reducing GHG emissions by capturing CO_2 at the site of the power plant, transporting it to an injection site, and sequestrating for long-term storage in suitable formations. Installation of a CCS unit at thermoelectric plants can efficiently capture about 85%–95% of the CO_2 processed in a capture plant (Eldardiry and Habib, 2018). However, the CCS is costly and substantially increases capital, operating and maintenance costs of power plants. The increase in costs depends on the methods used and the percentage of the carbon to be captured by the CCS system.

Water is used as an integral element of CCS processes. It is used for cooling and emission scrubbing. Thus, deployment of CCS will potentially increase water withdrawals to meet the added needs for chemical and physical processes of capturing and separating large volumes of CO_2. Thus, the CCS technologies introduce additional stresses on the sustainability of water systems. In addition to water needs, a power plant with a CCS system would also need roughly 10%–40% more energy than a plant of equivalent output without CCS.

There are three main components of the CCS process: (a) capturing CO_2 arising from the combustion of fossil fuels, (b) transporting CO_2 to the storage site, and (c) storing CO_2 for a long period of time, rather than being emitted to the atmosphere. The three common technologies for CO_2 capture in CCS systems are the following: (a) post-combustion capture, (b) pre-combustion capture, and (c) oxy-fuel capture. In post-combustion capture, CO_2 is separated from the flue gases before they are discharged to the atmosphere. The most commercially common method, amine scrubbing, is based on using amine gas treating to remove CO_2 by aqueous solutions of amines. The CO_2 removed from the amine solvent is then dried and compressed to reduce its volume before being transported to a safe storage site. The pre-combustion capture of CO_2 is based on the ability to gasify all types of fossil fuels with oxygen or air and/or steam to produce a synthesis gas (syngas) or fuel gas composed of carbon monoxide and hydrogen. Additional water (steam) is then added, and the mixture is passed through a series of catalyst beds for the water–gas shift reaction to approach equilibrium, after which CO_2 can be separated to leave a hydrogen-rich fuel gas. This hydrogen can be sent to a turbine to produce electricity or used in hydrogen fuel cells of transportation vehicles. Although the energy requirements in pre-combustion capture systems may be of the order of half that required in post-combustion capture, the pre-combustion process requires more water for the water–gas shift reaction. In the oxy-fuel capture, pure oxygen is used for combustion instead of air and gives a flue gas mixture of mainly CO_2 and condensable water vapor, which can be separated and cleaned relatively easily during the compression process.

After the CO_2 is captured, it gets compressed to a supercritical fluid with properties between those of a gas and a liquid. It is then transported to a location suitable for long-term storage. Multiple factors are typically considered when selecting CO_2 storage sites. These factors are volume, purity, and rate of the CO_2 stream, proximity of the source and storage sites, infrastructure for the capture and delivery of CO_2, existence of groundwater resources, and safety of the storage site. Several options are available for the storage of CO_2, including injection of CO_2 into the ocean so that it gets carried into deep water, or more commonly by using geological formations as natural reservoirs accessed using wells drilled deep underground.

POWER PLANT TECHNOLOGY SPECIFICATIONS

Plant Output and Heat Rate
The output of the plant is specified in terms of net nominal capacity (in MW). The nominal capacity (also known as the nameplate capacity, rated capacity, or installed capacity) is the intended full-load sustained output of a power plant.

The heat rate is the amount of energy used by an electrical generator (or power plant) to generate 1 kWh of electricity. The U.S. Energy Information Administration (EIA) expresses heat rates in British thermal units (Btu) per net kWh of electricity generated.

Plant Building and Operational Costs
When comparing power plant costs, the following costs are considered during early decision making and planning phases of the plant development program.

1. Overnight capital costs (capital costs without considering interest and repayments on the capital) ($/kW)

2. Fixed operating and maintenance (O&M) costs ($/kW-year)
3. Variable operating and maintenance costs ($/MWh)
4. Fuel costs for fossil fuel and biomass sources ($/kWh)

Other information used includes:

1. Likely annual hours per year run (or load factor, which is hours actually run in a year divided by hours available to run in a year expressed in percentage. The load factor is also known as capacity utilization or capacity factor. It may be as low as 30% for wind energy, or as high as 90% for nuclear energy.
2. Emission/pollutant rate for NO_x, SO_x and CO_2 (lb/MMBtu)
3. Plant life (typically 25–50 years)

External costs such as connections costs and the effect of each plant on the distribution grid are considered separately as additional costs to the calculated power cost at the terminals.

Table 2.3 presents capital costs, fixed and variable maintenance costs, and amount pollutants emitted by 25 utility scale power plants using different technologies (EIA, 2020d). The data show that fossil-fueled plants with carbon capture capability, nuclear power plants, fuel cells, and concentrating plant have high capital costs (over about $5,800/kW). Whereas, the natural gas plants have the lowest capital costs around $1,000/kW. The nuclear and geothermal plants have high fixed overhead and maintenance costs (over about $120/kW). The fossil-fueled power plants and the biomass power plants also generate high levels of CO_2 (over $117 lb/MMBtu).

Capacity Factors

The net capacity factor of a power plant is the ratio of its actual output over a period, to its potential output if it were possible for it to operate at full nameplate capacity indefinitely. The capacity factor is also an indicator of reliability of power plant. It will vary as a function of time and level of power demand and will also depend upon the design characteristics and operating conditions of each power plant. Figure 2.10 presents capacity factors for several utility scale power plants from EIA (2,020g).

Environmental Impacts of Electricity Generation

Nearly all parts of the electricity generation systems can affect the environment, and the size of these impacts will depend on how and where the electricity is generated and delivered. In general, the environmental effects can include:

a. Emissions of greenhouse gases and other air pollutants, especially when a fuel is burned.
b. Use of water resources to produce steam, provide cooling, and serve other functions.
c. Discharges of pollution into water bodies, including thermal pollution (water that is hotter than the original temperature of the water body).
d. Generation of solid waste, which may include hazardous waste.

TABLE 2.3
Cost and Performance Utility-scale Electric Power Generating Technologies

Case No.	Technology	Description	Net Nominal Capacity	Net Nominal Heat Rate (Btu/Kwh)	Capital Cost ($/kW)	Fixed O&M Cost ($/kW-year)	Variable O&M Cost ($/MWh)	NOx (lb/MMBtu)	SO₂ (lb/MMBtu)	CO₂ (lb/MMBtu)
1	650 MW Net, ultra-supercritical coal w/o carbon capture—Greenfield	1×735 MW gross	650	8,638	3,676	40.58	4.5	0.06	0.09	206
2	650 MW Net, ultra-supercritical coal 30% carbon capture	1×769 MW gross	650	9,751	4,558	5,430	7.08	0.06	0.09	144
3	650 MW Net, ultra-supercritical coal 90% carbon capture	1×831 MW gross	650	12,507	5,876	59.54	10.98	0.06	0.09	20.6
4	Internal combustion engines (natural gas)	4×5.6 MW	21	8,295	1,610	35.16	5.69	0.02	0	117
5	Combustion turbines – simple cycle (natural gas)	2×LM6000	105	9,124	1,175	16.3	4.7	0.09	0	117
6	Combustion turbines – simple cycle (natural gas)	1×GE 7FA	237	9,905	713	7	4.5	0.03	0	117
7	Combined-cycle 2×2×1 (natural gas)	GE 7HA 02	1,083	5,370	958	12.2	1.87	0.0075	0	117

(Continued)

TABLE 2.3 (Continued)
Cost and Performance Utility-scale Electric Power Generating Technologies

Case No.	Technology	Description	Net Nominal Capacity	Net Nominal Heat Rate (Btu/Kwh)	Capital Cost ($/kW)	Fixed O&M Cost ($/kW-year)	Variable O&M Cost ($/MWh)	NOx (lb/MMBtu)	SO₂ (lb/MMBtu)	CO₂ (lb/MMBtu)
8	Combined-cycle 1×1×1, single shaft (natural gas)	H class	418	6,431	1,084	14.1	2.55	0.0075	0	117
9	Combined-cycle 1×1×1, single shaft, w/ 90% carbon capture (natural gas)	H class	377	7,124	2,481	27.6	5.84	0.0075	0	117
10	Fuel cell	34×300 KW Gross	10	6,469	6,700	30.78	0.59	0.0002	0	117
11	Advanced nuclear (brownfield)	2×AP1000	2,156	10,608	6,041	121.64	2.37	0	0	0
12	Small modular reactor nuclear power plant	12×50-MW Small modular reactor	600	10,046	6,191	95	3	0	0	0
13	50-MW biomass plant	Bubbling fluidized bed	50	13,300	4,097	125.72	4.83	0.08	<0.03	206
14	10% biomass co-fire retrofit	300-MW PC boiler	30	1.50%	705	25.57	1.9	0%–20%	–8%	–8%
15	Geothermal	Binary cycle	50	N/A	2,521	128.544	1.16	0	0	0
16	Internal combustion engines – landfill gas	4×9.1 MW	35.6	8,513	1,563	20.1	6.2	0.02	0	117

(Continued)

TABLE 2.3 (Continued)
Cost and Performance Utility-scale Electric Power Generating Technologies

Case No.	Technology	Description	Net Nominal Capacity	Net Nominal Heat Rate (Btu/Kwh)	Capital Cost ($/kW)	Fixed O&M Cost ($/kW-year)	Variable O&M Cost ($/MWh)	NOx (lb/MMBtu)	SO$_2$ (lb/MMBtu)	CO$_2$ (lb/MMBtu)
17	Hydroelectric power plant	New stream reach development	100	N/A	5,316	29.86	0	0	0	0
18	Battery energy storage system	500 MW I 200 MWh	50	N/A	1,389 (347 $/kWh)	24.8	0	0	0	0
19	Battery energy storage system	500 MW I 100 MWh	50	N/A	845 (423 $/kWh)	12.9	0	0	0	0
20	Onshore wind – large plant footprint: great plains region	200 MW I 2.82 MW WTG	200	N/A	1,265	26.34	0	0	0	0
21	Onshore wind – small plant footprint: coastal region	50 MW I 2.78 MW WTG	50	N/A	1,677	35.14	0	0	0	0
22	Fixed-bottom offshore wind: monopile foundations	400 MW I 10 MW WTG	400	N/A	4,375	110	0	0	0	0
23	Concentrating solar power tower	with molten salt thermal storage	115	N/A	7,221	85.4	0	0	0	0
24	Solar PV w/ single axis tracking	150 MWAC	150	N/A	1,313	15.25	0	0	0	0
25	Solar PV w/ single axis tracking + battery storage	150 MWAC solar 50 MW I 200 MWh storage	150	N/A	1,755	31.27	0	0	0	0

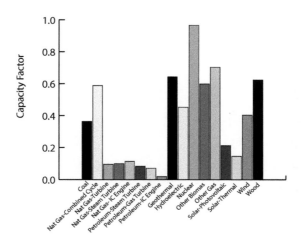

FIGURE 2.10   Capacity factors of different utility scale electric power plants.

   e. Land use for fuel production, power generation, and transmission and distri-
      bution lines.
   f. Effects on plants, animals, and ecosystems that result from the air, water,
     waste, and land impacts above.

Many of these environmental effects can also potentially affect human health, particularly if they result in people being exposed to pollutants in air, water, or soil.
These include:

   1. Effects of obtaining the fuels from mines, using the fuels, and dealing with
     the wastes following use of the fuel.
   2. Radiation during mining (uranium, coal, natural gas, emitted radon)
   3. Disposal and dispersion of waste generated from power generation (emitted
     gases, spent nuclear fuel, ashes)

Chapter 3 provides additional information on risks due to toxic substances.

Financial Incentives for Renewables and Energy Efficiencies

In order to promote development of renewable energy production, various incentives
are offered by different government agencies (federal, state, and local) in a form of
tax credits, low interest loans, rebates, and so forth. A brief summary of these incentives is provided below:

   a. Tax credits (usually by Federal Govt.) (e.g., 30% cost of renewable/energy
     saving equipment including installation)
   b. Sales tax incentives
   c. Property tax incentives
   d. Rebates (e.g., for energy efficiency equipment)

e. Loans/financing (low interest loans) (e.g., up to 10 basis points lower than standard interest rate)

f. Database of State Incentives for Renewables (DSIRE) and Efficiency (NC Clean Energy Technology Center, 2020) provides details on available incentives.

JEDI MODELS

The Jobs and Economic Development Impact (JEDI) models are user-friendly tools that estimate the economic impacts of constructing and operating power generation and biofuel plants at the local and state levels. Separate JEDI models are available for analyzing the energy impacts of wind, biofuels, concentrating solar power, geothermal, marine, and hydrokinetic power, coal, and natural gas power plants.

The models are created using Excel spreadsheets and are available for free download from the NREL's website (NREL, 2020) for the following types of power plants:

1. Biofuels
2. Coal
3. Conventional hydropower
4. Concentrating solar power
5. Geothermal
6. International
7. Marine and hydrokinetic power
8. Natural gas
9. Petroleum
10. Transmission line
11. Wind

The outputs of the JEDI models were used for cost–benefit analysis presented in Chapter 16.

U.S. MAP DATABASE WITH ENERGY SOURCES, TRANSMISSION, AND DISTRIBUTION

The U.S. energy mapping system (EIA, 2020e) is a comprehensive map-based database of all energy-related facilities such as energy resources (oil and gas wells, fossil fuel mines, uranium plants), power plants (using different technologies), fuel refining and processing plants, transmission lines, pipelines, transport and storage facilities, regions (federal lands, NERC regions, Indian lands, climate zones, and so forth). One can select the type of facility as a layer on the U.S. map, and the layer shows locations of all the facilities in the selected layer (e.g., coal power plants). Clicking on an individual facility, additional information on the type of facility and its characteristics (e.g., name, location, technology, capacity) is displayed. It is a very useful database to understand the distribution and locations of various energy-related facilities in the United States.

EPA's POWER PROFILER

EPA (2021) had provided a Power Profiler, which provides a mix of energy provided by different power sources in all eGRID subregions of the United States. The Emissions & Generation Resource Integrated Database (eGRID) is a comprehensive source of data on the environmental characteristics of almost all electric power generated in the United States. The database also provides CO_2, SO_2, and NO_x emission rates in lbs/MWh from the power sources in eGRID subregions and by U.S. postal zip codes.

FUTURE CHANGES IN ENERGY PICTURE

This section provides some data suggesting changes that can impact the energy picture in the future.

ANNUAL ENERGY OUTLOOK

The U.S. Energy Information Agency publishes Annual Energy Outlook (EIA, 2020i), which provides projections for a reference case and some special cases for 2020–2050. Some examples of projections included in the Annual Energy Outlook are listed below.

a. Energy production by sources
b. Energy consumption by sectors (residential, commercial, industrial, transportation, electrical power)
c. Energy gross trade (imports and exports)
d. Energy-related CO_2 emissions by sectors
e. Energy-related CO_2 emissions by fuels
f. Effect of critical drivers and uncertainty (e.g., high oil process, low oil prices, high oil and gas supply case, low oil and gas supply case, high economic growth case, low economic growth case, high renewable costs, low renewable costs)
g. Electricity generation from fuel sources and renewable sources
h. Electricity growth rates
i. Overnight installed cost of electricity from renewable energy sources and energy generating fuels
j. Energy generating capacity additions and retirements by energy generating sources and renewables

The document provides annual projection curves for 2020–2050 period, which can be used as assumptions in estimating costs and conducting cost–benefit analyses (see Chapters 7 and 8).

INCREASINGLY STRINGENT EMISSIONS AND GREEN ENERGY REQUIREMENTS

The requirements on the greenhouse gas emissions are expected to get more stringent over the future years and the cost of producing renewable energy is also expected to drop in the future years. Thus, cleaner air and less polluting power plants and transportation vehicles (see Chapter 12) are expected to make significant contribution in reducing pollution (NHTSA, 2020 and EPA, 2020).

INITIATIVES OF GOVERNMENT AGENCIES TO COMBAT CLIMATE CHANGE

Many initiatives are developed by energy-related government agencies such as the DOE, EPA, and NHTSA. Chapter 11 provides more information of these government agencies. It covers the functions, initiatives, and regulatory responsibilities of the U.S. Department of Energy (DOE), the Environmental Protection Agency (EPA), and the National Traffic Safety Administration (NHTSA). For example, the EPA is responsible for ensuring that the environment, primarily air and water, is safe, i.e., it does not contain toxic or harmful substances above certain specified concentration levels for human activities. The DOE implements minimum efficiency standards for a wide range of appliances and equipment used in residential and commercial buildings. Similarly, the NHTSA and EPA have jointly created requirements on minimum allowable levels of corporate average fuel economy (CAFE) [in miles per gallon of fuel consumed (mpg)] and maximum allowable levels of emissions requirements [in equivalent grams of CO_2 per mile] on light automotive products (passenger cars and light trucks).

REDUCING CARBON EMISSIONS FROM NATURAL-GAS-FUELED POWER PLANTS

Green hydrogen is a term used to use hydrogen as a fuel in combustion turbines to reduce emissions that are emitted by burning natural gas in combustion turbine power plants (Blunt, 2020). The hydrogen gas can be produced by using electrolysis technique and power from renewable sources such as solar and wind. The hydrogen thus generated is still 3–5 times costlier than the natural gas currently used for power generation. However, in the future, the price of hydrogen is expected to drop lower. NextEra Energy Inc. and Dominion Energy Inc. are expected to commercialize this costly technology to provide steady carbon-free power. By 2025, they expect to use a 30% hydrogen and 70% natural gas mixture to run power plants to reduce carbon emissions. As the price of generating hydrogen drops in the future, higher proportion of hydrogen is expected to be used the hydrogen–natural gas mixture to run the power plants.

CONCLUDING REMARKS

The energy picture of the United States is changing steadily due to a number of changes, such as increased focus on the renewable energy, increasing availability of natural gas, replacement of older fossil fuel plants with natural gas power plants, anticipated increase in electric vehicles, and implementation of smart grid features. The power generation and distribution problems are also getting more complex due to increasing number of power sources and increasing consumer demands for comfort and conveniences through the use of use of many devices, such as smart phones, smart appliances in homes, autonomous features in vehicles, shared services (e.g., ride sharing). The integration of these increasing number of the power sources and consumer demands will create a number of challenges in developing new and improving existing energy systems. The energy systems designers and engineers will have to make a number of decisions to accommodate these changes. This book provides many methods to solve such future challenges.

REFERENCES

Blunt, K. 2020. Utilities look to switch to green hydrogen. *The Wall Street Journal*. September 8, 2020.

EIA. 2015. Residential Energy Consumption Survey (RECS). Websites: https://www.eia.gov/consumption/residential/data/previous.php and https://www.eia.gov/consumption/residential/data/2015/index.php?view=consumption#undefined (Accessed: September 5, 2020).

EIA. 2019a. U.S. Primary Energy Use by Sources. Website: https://www.eia.gov/energyexplained/us-energy-facts/ (Accessed: September 4, 2020).

EIA. 2019b. Energy Use for Transportation. Website: https://www.eia.gov/energyexplained/use-of-energy/transportation.php (Accessed: November 5, 2020).

EIA. 2020a. Total Energy Review. Website: https://www.eia.gov/totalenergy/data/browser/index.php?tbl=T01.01#/?f=A&start=1949&end=2019&charted=4-6-7-14 (Accessed: November 10, 2020).

EIA. 2020b. Hourly Electric Grid Monitor. Website: https://www.eia.gov/beta/electricity/gridmonitor/dashboard/electric_overview/US48/US48 (Accessed: September 5, 2020).

EIA. 2020c. Residential Energy Consumption Survey (RECS). Website: https://www.eia.gov/consumption/residential/data/previous.php (Accessed: September 5, 2020).

EIA. 2020d. Capital Cost and Performance Characteristic Estimates for Utility Scale Electric Power Generating Technologies. Contract No. 89303019CEI00022. SL-014940 | Project No. 13651.005. Prepared by Sargent & Lundy, L.L.C. ("Sargent & Lundy"). Website: https://www.google.com/url?sa=t&rct=j&q=&esrc=s&source=web&cd=&ved=2ahUKEwiXsuXBjtPrAhVUZc0KHauUAjoQFjAAegQIAhAB&url=https%3A%2F%2Fwww.eia.gov%2Fanalysis%2Fstudies%2Fpowerplants%2Fcapitalcost%2Fpdf%2Fcapital_cost_AEO2020.pdf&usg=AOvVaw3zcjjqD4mUUPXIGVfx6x1f (Accessed: September 4, 2020).

EIA. 2020e. U.S. Energy Mapping System. Website: http://www.eia.gov/state/maps.cfm?v=Fossil%20Fuel%20Resources (Accessed: September 5, 2020).

EIA. 2020f. Electricity Data Browser. Website: https://www.eia.gov/electricity/data/browser/#/topic/5?agg=0,1&geo=vvvvvvvvvvvvvo&endsec=vg&linechart=ELEC.SALES.US-ALL.A&columnchart=ELEC.SALES.US-ALL.A&map=ELEC.SALES.US-ALL.A&freq=A&ctype=linechart&ltype=pin&maptype=0&rse=0&pin= (Accessed: September 7. 2020).

EIA. 2020g. Electric Power Monthly -- Capacity Factors for Utility Scale Power Plants. Website: https://www.eia.gov/electricity/monthly/epm_table_grapher.php?t=epmt_6_07_b (Accessed: September 7, 2020).

EIA. 2020h. Biomass Explained. Website: https://www.eia.gov/energyexplained/biomass/ (Accessed: September 7, 2020).

EIA. 2020i. Annual Energy Outlook 2020 with Projections to 2050. Website: https://www.eia.gov/outlooks/aeo/ (Accessed: November 10, 2020).

EIA. 2021. New U.S. Power Plants Expected to be Mostly Natural Gas Combined-Cycle and Solar PV. Website: https://www.eia.gov/todayinenergy/detail.php?id=38612 (Accessed: May 3, 2021).

Eldardiry, H., and Habib, E. 2018. Carbon capture and sequestration in power generation: review of impacts and opportunities for water sustainability. *Energ Sustain Soc* 8(6). Doi:10.1186/s13705-018-0146-3.

Electropaedia. 2020. Energy Efficiency of Various Energy Sources. Website: https://www.mpoweruk.com/energy_efficiency.htm (Accessed: September 5, 2020).

EPA. 2020. *Air Quality Trends Show Clean Air Progress*. Washington, DC: Environmental Protection Agency. Website: https://gispub.epa.gov/air/trendsreport/2019/#highlights (Accessed: September 8, 2020).

EPA. 2021. *Power Profiler*. Washington, DC: Environmental Protection Agency. Website: https://www.epa.gov/egrid/power-profiler#/ (Accessed: April 10, 2021).

Fu, R., D. Feldman, R. Margolis, M. Woodhouse, and K. Ardani. 2017. *U.S. Solar Photovoltaic System Cost Benchmark: Q1 2017*, National Renewable Energy Laboratory, Golden, CO. September 2017.

IEA. 2014. *Solar Energy Could Dominate Electricity by 2050*. International Energy Agency. Website: https://www.reuters.com/article/us-solar-iea-electricity-idUKKCN0HO11K20140929 (Accessed: June 6, 2021).

International Hydropower Association. 2018. 2018 Hydropower Status Report. Website: https://www.hydropower.org/publications/2018-hydropower-status-report (Accessed: September 7, 2020).

Matek, B. 2015. *2015 Annual U.S. & Global Geothermal Power Production Report*. Geothermal Energy Association. Washington, DC. February 2015.

National Highway Traffic Safety Administration. 2020. Website: https://one.nhtsa.gov/cafe_pic/CAFE_PIC_fleet_LIVE.html (Accessed: September 8, 2020).

National Renewable Energy Laboratory. 2020. NREL's JEDI Models Website: https://www.nrel.gov/analysis/jedi/models.html (Accessed: September 7, 2020).

NC Clean Energy Technology Center. 2020. Database of State Incentives for Renewables & Efficiency (DSIRE). Website: https://www.dsireusa.org/ (Accessed: September 7, 2020).

Union of Concerned Scientists. 2012. UCS EW3 Energy-Water Database V.1.3. Website: www.ucsusa.org/ew3database (Accessed: November 6, 2020).

World Nuclear Association. 2013. *Cooling Power Plants*. London. Website: https://world-nuclear.org/our-association/publications/technical-positions/cooling-of-power-plants.aspx (Accessed: November 6, 2020).

3 Risk Assessment Methods

INTRODUCTION

WHAT IS RISK?

A risk is generally associated with an occurrence of an undesired event, such as a financial loss and/or an injury resulting from an entity (e.g., product or equipment used in a plant or a facility) related failure. The risk can be measured in terms of the magnitude of the consequence due to the occurrence of an undesired event. The magnitude of the consequence can be measured by costs associated with undesired events, such as loss due to product/equipment failures, accidents, loss due to interruption of work, loss of revenue, loss of reputation, and so forth.

Risk is the combination of the predicted frequency of an occurrence of the undesired (initiating) event and the predicted consequence (magnitude or severity) of the damage such an event might cause if the ensuing follow-up events were to occur. The risk (in dollars) can be computed as:

$$\text{Risk}(\$) = [\text{Probability of occurrence}] \times [\text{Consequence of the undesired event}(\$)]$$

Two examples of the application of the above definition of risk can illustrate the concept and its computation. Let us assume that if a step-up transformer at an electric transmission station explodes due to higher temperature, the cost to repair the damage caused by the explosion would be \$150,000 and the probability of the explosion would be 0.0001 per year of operation. Then the risk would be \$150,000 × 0.0001 = \$15. In another example, if a blowout prevent valve in an oil well explodes and spills the oil, then the loss of the oil plus the damage due to the explosion is estimated to be \$300,000, and the probability of occurrence of the blowout valve is estimated from historic data to be 0.01 per year. Then, the risk in explosion of the blowout preventer valve will be \$300,000 × 0.01 = \$3,000. Thus, the risk due to the explosion of the blowout preventer valve of \$3,000/year would be much larger than the \$15/year risk due to explosion of the step-up transformer.

RISK ANALYSIS

Risk analysis is the process of defining and analyzing the dangers to individuals, businesses, and government agencies or their operations posed by potential natural and human-caused adverse events. Risk analysis is used to identify and assess factors that may jeopardize the success of a project or achieving a goal. This technique also helps to define preventive measures to reduce the probability of these factors from occurring

DOI: 10.1201/9781003107514-4

and identify countermeasures to successfully deal with these problems when they develop to avert possible negative effects on the competitiveness of the organization.

A risk analysis can be defined a decision-making exercise conducted to determine the next course of action after a potential undesired event has been identified and the magnitude of the consequence and relevant characteristics of the undesired event have been estimated.

The phase of identification of undesired event can be called as the "Risk Identification" phase, and the phase of estimation of the magnitude of the consequence due to the undesired event can be called as the "Risk Assessment" phase.

Some commonly used methods for Risk Identification, Risk Assessment, and Risk Analysis are given below.

1. *Risk identification methods*: Brainstorming, interviewing experts, use of checklists and historic data (e.g., past records of product defects, warranty problems, customer complaints), hazard analysis (see Chapter 5), failure modes and effects analysis (FMEA) (see later pages of the chapter)
2. *Risk assessment methods*: Estimation of probability (or frequency) of occurrence of undesired events, magnitude of the consequence (or severity) of the undesired event, and probability of detection of the undesired event can be obtained by using brainstorming, interviewing experts, safety analysis (e.g., fault tree analysis [see Chapters 5] and FMEA, and the historic data from company/organization records, e.g., costs of past accidents or failures)
3. *Risk analysis methods*: Risk matrix, risk priority number (RPN), nomographs, applications of existing design and performance standards, and specialized risk models (Floyd et al., 2006).

After assessing the risk, the decision needs to be made on future actions. Alternatives include accepting the risk, mitigating or eliminating the risk, transferring the risk, and avoiding the risk (Conrad et al., 2017).

Accept the risk: Some risks may be accepted. In some cases, it is cheaper to leave an asset unprotected due to a specific risk, rather than make the effort and spend the money required to protect it. This cannot be an ignorant decision; all options must be considered before accepting the risk. Low likelihood/low consequence risks are candidates for risk acceptance. High and extreme risks cannot be accepted. There are situations where accepting the risk is not an option, such as data protected by laws or regulations, and risk to human life or safety.

Mitigating risk: Mitigating risk means lowering the risk to an acceptable level. Lowering risk is also called risk reduction, and the process of lowering risk is also called reduction analysis.

Eliminating the risk: In some cases, it is possible to remove specific risks entirely; this is called eliminating the risk.

Transferring risk: The insurance model depicts transferring risk. Most homeowners do not assume the risk of fire for their houses; they pay an insurance company to assume that risk for them. The insurance companies are experts in risk analysis and buying risk is their business.

Risk avoidance: A thorough risk analysis should be completed before taking on a new project. If the risk analysis discovers high or extreme risks that cannot be easily mitigated, avoiding the risk (and the project) may be the best option.

PERCEPTION OF RISK

Different individuals will perceive the risk in any given undesired situation differently. Thus, level of risk perceived by an individual is very subjective. We voluntarily assume certain risks, e.g., flying, driving, driving on snow covered roads, or driving at night. There are also some involuntary risks that we take in situations for which we have no control. Some risks can be "Distributed" (i.e., spread over many incidences, e.g., accidents occurring at many locations with few casualties at each location), while other risks may be considered to be "acute" [concentrated] (e.g., one catastrophic accident).

The following factors have different and even opposite effects while accepting an event or situation as a risk.

a. assumed voluntarily vs. incurred involuntarily
b. consequences occur immediately vs. delayed
c. consequences reversible vs. irreversible
d. consequences occur over short term vs. long term
e. no alternatives available vs. many alternatives available
f. small uncertainty vs. large uncertainty
g. common hazard vs. unknown or "dreaded" hazard
h. exposure is necessary vs. exposure is optional
i. incurred occupationally vs. incurred non-occupationally
j. incurred by other people vs. not incurred by other people

Thus, risk of developing lung cancer due to living close to a coal-fired power plant can be considered to be voluntary (if you can afford to live away from the plant), long term (exposed to flue gases over many years), alternatives available (i.e., other alternatives available to reduce exposure of carcinogens), living near known hazard and non-incurred by other people (i.e., presence of other people does not affect causation of the cancer). Similarly, coalminers were working in coal mines by taking risks voluntarily due to their occupation; it was a commonly known hazard of developing a lung disease by working over many years (long term) in a coal mine. Whether the lung disease can be reversible after the working in coal mines was terminated sooner, or irreversible if the miner worked in the mining industry for many years, would depend upon factors such as the level of exposures (e.g., dosage and exposure time) and the composition of the contaminants.

Some Examples of Risks and Associated Causes

1. *Risks in power plant building*: The power plant may be completed late (schedule risk) and/or may be over-budget (financial risk). The actual output

of the power plant may be well below its design capacity (i.e., the new plant would be a disappointment – it cannot produce output at the design level – i.e., production risk). The operating and maintenance costs of the plant may be much higher than many other similar plants (financial risk).

2. *Risks in developing a new car program*: Many attributes of the vehicle such as its acceleration performance, fuel/energy consumption, braking capability may be well below the target levels specified during its early planning phase. Thus, many customers of the vehicle will not be satisfied, and future sales may be low (i.e., risk of product acceptance).

3. *Risk in creating a new drug for a disease*: The drug may not be effective, i.e., the effect may not last over its specified time duration (performance risk). The cost of developing the drug may be too high (financial risk).

4. *Risks in deep-water drilling*: The deep-water drilling project took a much longer time to complete due to several unexpected clogging incidences with pipelines (schedule risk). The oil output was well below initial estimates (production risk).

5. *Risks in using high-sulfur fuels in cars*: The high-sulfur fuel causes higher levels of pollutants resulting in higher number of cases involving respiratory illnesses among people living close to the roadways (heath risk).

6. *Risk in building wind turbines*: Numerous accidents during construction of wind turbines (e.g., accidents during transporting long blades and tower assembly, tower collapse/failures, slips and falls from higher work heights) and operation (e.g., blade breakage due to ice buildup, fires in power generating room and transformers) have caused injuries and losses (safety risks).

7. *Risk in installing solar panels on roofs*: Lower power output and poor durability of solar panels due to snow and wind damage and sandstorms requiring unscheduled stoppages for maintenance are examples of performance risks. Slips and falls (accidents) during solar panel installation and maintenance are safety-related risks.

8. *Risk in building dams*: Cracks and water leakage in dams and clogged intake water pipes reduce output levels of power generators located on the downstream side of the dam are examples of performance risks. Flooding of downstream areas due to heavy rains causing major disaster in low lying communities is an example of safety risk.

TYPES OF RISKS

Many different types of undesired events can occur. All major and minor types of risks should be identified, and plans must be initiated for managing the risks. The risks can be classified by using many different considerations. Some categories of risks are presented below. (Note that a given risk may fall into more than one category.)

1. *Revenue/Financial risk*: The company could not generate the required level of revenue to sustain profitability (e.g., income from energy sales is lower than energy generation and distribution costs). Financial/fiscal risks reflect

the uncertainty in financing of the power generation investment, changes in taxes, interest rates, sales and revenues, and the resulting variations in the profitability of the investment.

2. *Production risk*: Output quantity and quality are not at a level demanded by its customers. For example, a power plant may not be able to generate the required level of output electricity at the reliability level (absence of blackouts or brownouts) over a given period. This problem could be due to a technical problem (e.g., plant efficiency below estimated level when the plant was proposed) due to reasons, such as low pressures and temperatures in a boiler or combustion/steam turbines, or shortage of fuel and/or feed water.

3. *Demand risk*: The level of demand of the outputs (electricity consumed by the customers) may not be at the level required by a power company to generate the revenues needed to sustain profitability.

4. *Political risk*: Changes in national policy can affect regulations on future types of technologies to be used in the power generation plants. Some examples are: (a) climate change policy risks depend upon CO_2 reduction targets and fluctuations in CO_2 capture and sequestration costs, (b) changes in the climate change policy schemes (e.g., changes in renewable energy sources subsidy/promoting policies), and (c) forced retirements of fossil fuel power plants due to government actions.

5. *Safety risks*: Safety-related risks can be measured by estimating costs of accidents and safety programs implemented for prevention of the accidents. Thus, the two safety risks can be described as follows:
   a. *Accident cost risks*: Costs incurred due to accident occurrences within the organization's operations (e.g., medical costs to treat injured persons, costs of disabling injury accidents, costs to repair or replace plant/equipment for power generation, power transmission, and distribution).
   b. *Accident prevention cost risks*: Costs of installing and/or maintaining safety equipment (e.g., fire protection equipment, burn prevention heat shields, and/or insulations, high pressure and temperature alarms).

6. *Overhead risks*: Costs due to additional unplanned maintenance/repairs, cleanups of spills, and so forth.

7. *Insurance costs risks*: Unexpected increase in insurance premiums or changes in insurance coverages.

8. *Liability risks*: Costs of defending lawsuits initiated by customers, contractors, and suppliers related to inability of the company in meeting contractual agreements, negligence in conducting safety analyses and implementing safety countermeasures, providing cleaner/healthful working conditions for employees, and so forth.

9. *Health risks*: Costs due to health-related incidences (e.g., cardiovascular and respiratory diseases and other chronic diseases such as asthma). A new power plant fails to meet stringent pollution requirements.

The costs associated with many of the risk-related consequences depend upon medical costs in treating people involved in (a) accidents resulting in injuries or deaths, and/or (b) longer-term exposures of pollutants. The medical costs associated due to hospitalization, costs due to lost workdays (due to disabling injuries), and costs associated with loss of human lives can be estimated. Costs of various types of accidents can be estimated based on type and severity level of the injuries, and even deaths can be estimated based on historic data and record keeping (NSC, 2017). For example, NATA air quality database also provides estimates of probability of death due to a number of toxins and air quality measures for most counties within the United States (EPA, 2014). The increases or decreases in the costs related to the deaths and the medical cases can be used as estimates of health-related risks in conducting cost–benefit analyses (see Chapter 8).

10. *Global warming and climate change-related risks*: Greenhouse gases emitted by many fossil-fuel-fired power plants have steadily increased risks due to global warming and climate change. Recent declining trend in fossil fuel consumption by the power sector has been driven by a decrease in the use of coal and petroleum with an offsetting increase in the use of natural gas. Changes in the fuel mix and improvements in electricity generating technologies (e.g., renewable sources) have also led the power sector to produce electricity while consuming fewer fossil fuels. These rapid changes have added financial risks in the utility industry due to premature retiring of many fossil-fuel-fired power plants and meeting new government requirements, such as to require purchasing excess electric energy generated by many newer renewable power plants at higher prices.

11. *Cyber threats*: As the power industry is implementing smart grid (digital communications between smart devices/equipment, monitoring, and control of power demand and loads), higher number of hacker attacks are expected. The attackers can not only attempt to reduce their electricity bills, but they can also create unexpected changes in power demands and power shutoffs. For example, on May 7, 2021, Colonial Pipeline Co. experienced an online ransomware attack that forced the closure of the largest U.S. fuel pipeline. The 5,500-mile pipeline carries gasoline and other fuels from the Gulf Coast to the New York Metro area.

ENVIRONMENTAL RISKS

CANCER AND RESPIRATORY PROBLEMS

Air pollution is now the biggest environmental risk for early death, responsible for as many as 5 million premature deaths each year from heart attacks, strokes, diabetes, and respiratory diseases. That is more than the deaths from AIDS, tuberculosis, and malaria combined (Heath Effects Institute [HEI], 2019).

Poor air quality causes people to die younger as a result of cardiovascular and respiratory diseases, and it also exacerbates chronic diseases such as asthma, causing people to miss school or work, and eroding quality of life. Air pollution consistently ranks among the top risk factors for death and disability worldwide. Breathing polluted air has long been recognized as increasing a person's chances of developing heart disease, chronic respiratory diseases, lung infections, and cancer. In 2017, air pollution was the fifth highest mortality risk factor globally and was associated with about 4.9 million deaths and 147 million years of healthy life lost (Heath Effects Institute, 2019).

Two main pollutants considered key indicators of ambient, or outdoor, air quality are (a) fine particle pollution – airborne particulate matter measuring less than 2.5 μm in aerodynamic diameter, commonly referred to as PM2.5, and (b) ground-level (tropospheric) ozone. The air-pollutants are briefly described below.

Particulate matter (PM$_{10}$, PM$_{2.5}$): Particulate matter (PM) is made up of small airborne particles such as dust, soot, and drops of liquids. The majority of PM in urban areas is formed directly from burning of fossil fuels by power plants, automobiles, non-road equipment, and industrial facilities. Other sources are dust, diesel emissions, and secondary particle formation from gases and vapors.

Coarse particulate matter (PM$_{10}$, particles less than 10 μm in diameter) is known to cause nasal and upper respiratory tract health problems. Fine particles (PM$_{2.5}$, particles less than 2.5 microns in diameter) penetrate deeper into the lungs and cause heart attacks, strokes, asthma, and bronchitis, as well as premature death from heart ailments, lung disease, and cancer. Studies show that higher PM$_{2.5}$ exposure can impair brain development in children.

Fine particle air pollution comes from vehicle emissions, coal-burning power plants, industrial emissions, and many other human and natural sources. While exposures to larger airborne particles can also be harmful, studies have shown that exposure to high average concentrations of PM$_{2.5}$ over the course of several years is the most consistent and robust predictor of mortality from cardiovascular, respiratory, and other types of diseases.

The sources responsible for PM$_{2.5}$ pollution vary within and between countries and regions. Dust from the Sahara Desert contributes to the high particulate matter concentrations in North Africa and the Middle East, as well as to the high concentrations in some countries in western sub-Saharan Africa. A recent analysis by HEI found that major PM$_{2.5}$ sources in India include (a) household burning of solid fuels, (b) dust from construction, roads, and other activities, (c) industrial and power plant burning of coal, (d) brick production, (e) transportation, and (f) diesel-powered equipment. The relative importance of various sources of PM$_{2.5}$ in China was quite different, with a separate study identifying the major sources as (a) industrial and power plant burning of coal and other fuels, (b) transportation, (c) household burning of biomass, (d) open burning of agricultural fields, and (e) household burning of coal for cooking and heating.

Black carbon (BC): BC is one of the components of particulate matter and comes from burning fuel (especially diesel, wood, and coal). Most air pollution regulations focus on PM$_{2.5}$, but exposure to BC is a serious health threat as well. Populations with higher exposures to BC over a long period are at a higher risk for heart attacks and stroke. In addition, BC is associated with hypertension, asthma, chronic obstructive pulmonary disease, bronchitis, and a variety of types of cancer.

Nitrogen oxides (NO and NO₂): Nitrogen oxide (NO) and nitrogen dioxide (NO_2) are produced primarily by the transportation sector. NO is rapidly converted to NO_2 in sunlight. NO_x (a combination of NO and NO_2) is formed in high concentrations around roadways and can result in development and exacerbations of asthma and bronchitis and can lead to a higher risk of heart disease.

Ozone (O₃): Ozone is a gas generated with both natural and human sources. When it is high up in the atmosphere (in the stratosphere), ozone plays a protective role, shielding Earth from harmful rays and ultraviolet radiation. Ozone is formed in the atmosphere through reactions of volatile organic compounds and nitrogen oxides, both of which are formed because of combustion of fossil fuels. Ozone concentrations are measured in parts per billion (ppb). When assessing exposure to ozone, scientists focus on measurements taken in the warm season in each region, when ozone concentrations tend to peak in the mid-latitudes (where most epidemiological studies have been conducted), rather than on annual averages.

When it is near ground level (in the troposphere), it acts as a greenhouse gas and a pollutant, with harmful effects on human health. Most ground-level ozone pollution is produced by human activities (for example, industrial processes and transportation) that emit chemical precursors (principally, volatile organic compounds and nitrogen oxides) to the atmosphere, where they react in the presence of sunlight to form ozone. Exposure to ground-level ozone increases a person's likelihood of dying from respiratory disease, specifically chronic obstructive pulmonary disease. Short-term exposure to ozone can cause chest pain, coughing, and throat irritation, while long-term exposure can lead to decreased lung function and cause chronic obstructive pulmonary disease. In addition, ozone exposure can aggravate existing lung diseases.

Sulfur dioxide (SO₂): SO_2 is emitted into the air by the burning of fossil fuels that contain sulfur. Coal, metal extraction and smelting, ship engines, and heavy equipment diesel equipment burn fuels that contain sulfur. Sulfur dioxide causes eye irritation, worsens asthma, increases susceptibility to respiratory infections, and impacts the cardiovascular system. When SO_2 combines with water, it forms sulfuric acid; this is the main component of acid rain.

CLEANER AIR AND AIR QUALITY STANDARDS

The Clean Air Act (CAA) (42 U.S.C. 7401 et seq.) is a comprehensive Federal law that regulates all sources of air emissions (EPA, 2020). The 1970 CAA authorized the U.S. Environmental Protection Agency (EPA) to establish National Ambient Air Quality Standards (NAAQS) to protect public health and the environment. Under the CAA, the EPA is required to regulate emission of pollutants that "endanger public health and welfare." State and local governments also monitor and enforce CAA regulations, with oversight by the EPA.

RISKS IN PRODUCT DEVELOPMENT AND PRODUCT USES

RISKS DURING PRODUCT DEVELOPMENT AND MANUFACTURING

The operations of the energy companies (i.e., companies involved in the business of generating and selling energy) are affected by defects involved in designing and manufacturing of many products used in the energy industry such as pumps, boilers, turbines, generators, transformers, switches, transmission lines, and so forth. Thus, risk faced due to the defects during design and manufacturing of the products has a direct effect on the business risks faced by the energy companies.

The product programs involve many risks (Bhise, 2014). A risk is present when an undesired event (which generally incurs substantial loss) is probable (i.e., likely to occur with some level of probability). Risks associated with failures of the products are possible anytime during or after the product development processes. If a product producer takes too little risk by overdesigning (or using too high a safety factor), the product will be more costly and the extra cost most likely will be wasted. On the other hand, if the product designer takes too much risk by underdesigning (e.g., product has insufficient strength, or use of cheap low-quality materials and/or components), then the product program will be too costly due to high costs from product failures and repairs under warranty. The product failures can cause accidents, which can incur additional costs due occurrences of (a) injuries, (b) property damages, (c) loss of income, (d) interruptions or delays in work situations, and (e) product liability cases.

DEFINITION OF RISK AND TYPES OF RISKS IN PRODUCT DEVELOPMENT

A risk is generally associated with an occurrence of an undesired event such as a financial loss and/or an injury resulting from a product-related failure. The risk can be measured in terms of the magnitude of the consequence due to the occurrence of an undesired event. The consequence due to a risk can be measured by costs associated with customer dissatisfaction, loss due to product defects or resulting accidents, loss due to interruption of work, loss of revenue, loss of reputation, and so forth.

The risk is generally assessed by consideration of the following variables: (a) probability of occurrence of the undesired event, (b) the consequence (or severity) of the undesired event (e.g., amount of loss or severity of injuries), (c) probability of detection of the undesired event before or when it occurs, and (d) preparedness of the risk fighting unit (e.g., fire department, emergency response units, police, etc.) that can attempt to contain the severity of the loss or injury.

The risks during the product development process can be categorized as follows:

1. *Technical risk*: This type of risk occurs due to one or more technical problems with the design of the product. For example, a design flaw discovered during testing of an early production component. Such a problem may prevent the product to achieve required technical capability or performance. To eliminate the technical problem or the flaw, additional analyses, engineering changes and/or technology changes, and testing may be needed.

These additional tasks usually result in an increase in the costs and delays in the project schedule. Adoption of new technologies before adequate developmental work often leads to serious delays (e.g., problems in developing carbon fiber components and manufacturing problems with light-weight materials to increase strength and reduce weight of wind turbine blades).

2. *Cost risk*: This risk is associated with the cost overruns due to technical problems and resulting delays in the project schedule. The risk also may be due to underbudgeting caused by assuming optimistic estimates or under-estimation of required tasks, time, and costs (e.g., not providing sufficient allowance for rework).

3. *Schedule risk*: This risk is related to not able to meet the project schedule due to delays from a number of possible reasons (e.g., parts not delivered by suppliers on time, late changes made in the design due to failures uncovered in testing, or planned schedule may be too optimistic).

4. *Programmatic risk*: This risk is associated with the product development program (e.g., being over-budget, delayed, modified, or even cancelled due to several reasons). Since most of the complex products have many components that are made and supplied by various suppliers, selection of suppliers with unproven or low technical capabilities often leads to program delays, lower quality, and cost overruns.

The above four categories of risks are generally interrelated, i.e., a risk in any one of the categories also affects the risks associated in other categories. The risks also cause backward cascading effects of the problems in the work completed in the early phases but discovered in the later phases. These problems affect the progress in the succeeding phases due to factors such as redesign, rework, retests, delays, cost over-runs, and so forth.

TYPES OF RISKS DURING PRODUCT USES

The risk after the product is introduced in the market and used by end users can be categorized as follows:

1. *Loss of user confidence in the product*: The end users may be afraid to use a product because of a defect in the product. The defect may be caused due to a design or manufacturing defect or some "hidden-danger" that can cause an undesired event (e.g., sudden loss of control, a fire or an explosion, an accident, and exposure of toxic substances).

2. *Loss in future sales*: The likelihood of an undesired event can cause loss in the reputation of the producer and thus can affect future sales.

3. *Excessive repair or recall costs*: The producer will need to fix the product problem by repairing under warranty or by initiating a product recall.

4. *Product litigation costs*: The costs of defending the product in product lia-bility cases and costs related to settlements before the court trials or pay-ments of penalties, fines, and so forth.

METHODS FOR RISK ANALYSIS

Some commonly used methods for Risk Analysis are given below.

RISK MATRIX

The risk matrix involves simply creating a matrix with combinations of relevant variables associated with the degree of risk and the undesired outcomes. A risk matrix is a simple graphical tool. It provides a process for combining (a) the probability of an occurrence of an undesired event (usually an estimate) and (b) the consequence if the undesired event occurred (usually cost estimates in dollars).

The risk (in dollars) is computed as:

$$\text{Risk}(\$)=[\text{Probability of occurrence}]\times[\text{Consequence of the undesired event}(\$)]$$

Figure 3.1 shows a plot of the above relationship of probability of occurrence and consequence (loss in dollars) to the risk (expected loss in dollars) due to an undesired event. The plot is made by using log scales (logarithm to the base of 10) for both the axes. Thus, the expected loss, which is the result of the summation of the logarithms of the two values, is represented by a slanting line of risk on the plot with logarithmic axes. The magnitude of the risk is the same on any given risk line. The risk lines for $1–$1,000,000 are shown in Figure 3.1.

A simplified form of the above relationship between the probability of occurrence and the magnitude of the consequence can be presented in a matrix format. Figure 3.2 presents an example of a risk matrix. The cells of the matrix represent different risk levels increasing from low risk to high risk from the lower left corner

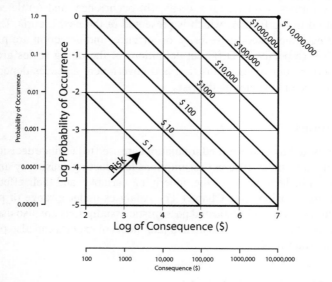

FIGURE 3.1 Relationship of risk to probability of occurrence and consequence of occurrence of an undesired event.

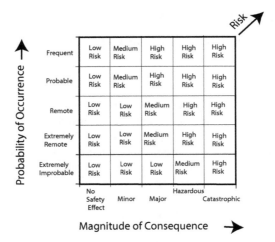

FIGURE 3.2 An example of a risk matrix. (Redrawn from: Federal Highway Administration, 2007.)

of the matrix to the top right corner of the matrix. The risk matrix thus allows for a quick assessment of the risk level after the occurrence probability and the magnitude of the consequence due to an undesired event are estimated.

Risk Priority Number (RPN)

RPN is another method used to assess the level of risk. It is based on multiplication of three ratings, namely: (a) severity, (b) occurrence, and (c) detection. This method is used in the FMEA, which is presented in the next section. Examples of 10-point rating scales used for severity, occurrence, and detection are presented in Tables 3.1–3.3, respectively. Different definitions of the rating scales are generally used in different companies, industries, and government agencies depending upon the entity involved in evaluating the risk.

Other Methods

Other methods such as modeling and simulations are used to facilitate decision-making. Exercising models under different assumptions (conducting sensitivity analysis) can provide a good understanding of the underlying variables and their effects on risks and subsequent decisions (see Chapter 17). Analyses under a range of possibilities with different levels of optimistic and pessimistic assumptions are also useful to estimate the limits of risks. Historic data and judgments of experts can also play a major role in the decision-making.

Failure Modes and Effects Analysis

The FMEA method was first used in the 1960s as a systems safety analysis tool. It was used in the early days in the defense and aerospace systems design to ensure that

TABLE 3.1

An Example of a Rating Scale for Severity

Rating	Effect	Criteria: Severity of Effect
10	Hazardous – without warning	Very high severity rating when potential failure mode affects safe product operation and involves noncompliance with government regulations without warning
9	Hazardous – with warning	Very high severity rating when potential failure mode affects safe product operation and involves noncompliance with government regulations with warning
8	Very High	Product inoperable with loss of primary function
7	High	Product operable with reduced level of performance. Customer dissatisfied.
6	Moderate	Product operable but usage with reduced level of comfort or convenience. Customer experiences discomfort.
5	Low	Product operable but usage without comfort or convenience. Customer experiences discomfort.
4	Very Low	Minor product defect (e.g., noise, vibrations, poor surface finish) only noticed by most customers
3	Minor	Minor product defect only noticed by average customer
2	Very Minor	Minor product defect only noticed by discriminating customer
1	None	No effect

TABLE 3.2

An Example of a Rating Scale for Occurrence

Rating	Probability of Failure	Possible Failure Rates
10	Very High: Failure is almost inevitable	≥ 1 in 2
9		1 in 3
8	High: Repeated failures	1 in S
7		1 in 20.
6	Moderate: Occasional failures	1 in BO.
5		1 in 800
4	Low: Relatively low: failures	1 in 2,000
3		1 in 15,000
2	Remote: Failure is unlikely	1 in 150.000
1		≤ 1 in 1500,000

the product (e.g., an aircraft, spaceship, or a missile) was designed to minimize probabilities of all major failures by brainstorming and evaluating all possible failures that could occur and then acting on the resulting prioritized list of the corrective actions.

For over the past 20 years, the method has been routinely used by the product design and process design engineers to reduce the risk of failures in the designs of products and processes used in the many industries (e.g., automotive, aviation,

TABLE 3.3

An Example of a Rating Scale for Detection

Rating	Detection	Criteria: Likelihood of Detection by Design Control
10	Absolutely uncertain	Design control will not and/or cannot detect a potential cause or mechanism for the failure mode; or there is no design control.
9	Very remote	Very remote chance that the design control will detect a potential cause/mechanism for the failure mode.
8	Remote	Remote chance that the design control will detect a potential cause/mechanism for the failure mode.
7	Very low	Very low chance that the design control will detect a potential cause/mechanism for the failure mode.
6	Low	Low chance that the design control will detect a potential cause/mechanism for the failure mode.
5	Moderate	Moderate chance that the design control will detect a potential cause/mechanism for the failure mode.
4	Moderately high	Moderately high chance that the design control will detect a potential cause/mechanism for the failure mode.
3	High	High chance that the design control will detect a potential cause/mechanism for the failure mode.
2	Very high	Very high chance that the design control will detect a potential cause/mechanism for the failure mode.
1	Almost certain	Design control will almost certainly detect a potential cause/mechanism for the failure mode.

utilities, and construction). The FMEA conducted by product design engineers is typically referred as DFMEA (the first letter "D" stands for design). And the FMEA conducted by the process designers is referred as PFMEA (the first letter "P" stands for process). In many automotive companies, product (or process) design release engineers are required to perform the task of creating the FMEA chart and to demonstrate that all possible failures with the RPN over a certain value are prevented.

FMEA is a proactive and qualitative tool used by quality, safety, and product/process engineers to improve reliability (i.e., eliminate failures – thus, improve quality and customer satisfaction). Development of an FMEA involves the following basic tasks:

a. Identify possible failure modes and failure mechanisms
b. Determine the effects or consequences that the failures may have on the product and/or process performance
c. Determine methods of detecting the identified failure modes
d. Determine possible means for prevention of the failures
e. Develop an action plan to reduce the risks due to the identified failures

The FMEA is very effective when performed early in the product or process development and conducted by experienced multifunctional team members as a team exercise.

The method involves creating a table with each row representing a possible failure mode of a given product (or a process) and providing information about the failure mode using the following columns of the FMEA table:

1. Description of a system, subsystem, or component
2. Description of a potential failure mode of the system, its subsystem, or component
3. Description of potential effect(s) of the failure on the product/systems, its subsystems, components, or other systems
4. Potential causes of the failure
5. Severity rating of the effect due to the failure
6. Occurrence rating of the failure
7. Detection rating of the failure or its causes
8. RPN. It is the multiplication of the three ratings in items 5, 6, and 7 above.
9. Recommended actions to eliminate or reduce the failures with higher RPNs
10. Responsibility of the persons or activities assigned to undertake the recommended actions and target completion date
11. Description of the actions taken
12. Resulting ratings (severity, occurrence, and detection) and RPNs (after the action is taken) of the identified failures in item 2 above

Examples of rating scales used for severity, occurrence, and detection are presented in Tables 3.1–3.3, respectively. The definitions of the scales generally vary between different organizations depending upon the type of industry, product or process, nature of the failures, associated risks to humans, and costs due to the failures.

An Example of a FMEA

A study of accidents during construction and operation of wind turbines was conducted by students in the author's course on "Risk Analysis in Energy Systems."

Many different types of accidents are possible during the construction and maintenance phases of wind turbine operations. Any time a structure as large as a wind turbine is being assembled, the transportation and proper placement of huge components such as the blades take careful planning and present great risks. Further, any time a worker must climb to the nacelle of a turbine for inspections or repair, a fall from that height of the system leads to the threat of personal injury or death.

The two most common accident varieties seen in Table 3.4, blade failure and fire, both have been seen to occur during the construction phase as well as the day-to-day operation of wind farms. These accidents deal with failures of the wind turbine system, and each accident can result from multiple causes. When blade failure occurs, pieces of a turbine blade or the entire blade separates itself from the nacelle and can fly great distances from the turbine's base. The threat of blade failure is one of the primary reasons that regulations exist as to the minimum distance between wind farms and residences. Almost any of the causes in Table 3.4 can lead to blade failure, from general wear and tear of the blade to a material defect that weakened its

TABLE 3.4

An Example of a FMEA

Item / Function	Potential Failure Mode	Potential Effect(s) of Failure	S = Severity Rating	Potential Cause(s)/ Mechanism(s) of Failure	O = Occurrence Rating	Current Design Controls	D = Detection Rating	RPN = Risk Priority Number	Recommended Actions	Responsibility and Target Completion Dates	Action Results				
											Actions Taken	SEV	OCC	DET	RPN
Rotor blades	Fracture: Separation of a part of the blade from the rest of the blade or the entire blade from the turbine head	Pieces of the blades / the whole blade acts as a projectile, jeopardizing the safety of those within a 1- mile radius	10	Corrosion	5	Anti-corrosion coating system, design rules regarding edges and weld seams, regulations around surface preparation before painting, routine regular maintenance	3	150	Conduct bench-level corrosion testing to refine preventative maintenance schedules	G. Jones/ Dec 2021	New fiber-glass material for blades	3	2	3	18
				Cyclical loading in the fasteners, resulting in fatigue failure	2	Industry standards for wind turbine fasteners, Computer- Aided Engineering durability analyses, prototype Design Verification testing	1	20	None						
				Manufacturing defect	2	Quality inspections, manufacturing standard processes, engineering sign-off stages, milestone and gateway reviews	1	20	None						

(Continued)

TABLE 3.4 (Continued)
An Example of a FMEA

Item / Function	Potential Failure Mode	Potential Effect(s) of Failure	S = Severity Rating	Potential Cause(s)/ Mechanism(s) of Failure	O = Occurrence Rating	Current Design Controls	D = Detection Rating	RPN = Risk Priority Number	Recommended Actions	Responsibility and Target Completion Dates	Actions Taken	Action Results			
												SEV	OCC	DET	RPN
				Lightning strike	2	Lightning receptors on the turbine blades, lightning grounding strategy built into turbine, surge protection on electronic systems	1	20	None						
	Ice throw: due to centrifugal forces, ice is thrown from the rotor blades and acts as a projectile	Pieces of ice seriously endanger anyone within close proximity to the wind turbine, damage to other turbine blades, damage to turbine head	10	Ice storm	5	Rotational speed sensors and integrated control systems, ice-phobic coatings for turbine blades, built-in blade heating systems (ex. electric foil)	3	150	Ensure all land within a 1000-ft radius of the turbine is not occupied by residential property	J. Green/ June 2020	Electrical deicing	6	4	2	48

(Continued)

TABLE 3.4 (Continued)
An Example of a FMEA

Item / Function	Potential Failure Mode	Potential Effect(s) of Failure	S = Severity Rating	Potential Cause(s)/ Mechanism(s) of Failure	O = Occurrence Rating	Current Design Controls	D = Detection Rating	RPN = Risk Priority Number	Recommended Actions	Responsibility and Target Completion Dates	Actions Taken	S E V	O C C	D E T	R P N
	Environmental failure: Rotor blade path interrupts native species' flight patterns	Protected species of birds are struck and killed by the wind turbine	7	Birds flying in / through the path of the rotor blade	10	Careful planning of the turbine's location, brighter blades, GPS tags on birds and corresponding shut-down systems on turbines, new blade shapes, lighting on the turbine at night	3	210	Increase utilization of GPS systems, particularly with endangered and protected birds						
	Transportation failure: damage to turbine / other property during transportation	Property damage, personal injury, lawsuits	8	Improper transportation route selected, driver not properly trained, hazardous road conditions	4	Government involvement in route planning, road closures to accommodate transportation, driver training, logistics teams	2	64	None						
Tower	Structural Failure: Tower uproots / collapses to the ground	Falling debris from tower seriously endanger anyone within close proximity to the tower	10	Vibration fatigue	2	Sensors and controls in place to detect and monitor vibration in main, yaw, and slew bearings, output shaft, and other rotating components	2	40	None						

(Continued)

TABLE 3.4 (*Continued*)
An Example of a FMEA

Item / Function	Potential Failure Mode	Potential Effect(s) of Failure	Potential Cause(s)/ Mechanism(s) of Failure	S=Severity Rating	O=Occurrence Rating	Current Design Controls	D=Detection Rating	RPN=Risk Priority Number	Recommended Actions	Responsibility and Target Completion Dates	Actions Taken	S E V	O C C	D E T	R P N
												Action Results			
			Earthquake		1	Foundational reinforcement, careful location planning	2	20	None						
			Installation error / defect		2	Turbines are heavily inspected prior to debut, regulations around physical installation methods	1	20	Ensure corroboration between any R&D projects into novel installation techniques against industry standards and regulations	J. Green/ Sept 2020					
			Unbalance rotor / hub assembly		2	Turbine blades are precision-balanced to ensure no centrifugal forces act on the turbine head	2	40	None						

(Continued)

TABLE 3.4 (*Continued*)
An Example of a FMEA

Item / Function	Potential Failure Mode	Potential Effect(s) of Failure	S = Severity Rating	Potential Cause(s)/ Mechanism(s) of Failure	O = Occurrence Rating	Current Design Controls	D = Detection Rating	RPN = Risk Priority Number	Recommended Actions	Responsibility and Target Completion Dates	Actions Taken	S E V	O C C	D E T	R P N
				Braking mechanism failure, resulting in excessive speeds, causing the head to come unhinged and ultimately collapsing the entire structure	2	Built-in redundancy between the rotor-lock system, electrical brakes, and mechanical brakes	1	20	Include brake inspections for all braking systems (including redundant systems) in preventative maintenance schedules						
Generator	Electrical Failure: Generator overloaded	Overheating of the generator, leading to permanent generator	7	High winds	3	Turbines specifically sized to their intended location	1	21	None						
		Overheating of the generator, leading to generator explosion	9	High winds	3	Turbines specifically sized to their intended location	1	27	None						

(Continued)

TABLE 3.4 (Continued)
An Example of a FMEA

Item / Function	Potential Failure Mode	Potential Effect(s) of Failure	S = Severity Rating	Potential Cause(s)/ Mechanism(s) of Failure	O = Occurrence Rating	Current Design Controls	D = Detection Rating	RPN = Risk Priority Number	Recommended Actions	Responsibility and Target Completion Dates	Actions Taken	Action Results S E V	O C C	D E T	R P N
Turbine Head	Mechanical Failure: Bearing malfunction	Gearbox / generator failure	7	Unbalanced shaft	2	Bearing data available for estimated lifetime under varying loading conditions; shaft precision-balanced; sensors monitor bearings	1	14	None						
				Poor lubrication	2	Oils / lubricants replaced regularly; sensors used to monitor oil / lubricants	1	14	None						
	Material Failure: Fire generated in the turbine head	Turbine head explodes, endangering anyone within close proximity	10	Faulty electrical wiring	2	Sensors monitor voltage signals; ground fault detection	1	20	None						

structure. Turbine fires also present one of the most common accidents in the wind energy industry. Lightning, failed electrical components, and faulty maintenance are all possible causes for fire taking hold in the nacelle. Modern wind turbines are generally outfitted with self-contained fire-fighting systems, but because of the concentration of flammable materials and constant air (oxygen) flow, flames are hard to contain once they have begun.

Table 3.4 presents an FMEA conducted to improve safety during operation of wind turbines.

FAILURE MODES AND EFFECTS AND CRITICALITY ANALYSIS

The Failure Modes and Effects and Criticality Analysis (FMECA) is very similar in format and content to the FMEA described above. It contains an additional column of criticality. The criticality column provides a rating illustrating level of criticality of the failure (in each row) in accomplishing the major goal (or mission) of the product. The technique is also called Failure Modes and Criticality Analysis (Hammer, 1980).

Criticality can be rated by using different scales for different products. The criticality ratings typically cover range from low criticality involving stoppage of equipment (requiring minor maintenance) to high criticality levels involving failures resulting in potential loss of life.

PROBLEMS IN RISK MEASUREMENTS

Assessing the risks involves identifying the hazards, assessing the potential consequences and the occurrence probability of such consequences. Identification of hazards is particularly difficult when both the potential customers and product (or system) uses are difficult to predict. Products or systems involving new technologies are also difficult to evaluate because very little failure data are generally available. It is especially hard to predict the risks during the early stages of product development when the product concept is also not fully developed.

The problems in risk measurements occur due many reasons. Most problems occur due to (a) lack of data on different types of hazards and risks, (b) subjectivity involved in identification and quantification of the data, and (c) differences in assumptions made during the design phases about how the customers or users will use the product vs. the actual uses of the product. The risk assessment models used in this area are therefore not precise. But they can be used as guides along with the recommendations of multiple experts and discussions between the decision-makers and the experts.

Subjective assessments of the three ratings (occurrence, severity, and detection) are also difficult and subject to several questions such as: Who would collect the data and conduct evaluations? Should the evaluations be conducted by experts, product safety advisory boards, teams, or individuals involved in the design process? Further, the level of understanding and awareness of risks varies considerably between different evaluators. Costs is also another problem in collecting failure-related data as the product tests are generally costly and funds are usually limited to undertake costly data collection studies. Failure testing is all very time-consuming as failure rate is generally very low.

There are trade-offs in the application of risk assessment methods between the consistency in the data, the level of details related to the outcomes, and the time and resources (particularly human and financial) required for the analyses. Apparently simple methodologies may contain implicit weightings that may not be appropriate for every product being assessed. Judgments may be intuitive, based on implicit assumptions, especially in relation to the boundaries between categories (or ratings). Taken together, these factors can result in a high degree of subjectivity in risk assessment, although the subjectivity can be reduced by the extent of guidance provided to the assessors in applying the various scales and ratings. In general, the potential for inconsistency in the results will be directly related to the amount subjectivity involved in the risk measurement process.

CONCLUDING REMARKS

Conducting risk analyses before undertaking a project is important as the analyses uncover potential risks and force the decision-makers to find ways to reduce risks and costs associated with them. The risk analyses also educate the technical and management teams about possible risks and prepare them to deal with the issues before they occur. It also improves the reliability of the system being developed and thus will improve customer satisfaction with the system after it is made operational. Including the results of the risk analyses in the financial analyses reduces surprises and provides the decision-makers in considering more realistic assessment of costs and prediction of financial results.

REFERENCES

Bhise, V.D. 2014. *Designing Complex Products with Systems Engineering Processes and Techniques.* ISBN: 978-14665-0703-6. Boca Raton, FL: CRC Press.

Conrad, E., Feldman, J. and S. Misenar. 2017. *Eleventh Hour CISSP®.* 3rd Edition. ISBN-13: 978-0128112489. Cambridge, MA: Elsevier.

Environmental Protection Agency (EPA). 2014. National Air Toxic Assessments. Website: https://www.epa.gov/national-air-toxics-assessment/2014-nata-assessment-results#emissions (Accessed: September 13, 2020).

Environmental Protection Agency (EPA). 2020. Clean Air Act (CAA). Website: https://www.epa.gov/laws-regulations/summary-clean-air-act (Accessed: September 14, 2020).

Federal Highway Administration. 2007. Risk Management. Website: https://www.fhwa.dot.gov/cadiv/segb/views/document/sections/section3/3_9_4.cfm (Accessed: September 14, 2020).

Hammer, W. 1980. *Product Safety Management and Engineering.* Englewood Cliffs, NJ: Prentice-Hall, Inc.

Heath Effects Institute. 2019. *State of Global Air 2019.* Boston, MA: The Health Effects Institute and the Institute for Health Metrics and Evaluation's Global Burden of Disease Project, Special Report. ISSN 2578-6873.

National Safety Council. 2017. *Injury Facts.* 2017 Edition. Itasca, IL: National Safety Council.

4 Systems Engineering in Energy Systems Projects

INTRODUCTION

WHY SYSTEMS ENGINEERING?

Most energy systems such an electric power generating plant are complex systems. Complex systems have many systems, and the systems have many lower-level systems such as subsystems and sub-subsystems and components. For example, a fossil-fuel-fired power plant system will have lower-level systems such as boiler, water injecting system, fuel loading system, combustion system, steam turbine, generator, plumbing system, wiring system, step-up transformer, and so forth. Systems Engineering (SE) is essential for efficient designing of complex systems (or products) that involve many systems. Implementation of systems engineering during the life cycle (from system conceptualization to its disposal) of any complex system helps in reducing delays and cost overruns and in creating the "right" system for the customers.

Design of many entities (systems, subsystems, and components) and interfaces (e.g., joints between entities) requires applications of SE techniques and processes involving many steps and iterations. For example, in designing a fossil-fuel-fired power plant, it is essential that the team designing the boiler constantly communicates with the team designing the steam turbine. Otherwise, the output steam generated by the boiler may not be delivered to the turbine at the optimal flow rate and pressure. The SE work thus involves coordinating the work of many multidisciplinary design teams involved in designing all the entities (subsystems and components) to ensure that the resultant whole system will produce the highest possible performance (i.e., greatest kW of power at the highest possible efficiency [kW/lbs of fuel] level). Thus, without the coordination between different design teams designing different subsystems and components, there could be chaos in the design office resulting into substantial delays to decide on appropriate trade-offs between different characteristics of subsystems in the whole system (i.e., the power plant).

The SE work thus involves:
a. Developing the whole system (or product) level requirements first by understanding customer needs, business needs, and government regulations
b. Maintaining constant focus on customer satisfaction
c. Developing an overall concept of the whole system
d. Determining functions of the whole system and cascading (i.e., allocating) requirements of the whole system level to its lower-level systems and components

DOI: 10.1201/9781003107514-5

e. Designing interfaces (i.e., joints and connections) between all systems and components to ensure functionality of the whole system
f. Conducting extensive design reviews of all entities within the whole system and trade-offs between attributes (e.g., capacity, space, configuration, fuel needs, and pollution potential) of the whole system
g. Conducting verification tests at each lower to higher-level systems (i.e., from components to sub-subsystems to subsystems to the whole system)
h. Conducting validation tests on the whole system to confirm that the designed system is the "right" system for its customers.
i. Providing technical and management support throughout the life cycle of the whole product

The systems engineers help manage the above work by creating a systems engineering management plan (SEMP). The SEMP is a document that describes how different design activities involving integration of many considerations and tasks (e.g., trade-offs between attributes, design reviews, verification testing) must be done iteratively to ensure that a balanced product is created while maintaining the required system configuration, project schedule, and spending in line with its budget.

WHAT IS SYSTEMS ENGINEERING?

SE is a multidisciplinary engineering decision-making process involved in designing complex systems or products. The SE activities involve both the technical activities and management activities from an early design stage of a system to the end of the life cycle of the system.

SE begins with understanding customer needs, developing customer requirements, and developing the concept of the system. SE is an iterative process. SE focuses on defining customer needs and required functionality early in the development cycle, documenting requirements, then proceeding with design synthesis and system validation while considering the complete problem.

SYSTEMS ENGINEERING DEFINITION

SE is the treatment of engineering design as a decision-making process involving people, organizations, processing, and technology. Like all other specialized engineering fields, systems engineers are constantly required to make many decisions regarding the system that they are involved in designing, improving, and/or operating. They work in teams and coordinate activities of team members from various different specialized disciplines such as electrical engineering, mechanical engineering, civil engineering, chemical engineering, environmental engineering, mining and drilling engineering, financial analysts, and so forth, so that the required technical activities are conducted at the right time to ensure that the right systems with the right characteristics are designed to satisfy the customers.

SE is both a technical and a management process. The technical process is the analytic effort necessary to transform the customer needs into specifications and requirements for the desired system and then designing the system of the proper size and configuration. The management process involves assessing costs and risks, integrating the engineering specialties and design groups, maintaining configuration control, and continuously auditing the effort to ensure that the costs, schedule, and technical performance objectives are satisfied to meet the original operational need.

Depending up on the amount of the engineering work and the overall cost of the project, budgets are allocated for various engineering activities. The budget of the SE activities will therefore affect the level of details such as the tasks to be accomplished, their schedule, and team members to be required to participate in each of the tasks. And hence, SE may be implemented differently in different organizations and in different system development programs.

SYSTEMS ENGINEERING APPROACH AND PROCESS

The SE approach in solving a problem, such as to develop a new power plant, a power distribution network, a wind turbine, or a solar plant, will begin with systematic implementation of the SE process (see Figure 4.1) based on the following SE considerations:

1. *Customer focus*: The SE work begins with understanding the needs of the customers to be satisfied by developing the system (product), and it ends with obtaining feedback on how well they liked and are satisfied with the system after they begin using the system.
2. *Requirements analysis*: The systems engineers help translate the customer needs into requirements on the system to be developed. The requirements are developed based on careful understanding and analysis of the customer needs. The customers to be satisfied are a) the external customers who purchase the outputs of the designed system, (b) government organizations whose requirements on the system must be complied, and (c) company management who decides and approves the program to develop the proposed system. The requirements should be also developed by considering inputs and design considerations of all the team members representing different disciplines (departments representing different specialized activities). It is very important to realize that the system development work cannot and should not begin without a carefully documented specifications and requirements on the proposed system. Otherwise, there will not be an agreement between different departments or disciplines on what system to develop, the characteristics (or attributes) of the system, and performance levels to be achieved by the system.
3. *Functional analysis and allocation*: As the requirements on the proposed system to be developed are being analyzed, the next tasks are to determine functions to be performed by each system and allocation of the functions to be performed by each of the lower-level systems (such as the subsystems of the system, sub-subsystems of the subsystems, and so forth, until the component level). These functional analysis and requirements allocation tasks to lower-level system are called the "cascading of the requirements from higher level system to lower level system." The requirements cascade ensures that lower-level systems are developed only to satisfy the requirements on its higher-level system. And the requirements on each level of

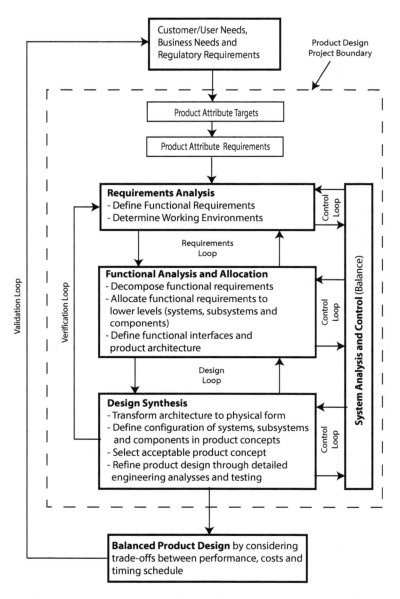

FIGURE 4.1 Systems engineering process showing the basic steps and five feedback loops.

system are then used to develop the systems at that level. Thus, the design process is strictly requirements driven at each system level. To ensure that many systems (or entities) work (i.e., function) with other systems or depend upon the operation of other systems, requirements on interfaces (e.g., joints, couplings, links, transfer of forces/pressures/energy, fluids, gases, data, signals, etc.) between the systems must be also developed.

4. *Design synthesis*: Design of the system and all lower-level entities of the system must be developed by meeting all the requirements developed earlier (in requirements analysis step). It involves synthesis of all design and engineering ideas and considerations and transforming them into possible configurations (or concepts) of the whole system with selected location and space allocated to each entity within the whole system. The synthesis work of integrating all requirements and translating those into a physical form of the system that will work and produce the required system output is very challenging and creative work. The work is done iteratively by exploring many possible alternate configurations of various higher and lower-level entities within the whole system. The output of the synthesis work is considered to be a "balanced design" of the whole system in which trade-offs between different designs, their attributes, configurations, functions, requirements, and functional outputs are carefully analyzed by specialists and experts from different disciplines all working together and constantly communicating their specialized needs and requirements.

5. *Feedback loops*: To ensure that the outputs of above-described activities provide a feasible and balanced configuration and design of the proposed system, the following five types of feedback loops are conducted simultaneously during the development process:

   a. *Requirements loop*: The requirements loop ensures that developed requirements on each entity are iteratively evaluated and modified to ensure that they are nonconflicting, feasible and produce an overall superior design.

   b. *Design loop*: The design loop ensures that the design is improved iteratively to come up with the best possible combinations of technological changes and efficient manufacturing and assembly considerations.

   c. *Control loops*: The control loops facilitate in taking advantage of continuous improvements in requirements, functional analysis and allocation, and design synthesis through systems analysis and control activities throughout the system development process.

   d. *Verification loop*: The verification loop ensures that tests (simulations and/or physical) are conducted to verify that the design and entities assembled from early production components meet their stated requirements.

   e. *Validation loop*: The validation loop allows testing of the whole system before final release (i.e., after production and construction) to evaluate if the right system was developed for its customers. Therefore, the validation tests should involve end users or customers to participate in the evaluation process.

AN ENERGY SYSTEM EXAMPLE: A WIND FARM

A wind farm system has the following subsystems: (a) wind turbines, (b) transformers, (c) internal access roads, (d) transformer station, and (e) transmission system connecting the facility to the national grid. Each of these subsystems has many sub-subsystems. For example, the wind turbine subsystem will have the following

sub-subsystems: tower, electrical wiring, transformer, turbine, shafts, gears, generator, computer control system, wind speed and direction sensing system, and so forth.
The sub-subsystems would be interfaced to ensure functioning of the turbine, i.e., the
energy captured by the turbine is transferred to the generator through the shafts and
gears in the gear box.

Design work begins with development of early specifications of the system such
as the desired level of power output, possible locations, wind direction and speed
surveys, followed with the steps such as allocation of functions and requirements.

Designing the complex system requires multifunctional team involving civil engineers, structural engineers, mechanical engineers, materials engineers, electrical/
electronics engineers, aerodynamic engineers, and so forth – all involved in tasks
such as designing the turbines, computer and control systems, wind speed and direction sensors and controlling the orientation of the turbine and the pitch of its blades,
transformers, and transmission lines.

Many factors need to be considered during the concept design of a new wind
farm. They include technical factors, economic factors, environmental factors, and
social factors. These factors are briefly described below.

The technical factors that affect the decision-making on-site selection include
wind speed, land topography and geology, grid structure, turbine size and spacing
between turbines. Economic factors to be considered are capital costs, land costs,
operation and maintenance costs, electricity market, incentives, and jobs creation.
Other environmental factors include visual impact, electromagnetic interference,
wildlife and endangered species, and noise impact on the surrounding communities. As a renewable energy source, wind farm does not cause reduction in natural
resources. As a result of having no input other than wind, there is no formation of
emission during the energy generation process. Wind turbines can generate noise
while they are working, and their image can be incompatible with the general view of
the region. Wildlife and endangered species could be disturbed during the construction of wind farm.

Social factors that affect the selection of a site include public acceptance, distance
from residential area, and alternative land use options of candidate wind farm site.
Some regulatory procedures may set restrictions or incentives to apply the wind farm
projects. Public acceptance is vital for the application of that kind of projects. Public
may oppose projects because of possible environmental or social effects. Distance
from residential area gains importance not to interfere with social life during wind
farm construction and operation. Other requirements resulting from regulatory
restrictions include some national or international-level regulations related with the
construction and operation of wind farms. These regulations must be explored before
evaluating the sociopolitical position of a wind farm project.

During the concept development phase, many design iterations would be conducted to ensure that all lower-level systems (e.g., tower structure, wind turbine
blades, shafts and gears, generator, transformers, and wiring systems) of the wind
turbine are developed to ensure that they meet the requirements of the whole wind
farm. The SE "V" model presented in the next section (Figure 4.2) shows how the
top-down design from concept level down to components can be designed to ensure
that the design synthesis shown in Figure 4.1 is accomplished.

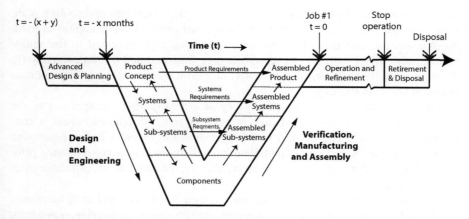

FIGURE 4.2   Systems engineering model showing life cycle activities of a new wind turbine from the advanced design and planning activities to the disposal of its production facilities.

SYSTEMS ENGINEERING "V" MODEL

The SE "V" model presents all important steps in the life cycle of a product or a system. The model is presented in Figure 4.2. The model is known as the SE "V" Model because the steps are arranged in a "V" shape with succeeding steps shown below or above the preceding steps and staggered in time (see Bhise (2014) and Blanchard and Fabrycky (2011) for more details). The model is described below in the context of development of a new wind turbine.

The model shows basic steps of the entire wind turbine program on a horizontal time axis, which represents time (t) in months before Job#1. "Job#1" is defined as the event when the first turbine is installed at the work site and begins producing electric power. The wind turbine program generally begins many months prior to the Job#1. The beginning time of the program depends upon the scope and complexity of the program (i.e., the changes in the new turbine as compared with the earlier version) and the state of management's approval to begin the turbine development process.

LEFT SIDE OF THE "V" – DESIGN AND ENGINEERING

In the early stages prior to the official start of the wind turbine program, an advanced design and product planning activity (which usually involves an advanced wind turbine planning department or a special new turbine planning team) determines the wind turbine specifications (i.e., characteristics and its preliminary architecture such as tower type, size, blade material and geometry, type of powertrain, performance characteristics, and so forth). It also provides a list of reference and competitors' wind turbines (used for benchmarking) that the new wind turbine may replace or compete with. A small group of engineers and designers (usually about 8–10) from the advanced design group are selected and asked to generate a few early wind turbine concepts to understand design and engineering challenges. A business plan including the projected sales volumes, the planned life of the new wind turbine, the wind turbine program timing plan, facilities and tooling plan, manpower plan, and financial plan (including estimates of costs, capital needed, revenue stream, and projected profits) is developed and presented to the senior management along with all other product programs planned by the company (to illustrate how the proposed program will fit within the overall corporate product plan and business strategy).

The wind turbine program typically begins officially after the approval of the business plan by the company management. This program approval event is considered to occur at x-months prior to Job#1 as shown in Figure 4.2. The figure also shows that the advanced design and planning activity begins at $(x+y)$ months prior to Job#1. (Depending upon the scope of the activity, the value of "y" can range from about 3 to 8 months.)

At minus x-months, the chief program manager is selected and each functional group (such as design, structural engineering, aerodynamics engineering, powertrain engineering, electrical engineering, packaging and ergonomics/human factors engineering, manufacturing engineering) within the product development and other related activities (e.g., engineers from the utility companies) are asked to provide personnel to support the wind turbine development work. The personnel are grouped into teams, and the teams are organized to design and engineer the wind turbine and its systems and subsystems.

The first major phase after the team formation is to create an overall wind turbine concept (labeled as "Product Concept" in Figure 4.2). During this phase, the designers and the package engineers work with different teams to create the wind turbine concept, which involves (a) creating early drawings or CAD models of the proposed wind turbine, (b) creating computer generated 3-D life-like images and/or videos of the wind turbine, and (c) physical scale model/mock-up of the wind turbine. The images and/or models of the proposed turbine are shown to prospective customers and to the management. Their feedback is used to further refine the product concept.

As the wind turbine concept is being developed, each engineering team decides on how each of the wind turbine systems can be configured to fit within the allocated space (defined by the exterior envelope and interior surfaces of the wind turbine concept – primarily by the nacelle, tower and turbine hub, and blades) and how the various systems can be interfaced with other systems to work together to meet all the functional, safety, and other requirements of the product. This step is shown as the "Systems" step in Figure 4.2. Any problems discovered during this phase may require iterating the process back (to previous phase) to refine or modify the product concept. (This feedback represents the up arrow from the systems to product concept shown in Figure 4.2.)

As the systems are being designed, the next phases involve a more detailed design of the lower level entities, i.e., design of subsystems of each system and components within each of the subsystems. These subsequent steps, straddled in time to the right, are shown as "Subsystems" and "Components."

The above steps form the left half of the "V" representing the time and activities involved in "Design and Engineering." The up arrows in the left half of the "V" indicate the iterative nature of the SE loops covered earlier in Figure 4.1.

Right Side of the "V" – Verification, Manufacturing, and Assembly

The right half of the "V," moving from the bottom to the top, involves manufacturing of components (or lower level entities) and testing to verify that they meet their functional characteristics and requirements (developed during the left half of the "V"). The components are assembled to form subsystems, which are tested to ensure that they meet their functional requirements. Similarly, the subsystems are assembled into systems and tested; and finally, the systems are assembled to create the whole system. At each of the steps, the corresponding assemblies are tested to ensure that they meet the requirements considered during their respective design steps (i.e., the assemblies are verified). These requirements are shown as the horizontal arrows from the left side to the right side of the "V" in Figure 4.2.

The right side of the "V" is thus labeled as "Verification, Manufacturing and Assembly." It should be noted that down arrows between various assembly steps in the right half of "V" are not shown in Figure 4.2. The down arrows would indicate failures in the verification steps. When failures occur, the information is transmitted to the respective design team to incorporate design changes to avoid repetition of such failures.

The engineers and technical experts assigned to various teams in the wind turbine program work throughout all the above steps and continuously evaluate the wind turbine design to verify that the needs of the turbine users are accommodated, and they will be able to use the wind turbine under all foreseeable usage situations. Early production wind turbines developed just before Job#1 are usually used for additional whole product evaluations for the product validation purposes.

Right Side of the Diagram – Operation and Disposal

After Job#1, the wind turbines are produced and transported directly to the customer designated installation site. The model diagram in Figure 4.2 also shows a period called "Operation and Refinement." During this period, the produced wind turbines are purchased, installed, maintained, and operated by the customers and serviced by the manufacturer (or other service providers). The wind turbines may also be refined with some changes (i.e., revised with minor changes during the existing model cycle or updated as a refreshed new model every few years) during its operational time. When the wind turbine design becomes old and outdated, it is pulled from the market. This marks the end of the production of the wind turbines. At that point, the assembly plant and its equipment are recycled or retooled for the next wind turbine model. As the wind turbines reach the end of their useful life, they are sent to the scrap yards where many of the components may be disassembled. The disassembled components are either recycled for extraction of the materials or sent to the junk yards.

The above SE "V" model thus shows the locations of the "Design and Engineering" and "Verification, Manufacturing and Assembly" tasks on the time axis along as the left and right sides of the "V" and "Operation and Refinement" and "Retirement and Disposal" tasks on the right side of the "V." The model also shows how the requirements developed during the design process are used for verification of the components, subsystems, systems, and the whole system, as they are assembled. The design process uses the "top-down" approach, i.e., it begins with the development of the product concept on the top left and ends at the top of the "V" on the right side with the assembled and product ready to be installed in a wind farm.

SYSTEMS ENGINEERING MANAGEMENT PLAN (SEMP)

The SEMP is a higher-level plan (not very detailed) for managing the SE effort to produce a final operational product (or a system) from its initial requirements. Just as a project plan defines how the overall project will be executed, the SEMP defines how the engineering portion of the project will be executed and controlled. The SEMP describes how the efforts of system designers, test engineers, and other engineering and technical disciplines will be integrated, monitored, and controlled during the complete life cycle of the product. In other words, the SEMP describes what each team (or department) needs to do and when to achieve the product program objectives. During the development for a complex system (or product) that has many systems and subsystems, various teams designing different systems need to be coordinated in terms of important events (e.g., milestones or gateways such as beginning and end phases, design reviews and management approval and sign-offs) during the timeline of the program. SEMP makes sure that all teams know upcoming events and communicate and resolve their design problems and issues with other teams to ensure that the product being developed meets all product-level as well as the systems-level requirements.

Figure 4.3 presents a flowchart illustrating the relationship of the SE management plan to project work and project management. The SEMP thus uses the information on project work and specifies all major engineering activities in terms of details such as what needs to be performed, how, and when.

For a small project, the SEMP might be included as part of the project plan document, but for any project or program of a greater size or complexity, a separate document is recommended. The SEMP provides the communication bridge between the project management team and the technical teams. It also helps coordinate work between and within different the technical teams (i.e., teams designing different systems of the product such as turbine, gear box, generator, control system – for a wind turbine). It establishes the framework to realize the appropriate work (or tasks to be performed) that meets the entry and success criteria of the applicable project phases. The SEMP provides management with necessary information for making SE decisions (e.g., when to proceed to the next phase of the program). It focuses on requirements, cascading of the product-level requirements down to lower-level entities, design, development (detailed engineering), test, and evaluation. Thus, it addresses traceability of stakeholder requirements and provides a plan to ensure that the right product (or system) will be developed during the entire project.

CONTENTS OF SEMP

The purpose of this section is to describe the activities and plans that will act as controls on the project's SE activities. For instance, this section identifies the outcomes of each SE activity, such as documentation, meetings, design reviews, and design

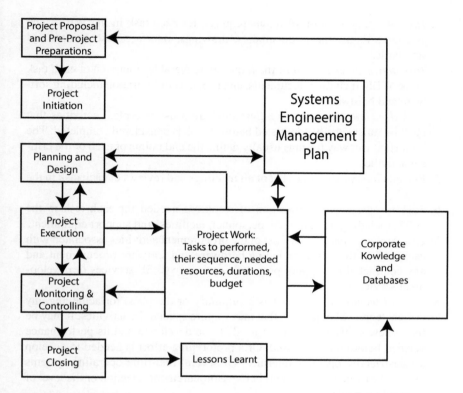

FIGURE 4.3 Relationship of the systems engineering management plan (SEMP) to project work and project management.

approvals. This list of required outcomes will control the activities of the team and thus will ensure the satisfactory completion of the activities. Some of these plans may be defined in detail in the SEMP. For other plans, the SEMP may only define the requirements for a particular plan. The plan itself may include detailed plan for an SE activity, e.g., a verification plan or a validation plan. Almost any of the plans described below may fall into either category. It all depends on the complexity of the project (or program) and the amount of up-front SE work that can be done at the time the SEMP is prepared.

The first set of required activities relates primarily to the successful management of the project. These activities are likely to have already been included in the project/ program plan but may need to be expanded in the SEMP (USDOT, 2007). Generally, they are incorporated into the SEMP but, on occasion, may be developed as separate documents. The items that can be included in the SEMP are listed below. The items and their descriptions provided in USDOT (2007) were modified to meet the needs of complex product development.

1. *Work breakdown structure* [WBS] consists of a list of all tasks to be performed on a project, usually broken down to the level of individually budgeted items.

2. *Task inputs* is a list of all inputs required for each task in the WBS, such as source requirements documents, drawings, interface descriptions, and standards.

3. *Task deliverables* is a list of the required deliverables (outputs) of each task in the WBS, including documents, and product configuration including software and hardware.

4. *Task decision gateways (or milestones)* are a list of critical activities that must be satisfactorily completed before a task is considered completed. The important gateway timings usually define the end points of each of the critical activities.

5. *Reviews and meetings* is a list of all meetings and reviews of each task in the WBS.

6. *Task resources* is identification of resources needed for each task in the WBS, including, for example, personnel, facilities, and support equipment.

7. *Task procurement plan* is a list of the procurement activities associated with each task of the WBS, including hardware and software procurement and any contracted or supplier provided services (e.g., SE services or development services).

8. *Critical technical objectives* is a summary of the plans for achieving any critical technical objectives that may require special SE activities. It may be that a new software algorithm needs to be developed and its performance verified before it can be used. Or a prototyping effort is needed to develop a user-friendly operator interface. Or several real-time operating systems need to be evaluated (verified) before a procurement selection or the level of assembly is made.

9. *Systems engineering schedule* a schedule of the SE activities that shows the sequencing and duration of these activities. The schedule should show tasks [at least to the level of the WBS], deliverables, important meetings, and reviews (including lists of key attendees and representation of departments), and other details (e.g., timings and requirements to be met) needed to control and direct the project. The SE schedule is an important management tool. It is used to measure the progress of the various teams working on the project and to highlight work areas that need management intervention.

10. *Configuration management plan* describes the development team's approach and methods to manage the configuration of the systems within the products and processes. It will also describe the change control procedures and management of the system's baselines as they evolve.

11. *Data management plan* describes how and which data will be controlled, the methods of documentation, and where the responsibilities for these processes reside. The data should include product design (e.g., CAD models, dimensional data, and manufacturing processes and materials selected for each component), schedules of different events (e.g., reviews, tests), results of tests, costs, communications, and so forth.

12. *Verification plan is always required*. This plan is written along with the requirements specifications. However, the parts of tests to be conducted can be written earlier (e.g., included in the systems design standards). The

verification procedures are generally developed by the core engineering experts, and they define the step-by-step procedures to conduct the verification tests.

13. *Validation plan* is required. It ensures that the product being designed is the right product and would meet all the customer needs. It describes the test procedures and data to be collected using customers/users as test subjects for the product.

The second set of plans can be designed to address specific areas of the SE activities. They may be included entirely in the SEMP, or the SEMP may give guidance for their preparation as separate documents. The plans included in the first set listed above are generally universally applicable to any project. On the other hand, some of the plans included in this second set are required on an as-needed basis. The unique characteristics of a project will dictate their need. For a complex product, such as an automobile, a power plant, or a wind turbine, many of these second set plans are required items. These items are described below. The items and their descriptions provided in USDOT (2007) were modified to meet the needs of complex product development.

1. *Software development plan* describes the organization structure, facilities, tools, and processes to be used to produce the project's software. It also describes the plan to produce custom software and procure commercial software products.
2. *Hardware development plan* describes the organization structure, facilities, tools, and processes to be used to produce the project's hardware. It describes the plan to produce custom hardware (if any) and to procure commercial hardware products.
3. *Technology plan* (if needed) describes the technical and management process to apply new or untried technology. Generally, it addresses technical risks, performance criteria, assessment of multiple technology solutions, and fallback options to existing technology.
4. *Interface control plan* identifies all important interfaces within and between systems (within the product and external to the product) and identifies the responsibilities of the organizations on both sides of the interfaces.
5. *Technical review plan* identifies the purpose, timing, place, presenters and attendees, topics, entrance criteria, and the exit criteria (resolution of all action items) for each technical review to be held for the project/program.
6. *System integration plan* defines the sequence of activities that will integrate various product chunks involving components (software and hardware), subsystems, and systems of the product. This plan is especially important if there are many subsystems, and systems are designed and/or produced by different development teams from different organizations (e.g., suppliers).
7. *Installation plan or deployment plan* describes the sequence in which the parts of the product are installed (deployed). This plan is especially important if there are multiple different installations at multiple sites. A critical part of the deployment strategy is to create and maintain a viable operational capability at each site as the deployment progresses.

8. *Operations and maintenance plan* defines the actions to be taken to ensure that the product remains operational for its expected lifetime. It defines the maintenance organization and the role of each participant. This plan must cover both hardware and software maintenance.
9. *Training plan* describes the training to be provided for both maintenance and operation personnel.
10. *Risk management plan* addresses the processes for identifying, assessing, mitigating, and monitoring the risks expected or encountered during a project's life cycle. It identifies the roles and responsibilities of all participating organizations for risk management.
11. *Other plans* that might be included are, for example, a safety plan, a security plan, and a resource management plan.

This second list is extensive and by no means exhaustive. These plans should be prepared when they are clearly needed. In general, the need for these plans becomes more important as the number of stakeholders and systems involved in the project increases.

The SEMP must be written in close synchronization with the project plan. Unnecessary duplication between the project plan and the SEMP should be avoided. However, it is often necessary to put further expansion of the SE effort into the SEMP even if they are already described at a higher level in the project plan.

CHECKLIST FOR CRITICAL INFORMATION

The USDOT (2007) guide also provides a checklist to assure that the SEMP includes the following:

1. Technical challenges of the project
2. Description of the processes needed for requirements analysis
3. Description of the design processes and the design analysis steps required for an optimum design
4. Identification and documentation of any necessary supporting technical plans, such as a verification, an integration, and a validation plan
5. Description of stakeholder involvement when it is necessary
6. Identification of all the required technical staff and development teams, and the technical roles to be performed by the system's owner, project staff, stakeholders, and the development teams
7. Description of the interfaces (or interactions) between the various development teams

ROLE OF SYSTEMS ENGINEERS

The role of the systems engineers assigned to the program is essentially to do what is needed to implement the SE process. A carefully developed SEMP will provide a clear roadmap for the systems engineers. They should work closely with all other team members, technical and program planning departments, to ensure that all basic SE steps are followed (see Figures 4.1 and 4.2).

The systems engineers will usually play the key role in leading the development of the product and/or system architecture, defining and allocating requirements, evaluating design trade-offs, balancing technical risk between systems, defining and assessing interfaces, and providing oversight on verification and validation activities. The systems engineers will usually have the prime responsibility in developing many of the project documents, including the SEMP, requirements/specification documents, verification and validation documents, certification packages, and other technical documentation (NASA, 2007).

SE is about trade-offs and compromises, about generalists rather than specialists. SE is about looking at the "big picture" and not only ensuring that they get the design right (meets requirements) but that they get the right design. Thus, a system engineer needs to perform the following tasks:

1. Understand customer and program needs
2. Obtain required data
3. Develop SEMP
4. Communicate the SEMP to program teams
5. Provide recommendations to program teams on SE tasks
6. Assist teams in conducting necessary trade-off analyses
7. Continuously communicate with program teams to perform above tasks

VALUE OF SYSTEMS ENGINEERING MANAGEMENT PLAN

A carefully developed and well-executed SEMP will enable proper implementation of SE during the program, i.e., all the SE steps from obtaining customer needs to the product validation in the product development and subsequent steps during the product operations and disposal stages are completed by the program teams in a timely manner.

The value of the SEMP can be summarized as follows:

1. It will facilitate reducing the risk of schedule and cost overruns and will increase the likelihood that the SE implementation will meet the user's needs.
2. It will engage right specialists at the right (needed) time (because they will know what needs to be done) and make sure that the design team members perform the right tasks (e.g., communicating their designs and issues to other team members, conducting analyses or tests); thus, resulting in improved stakeholder participation.
3. The product team will be more adaptable, and the developed products and systems will be resilient and meet customer needs.
4. All entities within the product will be verified for functionality, and thus the product should have fewer defects.
5. The experience gained and lessons learned during the implementation of the SEMP can be used to create improved SEMP documentation for the next program.

ADVANTAGES OF MANAGEMENT BY SYSTEM ATTRIBUTES

In SE applications, the product requirements are usually written in terms of attribute requirements. The product attributes are the characteristics of the product that the customers include in their wants and needs list. Meeting these product attribute requirements will help sell the product to its intended customers. Thus, many product design organizations that develop complex products develop lists of product attributes and their lower-level sub-attributes. They prepare detailed descriptions of requirements on each of the product attributes and cascade the product-level attribute requirements to sub-attributes requirements and apply them to systems and their lower-level entities (e.g., subsystems). This cascading of requirements ensures that the product possesses the required attributes, and the systems and lower-level entities of the product are designed to support all the functions that the product must possess.

Thus, the advantages of managing by product attribute requirements (and the cascading process described above) are as follows:

1. It helps everyone in the system development process to understand the traceability of the requirements on the system attributes (i.e., any requirement can be traced back to one or more product attributes).
2. People specialized in an attribute can be made responsible to ensure that the system is being designed to meet the attribute requirements.
3. It helps to ensure that all the system attributes are studied at every system development phase, and the compliance to the attribute requirements is reviewed at all major milestones in the system program.

The implementation of SE, thus, helps efficient development of complex products and complex systems through meeting all applicable requirements related to attributes such as performance, quality, safety, costs (budget), and timings. And most importantly, it helps creating the "right" (product or system) for its customers.

To reduce the workload of creating and implementing the attribute requirements, many product development organizations create and continuously update their system design specifications (design manuals), product development process documentation, and test procedures manuals. The information contained into these predeveloped manuals can be adapted quickly with minimum number of changes in supporting SE activities of new product programs.

CONCLUDING REMARKS

SE is a disciplined approach that begins with the understanding of the customer needs of the system being designed and ends with the disposal of the system. Thus, it considers the entire life cycle of the system. It also using the top-down approach from designing the system at its concept level and then designing lower-level systems through systematic decomposition of the whole system until all components of the system are developed. The SE approach also includes the bottom-up approach in assembly and verification of entities by ensuring that the requirements on each entity developed during the design are verified. Implementation of SE approach helps in creating the right product for its customers with reduced costs and time overruns.

REFERENCES

Bhise, V.D. 2014. *Designing Complex Products with Systems Engineering Processes and Techniques.* ISBN: 978-1-4665-0703-6. Boca Raton, FL: CRC Press, Taylor and Francis Group.

Blanchard, B. S. and W. J. Fabrycky. 2011. *Systems Engineering and Analysis.* Upper Saddle River, NJ: Prentice Hall PTR.

National Aeronautics and Space Administration (NASA). 2007. *NASA Systems Engineering Handbook.* Report no. NASA/SP-2007-6105 Rev1. NASA Headquarters, Washington, DC. 20546. http://ntrs.nasa.gov/archive/nasa/casi.ntrs.nasa.gov/20080008301_2008008500.pdf (accessed October 15, 2012).

U.S. Department of Transportation (USDOT)-Federal Highway Administration and Federal Transit Administration. 2007. *Systems* Engineering for Intelligent Transportation Systems: An Introduction to Transportation Professionals. Report no. FHWA-HOP-07-069, January 1, 2007. Website: https://rosap.ntl.bts.gov/view/dot/42529 (Accessed: September 14, 2021).

5 Safety Engineering in Energy Systems

INTRODUCTION AND BACKGROUND

DEFINITION OF SAFETY ENGINEERING

Safety Engineering is a specialized engineering field that deals with application of multidisciplinary concepts and techniques to design and evaluate products, systems, and processes with the primary objective of improving safety (i.e., preventing accidents and injuries) and providing healthful working environments (i.e., reducing harmful polluting substances).

SAFETY PROBLEMS IN ENERGY SYSTEMS

Accidents in energy systems field have a long history due to working in mines and oil fields and extracting of fuels, such as coal, oil, and natural gas. Many of the accidents inloved fires and explosions. Renewable energy sources also have their accidents, such as dam collapses at hydro electric plants, well blowouts at geothermal plants, fires inside wind turbines, and ice buildups on wind turbine blades causing blade breakages. Table 5.1 presents examples of accidents that have occurred during use of different energy sources.

Accidents are mostly caused by unsafe acts committed by people and/or unsafe working conditions at energy processing plants or distribution locations. The accidents can cause injuries and losses due to damages to the equipment and properties. The accidents can also lead to work stoppages and inefficiencies in work processes. Many research studies involving analyses of the accident data and accident causation factors have shown that most accidents are preventable (Heinrich, Petersen and Roos, 1980). Developers of systems, product/equipment manufacturers, and operators of the equipment incur safety costs due to (a) medical service for the injured persons, (b) repairing damaged equipment, (c) litigations and liabilities resulting from the accidents caused by their systems/products, (d) extra efforts undertaken to create safer systems/products through accident prevention actions (e.g., incorporation of safety features in the processing equipment or products, conducting special design reviews and tests during their development to ensure that the systems/products are safe), and (e) accident occurrences during the life cycle of the systems/products. The safety-related costs of a system developer or a product manufacturer can amount to about 5%–15% of the revenues generated by the products (i.e., adding all the costs associated with: conducting safety-related literature surveys, benchmarking systems and processes, safety analyses, safety reviews, purchasing, testing, and installing safety devices, defending litigations, settlement of accident cases, and so forth).

The need for safety engineering generally becomes obvious when any of the following problems occur: (a) increase in number of accidents and injuries, (b) increase in the costs

DOI: 10.1201/9781003107514-6

TABLE 5.1

Accidents During Use of Different Energy Sources

Energy Source	Accidents
Coal	Explosions and fires in underground coal mines. Coal dust and or Methane gas explosions. Collapse of roofs or walls in mines.
Oil	Fires and explosions from oil leaks in processing plants. Oil/petroleum tank/storage and pipelines explosions. Transportation accidents – fires, explosions from spills. Train disasters causing fuel tank-car explosions. Off-shore rig accidents. Oil spills at sea due to leaks or blowout from oil wells or tanker-ship wrecks and explosions.
Natural gas	Fires and explosions from gas leaks/blowouts.
Nuclear	Loss of coolant water. Nuclear meltdown. Accidents during shipment of radioactive waste. Radiation leakage due to accidents (e.g., tsunami). Radio-isotope releases.
Hydro	Flooding due to dam collapse. Building collapse due to flooding.
Geothermal	Well blowouts. Release of toxic gases.
Biomass	Fires due to inadvertent burning of biomass materials. Boiler explosions.
Wind	Blade failure due to ice buildup. Fires in nacelles. Tower collapse. Accidents during transportation and construction.
Solar-Photovoltaic	Release of toxic materials during photocell manufacturing.
Solar-Thermal	Release of toxic working fluid.

or losses resulting from the accidents, (c) employee turnover, (d) employee complaints, and/or (e) increased litigations due to defects and/or unsafe equipment and environment.

Safety engineers also work with environmental engineers to reduce harmful effects of pollutants and toxic substances emitted during construction and operation of energy systems and in providing healthful working and living environments for people as well as the wildlife in areas around the energy systems.

SAFETY ENGINEERING APPROACH

Figure 5.1 illustrates the safety engineering approach in dealing with existing or new systems or products. The safety problems are generally noticed with the mounting accident statistics obtained from the accident databases, changes in the safety regulations, increases in customer complaints, and/or product liability litigations. The data from these sources are reviewed by the product/systems designers and engineers to determine how various safety problems can occur. Many accident causation theories are considered, and accident situations are analyzed to determine the causes of the accidents. Several hazard analysis techniques are also applied to predict safety critical situations before occurrences of any accidents and/or unsafe situations. The accident and hazard analyses can uncover many potential causes for future accidents or unsafe situations. The accidents and unsafe situations can be eliminated by undertaking one or more accident preventive countermeasures.

Cost–benefit analyses are usually performed to determine the countermeasures that can provide higher ratios of the benefit-to-cost. The benefits can be due to reduction in costs due to accidents and the costs can be due to implementation of the countermeasures. Before and after the selected countermeasures are implemented, the safety performance is monitored (e.g., by maintaining control charts of the accident data) to determine the effectiveness of the countermeasures. If the countermeasures are found to be ineffective, then further changes in safety prevention strategies are considered by iterating the whole process as shown by the feedback loop in Figure 5.1.

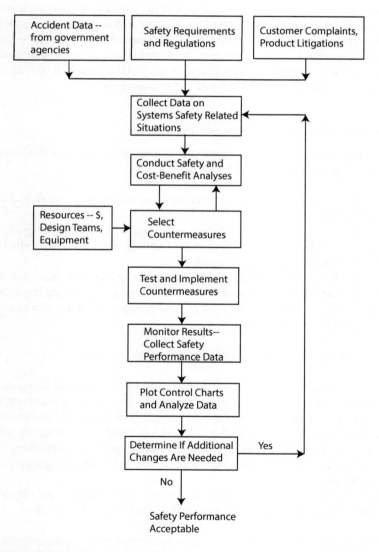

FIGURE 5.1 Safety engineering approach.

IMPORTANCE AND NEED OF SAFETY ENGINEERING

Safety engineering is an important field as it seeks to (a) design safer products and (b) build and operate safer systems. Safer products and systems are generally efficient to use as they tend to reduce costs and time losses due to accidents and health-related problems. Thus, safety engineering should be considered as one of the important disciplines during the entire product/system life cycle. It is especially useful to involve safety engineers during the early phases of product and system development.

Goetsch (1996) has pointed out that need for safety engineering becomes more acute when any one or more of the above problems are observed in an organization: (a) rapidly increasing safety related costs, (b) need to meet new safety regulations, (c) increasing litigations, (d) growing interest in ethics and corporate responsibility, (e) increased pressure from labor organizations and employees, (f) realization that safer products and systems are more productive, (g) realizing that safety and quality are closely related, (h) greater awareness and professionalization of health and safety, and (i) new hazards due to faster pace of technological changes.

3ES OF SAFETY ENGINEERING AND COUNTERMEASURES

In developing safe products and systems, engineers should always think about the "3Es" of safety engineering, which include:

1. *First "E" for Engineering*: Products/systems should be designed and engineered with safety in mind.
2. *Second "E" for Education*: Safety education will help the designers and users to understand the importance of safety. Thus, they will eliminate unsafe practices in work situations.
3. *Third "E" for Enforcement*: Safety requirements and safety practices must be enforced (through approaches such as training, incentives, inspections, audits, regulations, fines or penalties, and product/system recalls) to ensure that people act responsibly.

Thus, commonly considered safety countermeasures include:

1. Engineering Solutions (e.g., incorporation of fail-safe designs, lockouts, and alterations to products, systems, and processes to minimize occurrences of accidents, injuries, and exposures to harmful pollutants and substances)
2. Administrative Solutions (e.g., screening employees, limiting exposures to unsafe/toxic environments, rotating people, establishing policies, regulations/laws, practices/procedures, enforcement, training, and awareness)
3. Personal Protection (e.g., isolation of people from hazards, providing fire alarms, hard hats, safety goggles, masks, seat belts, and so forth).

Incorporating engineering solutions is generally the preferred countermeasure as they do not depend upon changing work practice and behavior of people in the workplaces.

METHODS USED IN SAFETY ENGINEERING

The techniques of safety engineering should be applied by engineers in designing complex products and systems to minimize the probability of safety-critical failures. The "Systems Safety Engineering" function helps to identify "safety hazards" in emerging designs and can assist with the techniques to "mitigate" the effects of potentially hazardous conditions that cannot be designed out of systems.

The methods used by safety engineers are listed below.

1. Critical Incident Technique
2. Behavioral Sampling
3. Checklists to identify hazards
4. Hazard Analysis (or Methods Safety Analysis)
5. Fish Diagram (Cause and Effect Diagram)
6. Failure Modes and Effects Analysis (FMEA) and Failure Modes, Effects, and Criticality Analysis (FMECA)
7. Logical Analyses and Fault Tree Analysis (FTA)
8. Reliability Analysis
9. Risk Analysis
10. Cost–Benefit Analysis
11. Accident Data Analysis (e.g., data gathering and statistical analysis of accident frequency, rates, severity, accident costs, and so forth.)
12. Accident Investigation (understand events that led to the accident and its causation)
13. Accident Reconstruction and Accident Simulation
14. Control Charts (plots of number of occurrences of different types of hazards, unsafe acts, unsafe conditions, and accidents as functions of time)
15. Experimental Studies (e.g., to determine effects of countermeasures on near-accident and accident rates by comparing "before" the countermeasures to "after" the countermeasures data)

The first ten of the methods listed above can be conducted without waiting for accidents to occur. The accident-data-based methods (#11, 12, and 13 above) can be applied after the accidents have occurred. The control charts and experimental studies can be based on measurements of non-accidents (e.g., unsafe acts, unsafe conditions, hazards, or near-accidents) and/or accidents. The above methods are described later in this chapter.

DEFINITION OF AN ACCIDENT

Several safety researchers have provided definitions to describe an accident, i.e., when would a situation be called an accident. The definitions also help in understanding the concepts of an accident and issues related to accident prevention. A few commonly referred accident definitions are provided below.

1. An accident is a set of complex events involving sequence, human actions/ behavior (unsafe acts), unsafe conditions, and some degrees of the following characteristics (Petersen and Goodale, 1980):

   a. *Degree of unexpectedness*: the less an event could have been antici-
      pated, the more it is likely to be called an accident.
   b. *Degree of avoidability*: the less the event could have been avoided, the
      more likely it is to be called an accident.
   c. *Degree of intention*: the less the event resulted from a deliberate action
      or lack of an action, the more likely it is to be called an accident.
   d. *Degree of warning*: the less warning, the more likely it is called to be an
      accident.
   e. *Duration of occurrence*: the more quickly it happens, the more likely it
      is to be called an accident.
   f. *Degree of negligence*: the more reckless or carelessness involved, the
      less likely it is to be called an accident.
   g. *Degree of misjudgment*: the more mistakes in judgment involved, the
      less likely it is to be called an accident.

2. An accident is any unplanned and uncontrolled event caused by human,
   situational, or environmental factors, or any combination of these factors,
   which interrupts the work process, which may or may not result in injury,
   illness, death, property damage, or other undesired events, but which has a
   potential to do so (Colling, 1990).
3. An accident is an unplanned and uncontrolled event in which the action
   or reaction of an object, substance, person, or radiation results in personal
   injury or probability thereof (Heinrich et al., 1980).
4. Accident is an unplanned, not necessarily injurious or damaging event,
   which interrupts the completion of an activity, is invariably preceded by an
   unsafe act and/or an unsafe condition, or some combination of unsafe acts
   and/or unsafe conditions (Tarrants, 1980).

ACCIDENT CAUSATION THEORIES

Many researchers have proposed theories to explain how accidents occur. It is important to understand the theories so that countermeasures can be generated to reduce
the occurrences of the accidents. These theories also help in undertaking accident
preventing actions during the development of safe products and processes. This section provides brief descriptions of several accident causation theories that are useful
in designing safe products. More detailed descriptions of the theories are provided
by Petersen and Goodale (1980).

   1. *Act of God (Demons or other supernatural forces)*: This theory is recog-
      nized in the legal literature and by the insurance industry that some acci-
      dents can be only explained as the "acts of god." The theory assumes that
      such acts are bad happenings outside an individual's control (or reasons
      such as the victims were presumably marked for punishment because of
      some unknown quality, the devil did his handiwork, and so on).
   2. *Accidents are "rare and random" events*: This theory recognizes that acci-
      dents are very low probability events, and they can happen and do happen
      to anyone. Here, an accident is considered as a "lottery" – an event whose

outcome seems to be determined by chance. In early 1900s, the accidents to a large extent were considered uncontrollable and unpredictable. Minimum thought was given to design of environments that could reduce probability of an accident or harm. It also suggests that repeated violations of common-sense safe practices eventually and invariably will lead to an accident.

3. *Accident-prone theory*: This theory assumes that accidents are caused by some invariant human characteristics identified as "accident proneness." Accident proneness may be defined as the continuing or consistent tendency of a person to have accidents as a result of his or her stable characteristics (or response tendencies). Such accident-prone people are also called accident "repeaters," i.e., they are involved in repeated number of accidents (or are "over-involved" in accident occurrences as compared with normal individuals).

   (Note: The accident proneness theory can be tested by analyzing accident data. Compare an "observed" distribution of number of individuals involved in x number of accidents in a population with a "fitted" distribution of expected number of individuals involved in x number of accidents assuming the Poisson distribution ($P[x]$, where x is the number of accidents occurring to an individual and $x = 0,1,2,3, \ldots$). Statistically significant difference between the "observed" and "fitted" (expected) distributions suggests the presence of accident proneness. This suggests that individuals with certain accident-prone characteristics are "over-involved" in accidents, i.e., accident-prone individuals have larger values of x than others with smaller values of x).

4. *Chain of multiple events*: This model assumes the existence of many factors influencing accidents rather than any key cause. The probability that an accident (P) will occur in each unit of activity (A) is assumed to be a function of a whole set of factors and conditions. If these factors are designated as $x_1, x_2, x_3, x_4, x_5, \ldots, x_k$, the probability of an accident in the activity would be a function of the factors (i.e., $P_A = f[x_1, x_2, x_3, x_4, x_5, \ldots, x_k]$).

5. *Energy exchange model*: Most accidents are caused by unplanned or unwanted release of excessive amounts of energy (e.g., mechanical, electrical, chemical, thermal, and ionizing radiation) or hazardous materials (e.g., carbon monoxide, carbon dioxide, hydrogen sulfide, methane, water, and so forth).(However, with a few exceptions, these accidents due to energy releases can be also explained by the unsafe acts theory where an unsafe act may trigger the release of large amounts of energy or a hazardous material, which in turn causes the accident.)

6. *Epidemiological model*: Epidemiology is the study of causal relationship between environmental factors and disease. The epidemiological theory suggests that the models used for studying diseases can be also used to study causal relationships between environmental factors and accidents. The model assumes that the key components are (a) predisposition characteristics (i.e., a susceptible host, a disease producing agent or a virus, and a hazardous environment), and (b) situational characteristics (e.g., not wearing sufficiently warm [low insulation] clothing in cold environment, risk taking,

and an untrained host). The predisposition characteristics are assumed to create like a disease producing condition or an accident-like condition (i.e., unexpected, unavoidable, or unintentional), which in combination with the situational characteristics results in an accident, which causes injuries or damage (see Figure 5.2).

7. *Domino theory*: An accident is caused by a sequence of events. Each event can be assumed to be represented by a domino.

   The domino sequence is as follows (see Figure 5.3):

   1. Faults of persons are created by environment or acquired by inheritance.
   2. Unsafe acts and conditions are caused by faults of persons.
   3. Accidents are caused by unsafe acts of persons and/or exposure to unsafe conditions.
   4. Injury results from an accident.

   The Domino theory thus suggests that the above four-event sequence involved in the causation of an accident can be broken by removing any one of the events (such as removal of a Domino in the chain) to prevent an accident from occurring.

8. *Unsafe acts theory*: Accidents occur primarily due to unsafe acts of people. The unsafe acts occur due to (a) misunderstanding of instructions, (b) lack of knowledge or training, (c) recklessness or violent temper, and/or (d) actions that exceed human capabilities and limitations (e.g., speed of response needed was beyond operator's capability to react).

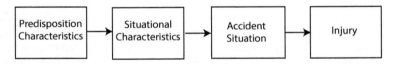

FIGURE 5.2 Sequence of events in the epidemiological model of accident causation.

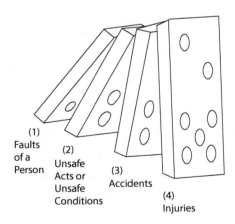

FIGURE 5.3 Sequence of dominos in the domino theory of accident causation.

9. *Human factors or human error models*: There are several human factors models that can be related to accident causation (Bhise, 2012, 2014). The models postulate that human failures occur due to reasons such as (a) task demand exceeds operator capabilities, (b) operator experiences information overload, (c) operator's attention is diverted/distracted, (d) operator is not consistent in his response (i.e., variability in human operator's output is too high), (e) operator is under stress, and (f) operator fails to get the right information at the right time [or fails to process the information] needed to make the decision and thus, does not make the right response needed to avert the accident. Most human failures can also be explained as a result of one or more human errors.

Many of the human failures are due to information processing errors where the human operator fails to make the right decision at the right point in time. Accidents can be caused by information processing failures. Some examples human errors caused by human information failures are (a) interpretational errors (i.e., errors in interpreting situations or signal interpretation. For example, the operator misunderstood the meaning of the "red" flashing light on his instrument panel.), (b) substitution errors (i.e., substituted a different action instead of the intended action. For example, the driver pressed the gas pedal instead of the brake pedal – substituted a wrong control), (c) reversal errors (i.e., operator responded with an action in the opposite direction instead of the intended direction. For example, turned a control clockwise instead of counterclockwise), (d) legibility errors (e.g., operator could not read a display due to small font size, poorly lit display at night, or sun-light reflection [glare] from the display lens), (e) forgetting errors or omission errors (e.g., forgot to perform an action. For example, operator turned off cooling water pump without reading the temperature gauge.), (f) commission errors (e.g., performed a task or step when not required or in a different sequence.), (g) other errors (e.g., control operational errors due to violation of one or more human factors principles in the design of a control (Bhise, 2012). Examples of such violations are (a) a control is not located at an expected location, (b) the direction of control motion violates its direction-of-motion stereotype, (c) the control not located with other controls of the same functional group, and (d) the control was not located close to its associated display.)

SAFETY PERFORMANCE MEASURES

Two types of measures (i.e., variables) are used to measure safety performance. They are (a) accident-based measures (i.e., based on accident data), and (b) non-accident measures (i.e., based on measurements of data from events other than accidents. For example, behaviors exhibited during the use of a product, unsafe acts committed, or errors made in operating equipment). The accident-based measures are more believable (or "hard") as compared with non-accident measures, which are regarded as "soft" measures of safety. The advantages of the non-accident measures are that they can be obtained without waiting for accidents to occur, and the non-accident events occur at much higher rates than accidents, which in general occur very rarely. On the other hand, the accident-based measures are "hard" or ultimate measures of safety and have higher face validity.

WHY MEASURE SAFETY PERFORMANCE?

Measurements are essential to determine the level of a problem or effects of changes in the design of a product or a system. The measurement of safety performance allows us to assess the following types of problems:

a. State of safety, i.e., accurately determining the level of safety in an operation or effectiveness of a product and/or process in achieving safety objectives
b. Assist in business planning and safety improvement activities
c. Allow us to evaluate, compare, or calibrate accident prevention initiatives
d. Provide feedback on past safety actions
e. Predict future safety costs

CURRENTLY USED ACCIDENT MEASURES

Some examples of currently used accident-based safety performance measures are provided below.

a. Number of accidents
b. Number of injury accidents
c. Number of persons injured in accidents
d. Number of disabling injury accidents (disabling means that the injured person is unable to perform normal activities due to the injury)
e. Number of fatal accidents
f. Number of fatalities in accidents
g. Incident rates (e.g., lost time [disabling] injury frequency rate, number of accidents per 200,000 work hours)
h. Lost workdays (number of workdays lost due to accidents)
i. Accident costs

The above measures are computed over a predefined exposure (e.g., time duration or number of product usage cycles). The accidents can be also categorized by using a number of classification criteria, e.g., accident type, accident severity, accident location, characteristics of a person involved in the accident (e.g., age, occupation), type of equipment, type of operation, or environment involved in the accident).

ACCIDENTS-BASED INCIDENT RATES

Incident rates are popularly used to measure safety performance in work-related industries (Tarrants, 1980; Goetsch, 2007; NSC, 2012). The commonly used incident rates based on incidents of injuries, illnesses, or fatalities are defined as follows:

$$IR = (N \times 200,000) / T$$

Where IR = Total injury and illness incident rate (incidences per 200,000 hours of work)

N = Number of injuries, illnesses, and fatalities resulting from the accidents
T = Total hours worked by all employees during the period in question
200,000 hours = 100 employees working 40 hours/week times 50 weeks in 1 year
The incident rates based on the 200,000 hours exposure are:

a. Injury rate
b. Illness rate
c. Fatality rate
d. Lost workday cases rate (accident cases where at least one workday was lost due to an accident)
e. Number of lost workdays rate
f. Specific hazard rate
g. Lost workday injury rate

The problems and issues with the use of the incident rates are as follows:

a. Since they are based on accidents, they are postmortem. Thus, one must wait for accidents to occur before the value of the measure can be computed. Thus, long periods of time need to elapse before reliable estimates of the safety measure can be obtained.
b. Since accidents are rare events, they are unreliable for small work organizations (i.e., they do not accumulate large number of work hours).
c. They are not very useful to predict effectiveness of safety countermeasures in shorter time periods.

ADVANTAGES AND DISADVANTAGES OF CURRENT ACCIDENT-BASED MEASURES

Some advantages of the currently used accident-based measures are as follows:

a. Quick acceptance (i.e., they are an "accepted standard") as compared to non-accident data, which may be regarded as questionable by many decision-makers.
b. Motivate management (i.e., they get management's attention and motivate them to take prompt actions).
c. Long history of use
d. Used by government agencies (e.g., U.S. Occupational Health and Safety Administration) and industry associations
e. Easy to calculate
f. Indicate trends in performance
g. Good for self-comparison

Some disadvantages of the currently used accident-based measures are as follows:

a. They can be only computed after the accidents have occurred (i.e., they are reactive or postmortem).
b. The numbers can be easily manipulated as many unreported accidents (intentionally or unintentionally) can cause under estimation of the safety problem.
c. They may be biased (due to management attitude to restricted work, doctor influence on reporting, worker attitude to light duties, compensation system, motivation to achieve safety awards, and competitions between organizations based on safety performance).
d. The measured number of accidents is typically low, making it difficult to establish trends (accidents, in general, are rare events).
e. The accidents differ in severity (i.e., the severity of injuries, amount of property damage, and losses are different in different accidents). Thus, comparisons based only on number of accidents can be misleading.
f. Some managers or safety specialists may regard an accident as a "once off" or a "freak" event. (Thus, may disregard the accident data.)

Non-Accident Measures

Unlike the accident-based measures, the non-accident measures are not standardized by the government agencies or industries. However, methods have been used (e.g., Behavior Sampling and Critical Incident Technique covered in a later section of this chapter) to measure unsafe acts, unsafe conditions, and errors. The frequency and occurrence rates of such incidences have been used to evaluate safety performance (Tarrants, 1980).

Safety costs have been routinely tracked by many organizations. They include (a) costs due to accidents and (b) accident prevention costs. The costs due to accidents are generally underestimated due to unreported or unaccounted accidents. Incidental costs of accidents have been estimated to be four times as great as the actual costs. The accident prevention costs include costs of safety analyses, engineering changes, evaluations/tests, reviews, training, protection devices, and so forth. The safety costs are covered in Chapter 7.

SAFETY ANALYSIS METHODOLOGIES

Two Possibilities: Accident vs. Hazard

The distinction between accident and hazard can be understood by considering the following two considerations:

1. *Accident*: Accident is an event in which damage to property or injury to personnel has occurred or occurring (i.e., accident cases and data are accumulating).
2. *Hazard*: Hazard is a real or potential condition that could cause damage to property or injury to personnel but has not occurred so.

The following two possibilities need to be considered prior to deciding on type of safety analysis and methods to use (Colling, 1990; Hammer 1980, 1989).

1. *An accident has occurred, or accidents are occurring*: This possibility leads to (a) accident investigation and (b) accident analysis. An accident investigation usually precedes an accident analysis.
2. *Not waiting for accidents to occur*: This possibility leads to conducting hazard identification and hazard analysis. The hazard analysis involves (a) hazard identification, (b) determining whether controls are in place to prevent occurrences of hazards, (c) formulate countermeasures, and (d) select the best countermeasures to implement to avoid future accidents.

ACCIDENT ANALYSIS METHODS

Accident analysis can be considered to include the following methods: (1) Accident Investigation, (2) Accident Analysis, and (3) Accident Data Analysis. The three methods are not distinctively different, and there is considerable overlap between their contents. The applications of the methods can also vary depending upon the accident researcher involved in performing the analyses. The three methods are described below.

1. *Accident investigation*: The accident investigation involves reading the accident report, visiting the accident site, talking to the witnesses, gathering all facts about a particular accident such as who was involved, how did the accident occur, what were the injuries and losses, and so on. A detailed accident investigation typically involves (a) reading individual accident reports, (b) sending independent accident investigators (or a team of multidisciplinary experts – in case of detailed investigation) to verify the details of the accident, (c) reconstructing the accident (i.e., describing how the chain of events led to the accident), and (d) preparing detailed report on each accident case.
2. *Accident analysis*: An accident analysis usually involves more than one accident of a given type (e.g., accidents involving a particular type of power plant under a certain type of situation, while performing a certain task or a maneuver). The accident analysis involves (a) collecting and analyzing accident data, (b) determining causes and circumstances of the accidents, (c) investigating possible chains of events that led to the accidents, and (d) creating a model of the accident situation to reconstruct and illustrate details about the behavior of various elements or events that led to the accidents.
3. *Accident data analysis*: The data accident analysis usually involves (a) securing access to one or more accident databases, (b) understanding variables and categories used in creating the database, (c) evaluating completeness of the data (i.e., understanding missing data or uncategorized variables that are generally categorized as "other" or "not available"), (d) creating tabular summaries of accident data based on relevant variables of interest, and (e) conducting statistical tests to determine if any of the differences due variables and their categories are statistically significant.

Hazard Analysis Methods

The methods used for hazard analysis also have some overlap in content and differences in formats and details depending upon the individuals or organizations conducting the analyses. The methods used for hazard analysis are listed below.

1. General Hazard Analysis
2. Detailed Hazard Analysis
3. Methods Safety Analysis (like operations analysis)
4. Job Safety Analysis/Job Hazard Analysis (to uncover hazards in a job)
5. FMEA
6. FMECA
7. FTA
8. Human Error Analysis
9. Human Reliability Analysis

The methods used for hazard analysis are described in the next sections of this chapter. More information on the above listed hazard analysis methods can be obtained from (Brown, 1976; Hammer, 1980; Goetsch, 2007; Roland and Moriarty, 1990; Colling, 1990).

Hazard Analysis

A hazard can be defined as an unsafe situation that can cause injuries and losses if allowed to remain without the removal of underlying causes of its occurrence. The occurrence probability of an unsafe situation can be generally estimated from historic information or by interviewing experts who have experienced or studied similar situations. Hazard analysis can be defined as a systematic method used by the safety engineers to study the causation of an unsafe situation by identifying its characteristics such as, unsafe conditions, unsafe acts or actions committed by people, or combinations of unsafe acts and conditions and determining their probabilities of occurrence. Many safety researchers such as Heinrich et al. (1980), Tarrants (1980), and Brown (1976) have suggested different formats for documentation of the hazard analysis with differing levels of details.

General Hazard Analysis

Brown (1976) suggested a simple hazard analysis involving a General Hazard Analysis Card. The General Hazard Analysis Card can be filled out by anyone involved or familiar with a process or a situation, which can involve a hazard. The person filling out the card is required to provide the following information: (a) hazard description, (b) location of the hazard in the workplace, (c) severity level (nuisance, marginal, critical, or catastrophic), (d) probability of accident (unlikely, probable, considerable, or imminent), and (e) possible costs due to an accident resulting from the hazard (prohibitive, extreme, significant, or nominal). The cards are then reviewed by a safety professional who categorizes each card into the following actions: defer, need more analysis, or immediate action needed.

The hazard analysis thus involves an overall look at the system under consideration and to identify those safety problems that require more detailed analysis. The sources of inputs for the general hazard analysis include (a) obvious hazards due to nature of the process (e.g., likelihood of a fire in processes involving flammable fluids and electrocution hazards with electrical systems involving high voltages), (b) hard recordkeeping data required by law or organizational policies (e.g., accident reports and injury records, violation of safety practices), and (c) other investigations that stress the cooperation of the line and staff personnel.

DETAILED HAZARD ANALYSIS

A detailed hazard analysis generally involves breaking down a task or a process into a series of steps (or subtasks) that may contain one or more hazards. The detailed hazard analysis is documented by creating a matrix. The steps are listed as rows of the matrix, and hazards are listed as columns of the matrix. The matrix is usually created by a safety professional. The cells of the matrix where a particular hazard element is present in each step are identified by entering "X" marks. The safety professionals evaluate each "X" marked cell to determine one or more possible countermeasures that could be applied to avoid the hazard. The effect of each countermeasure on all possible hazards in all rows (steps) is rated in terms of (a) R_i = the hazard can be reduced by the countermeasure to a degree indicated by the subscript i, (b) E = the hazard in the step can be eliminated, (c) I = the hazard would increase by the countermeasure, or (d) X = the hazard still exists in the step. The above determined ratings are placed in the corresponding cells to understand the effect of each countermeasure. The analysis is usually repeated by applying combinations of different countermeasures to select the best set of countermeasures. In addition, cost–benefit analysis can be performed on each alternative (or countermeasure) to determine the most cost-effective alternative that produces the highest benefit-to-cost ratio. The benefits can be defined as the reduction in costs due to elimination or reduction of the accidents. The costs are defined by adding all costs involved in the implementation of the countermeasures.

METHODS SAFETY ANALYSIS

A Methods Safety Analysis is also a hazard analysis technique, which involves breaking down a task or a process in a series of steps. Application of this technique involves a safety professional to analyze each step of the process (usually by discussing each step with the team members involved in the process) to identify possible hazards in each step and the causes of the hazards and propose improvements in the steps to eliminate the hazards. The improvements are incorporated in the process and a new improved method is developed. Thereafter, the team members are trained to follow the newly developed process. This technique is referred as the "Methods Safety Analysis" because it is like the traditional "Methods Analysis" (or Operations Analysis) technique used by the industrial engineers. In Methods Analysis, an industrial engineer analyzes the method involved in performing a task or an operation by systematically breaking down each step (or work motion) and then improving each step by asking questions such as can the step be eliminated, combined, or made more efficient by reducing unnecessary motions. In the Methods Safety Analysis, the emphasis is on improving safety. Thus, the methods safety analysis is primarily concerned with investigating hazards involved in the methods of an operation. The fundamentals of the Methods Safety Analysis are the same as the Methods Analysis. The steps in methods safety are (a) break downing the job or operation into its elemental tasks, (b) listing them in proper order, and (c) examining them critically to find opportunities to eliminate the hazards, and (d) developing a new safer and more efficient method.

The advantages of Methods Safety Analysis are as follows:

1. It maps out all the details of an operation so that they can be studied and restudied.
2. It is quick, simple, factual, and objective.
3. It permits comparison of current and proposed methods.
4. It presents a picture of the effect on production by the safety improvements.
5. It aids management in reviewing the benefits.
6. It permits engineer to improve productivity.
7. It facilitates analysis of safety potential of an operation before an accident occurs, i.e., it is proactive.
8. It assists in thorough investigations of methods or operations involved in the accident where the causes are obscured.

Checklists to Uncover Hazards

Determining causes for hazards and developing countermeasures require specialized safety professional with experience and knowledge in hazard recognition and accident causation theories. The hazard recognition abilities can be enhanced by creation and use of hazards checklists. For example, Hammer (1980) provided comprehensive set of checklists to analyze product designs. Hammer's checklists (with specific questions) included the following categories of hazards resulting from (a) acceleration (e.g., Does the product contain any spring loaded or cantilevered object or device that could be affected by acceleration or deceleration?), (b) chemical reaction (e.g., Is there any material present that will react with the oxygen in the air to produce a toxic, corrosive, or flammable material or one that will ignite spontaneously?), (c) electrical shock (e.g., Are the voltages and amperages high enough to cause arcing or sparking, which can ignite a flammable gas or combustible material?), (d) explosives and explosions, (e) flammability and fires, (f) heat and temperatures, (g) mechanical items (e.g., sharp points, edges), (h) pressure (e.g., high-pressure lines and vessels), (i) radiation, (j) toxic materials, (k) vibration and noise, and (l) other miscellaneous hazards. The product and system designers should construct specialized checklists containing issues related to hazards inherent in their product or system and use the checklists to identify all possible hazards.

Risk Analysis

Risk analysis is a methodology used to compute the level of risk in each product, process, or situation. The level of risk is generally expressed in terms of a risk priority number called the RPN. The computation of an RPN is similar to the approach used in the FMEA (see Chapter 3), which consists of multiplication of three rating values, namely severity rating (*S*), occurrence rating (O), and hazard detection rating (D). Thus, RPN = S × O × D. Many models of RPN with differing scales are used in the industry and by government agencies for risk analysis.

Figure 5.4 presents an example of a nomograph (also called nomogram) used to illustrate risk assessment. Three situations S1, S2, and S3 are shown by joining points on the three scales through a tie line between the first two scales. The first situation S1 involves a situation with moderate potential for injury, highly probable hazard occurrence, and improbable hazard recognition resulting in extremely

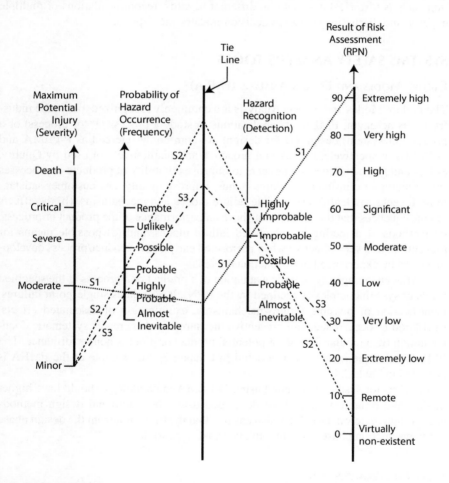

FIGURE 5.4 Example of a nomograph for risk assessment. (Redrawn from: Risk & Policy Limited. 2006. "Establishing a Comparative Inventory of Approaches and Methods Used by Enforcement Authorities for the Assessment of the Safety of Consumer Products Covered by Directive 2001/95/EC on General Product Safety and Identification of Best Practices". Final Report, February 2006 prepared for DG SANCO, European Commission.)

high risk. Whereas situation S2 involves minor potential for injury, probable hazard occurrence, and highly improbable hazard recognition resulting in remote to virtually nonexistent risk. The third situation S3 involves minor potential of injury, highly probable hazard occurrence, and improbable hazard recognition resulting in extremely low risk.

The problems with the RPN computation methodology are as follows: (a) the selection of values of the ratings is based on the subjective judgments of the analyst, (b) lack of objective data to determine values of the ratings, and (c) assumptions are made about consumer or user behavior. Thus, the RPN-based risk analysis models are not precise. But they can be used as guides along with other sources of information such as historical cost data on different hazards, recommendations of multiple experts, and discussions between decision-makers and experts.

SYSTEMS SAFETY ANALYSIS TOOLS

FAILURE MODES AND EFFECTS ANALYSIS (FMEA)

The Failure Modes and Effects Analysis is a commonly recognized tool in the industry by its acronym, FMEA. Its applications to study a process (P for process) or a product (or system) design (D for design) are commonly refereed as P-FEMA and D-FEMA, respectively. FMEA is a proactive and qualitative tool used by Quality, Safety and Product/Process engineers to improve reliability of products or processes (by reducing or eliminating failures – thus, improve quality, i.e., customer satisfaction). It seeks to identify (a) possible failure modes and mechanisms, (b) the effects or consequences that the failures may have on performance of the product or process, (c) methods of detecting the identified failure modes, and (d) possible means for prevention. It is very effective when performed early in the product/process development and by experienced multifunctional teams.

FMEA encourages systematic evaluation of a product or a process at the specified levels of system complexity (defined by the RPN). It postulates single-point failures, identification of possible failure mechanisms, examination of associated effects, likelihood of occurrence, and preventive measures. It also creates systematic documentation (in a tabular format) of potential product or process nonperformance. The FMEA technique is described in detail in Chapter 3. An example of the FMEA is provided in Table 3.4.

Products (or systems) designed using D-FEMA methodology should have higher quality and reliability than those developed using the traditional design methods (e.g., design reviews). D-FEMA also ensures that the transition from the design phase to the production phase is as smooth and rapid as possible.

FAULT TREE ANALYSIS (FTA)

Purpose
The purpose of a fault tree is to fully describe occurrence of an event placed at the top of the fault tree diagram. The top event is called the head event. The fault tree shows the branches (or paths) leading the occurrences of head event. Another

purpose of the fault tree is to determine the probability of occurrence of the head event.

Description

In a fault tree, all possible combinations of events that can cause the head event to occur are described as branches underneath the head event. The relationships between all events can be described in terms of a series of Boolean algebraic equations. The equations can be used to analyze occurrences of any of the events in the tree. The Boolean expressions can be used to determine probability of occurrence of any of the events by using the probabilities of the terminating events under each of the branches.

Fault tree is a logical tree diagram showing how the event shown at the top of the tree (head event) can occur. It describes all possible events that can lead the occurrence of the head event. A fault tree has only one head event. It is generally an "undesired" event. Various events that can lead to an event are shown by using logical operators called "gates." "AND" and "OR" gates (and other gates) are used to show how an event described at the top of the gate can occur due to occurrences of events shown below the gate.

Application of Boolean Algebra

Boolean algebra can be used to provide logical description of a fault tree and its branches.

It can also help in performing computational analyses for large complex fault trees using computers. A given Boolean expression defining the same head (output) event can be rewritten in various equivalent expressions for (a) restricting of the tree, (b) equation simplifications of the tree, and (c) determination of mutually exclusive branches or events. The probability of a head event can be computed from its Boolean expression defining the event. The fault trees of product (or system) failures can be also constructed in configurations involving different product design alternatives or countermeasures. And comparisons of quantitative evaluations of the alternate configurations of fault trees can be made to reduce the occurrences of the undesired events and to determine effectiveness of the countermeasures. The basics of Boolean algebra are described below.

Let us assume that $X, Y, Z, A, B, C, D, \ldots$ are Boolean algebra variables (i.e., logical variables). Each Boolean variable defines an event. Variable X can be defined to convey that event X exists, i.e., X is true. Then $\bar{X}$ (i.e., X-bar) denotes that event X does not exist, or X is not true, i.e., X is false.

The primary logical operators are (a) + (plus), which denotes "OR" gate situation, and (b) • (dot), which denotes "AND" gate situation.

Therefore, Boolean equations can be written as follows:

$$X + X = X$$

$$\bar{X} + X = 1$$

$$X \cdot X = X$$

$$X \cdot \bar{X} = 0$$

Where $0 =$ Null (no event exists) and $1 =$ All events coexist (i.e., the universe).

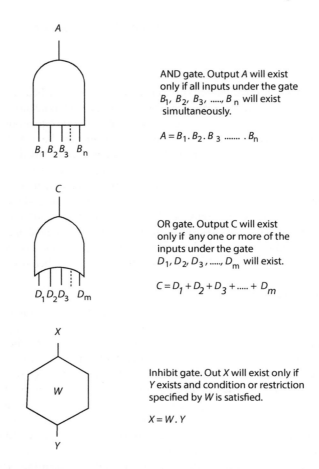

FIGURE 5.5 AND, OR, and INHIBIT gates used in fault trees.

Figure 5.5 illustrates AND, OR, and Inhibit gates. All the events above, below, and within the gates are defined by Boolean variables. Figure 5.5 thus illustrates the following:

a. The "AND" gate illustrates that $A = B_1 \cdot B_2 \cdot B_3 \ldots B_n$.
b. The "OR" gate illustrates that $C = D_1 + D_2 + D_3 + \cdots + D_m$
c. The inhabit gate illustrates that $X = W \cdot Y$.

The events denoted by Boolean variables are shown in the fault tree diagram by using different shaped boxes. The notations used for different types of events are shown in Figure 5.6. The description of each event is written inside the box in words or in terms of its Boolean variable (such as, *G. F, E, and H* in Figure 5.6). The output of any event box is on the top side and input is from the bottom side.

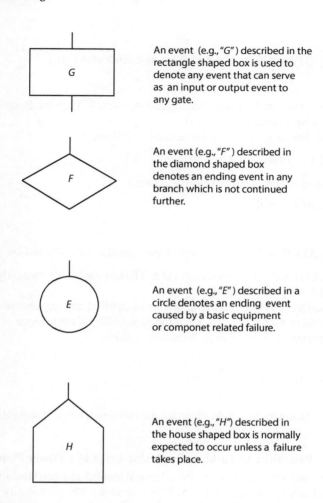

An event (e.g., "G") described in the rectangle shaped box is used to denote any event that can serve as an input or output event to any gate.

An event (e.g., "F") described in the diamond shaped box denotes an ending event in any branch which is not continued further.

An event (e.g., "E") described in a circle denotes an ending event caused by a basic equipment or componet related failure.

An event (e.g., "H") described in the house shaped box is normally expected to occur unless a failure takes place.

FIGURE 5.6 Types of events shown in different shaped boxes in fault trees.

"AND" Gate

Assume that events A and B are required to be present simultaneously to generate event X. Then the Boolean (logical) expression can be written as: $X = A \cdot B$

The probability of $X = P(X)$ can be defined as follows:

$$P(X) = P(A)P(B \mid A)\left(\text{If } A \text{ and } B \text{ are not independent events.}\right)$$
$$= P(B)\ P(A \mid B)$$

Note that $P(A|B)$ and $P(B|A)$ are conditional probabilities. Thus, $P(A|B)$ is defined as probability of A has occurred given B has occurred.

Otherwise,

$$P(X) = P(A)P(B)\,(\text{Note: If } A \text{ and } B \text{ are independent events.})$$

"OR" Gate

Assume that events A or B are required to generate event Y. Then the Boolean (logical) expression can be written as: $Y = A + B$

The probability of $Y = P(Y)$ can be defined as follows:

$$P(Y) = P(A) + P(B) - P(AB)$$

$$\big(\text{If } A \text{ and } B \text{ are not mutually exclusive events.}\big)$$

$$P(Y) = 1 - P(\overline{A}) \cdot P(\overline{B})$$

Otherwise,

$$P(Y) = P(A) + P(B)\,\big(\text{Note: If } A \text{ and } B \text{ are mutually exclusive events.}\big)$$

If $P(AB) = 0$. Events A and B cannot coexist. Thus, A and B are mutually exclusive. Therefore, $P(A + B) = P(A) + P(B)$.

This mutually exclusive requirement can be applied and generalized for any n mutually exclusive events (A_1 to A_n) and the probability of occurrence of any one or more of the events (A_1 to A_n) can be written as follows:

$$P(A_1 + A_2 + A_3 + \cdots + A_n) = \sum_{i=1}^{n} P(A_i)$$

The following examples will help illustrate the independence and mutually exclusive events.

An Example: Reliability of Turbine Generator Units in a Power Plant

Let us assume that a power plant has two identical natural-gas-fired turbine generator units (TGU). Each unit produces 100 MW of electricity. The probability that a TGU will work continuously over 24-hour period is 0.99. Compute the probabilities that at least one TGU unit will work in a 24-hour period and when both the TGUs will work in a 24-hour period. Assume that the two TGUs are independent (i.e., operation of one unit does not affect the other unit).

The first part of the problem can be solved by considering an "OR" gate where at least one TGU will work (i.e., TGU A or TGU B will work. See upper fault tree in Figure 5.7). The second problem is solved by considering an "AND" gate (see lower fault tree in Figure 5.7). Thus, the probability that the power plant will provide at least 100 MW of power over a 24-hour period (i.e., at least 1 TGU will work) is 0.9999; and the probability that the power plant will generate 200 MW of power over a 24 hour period (i.e., both the TGUs will work) is 0.9801.

Fault Tree Development Rules

Brown (1976) has provided the following three rules for fault tree development. The rules allow systematic development of a fault tree by using OR gates first (at the top

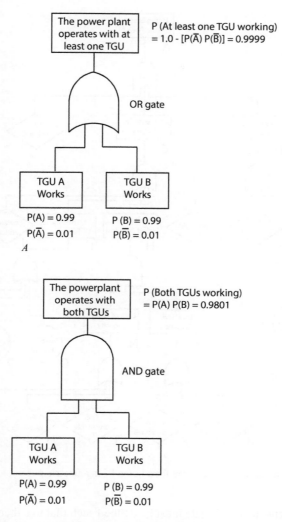

FIGURE 5.7 Illustration of application of OR and AND gates.

and every subsequent events) and organizing conditional events to reduce errors in computations of the probabilities.

Rule 1: Fault Tree Development Rule

For any event that requires further development, analyze all possible OR input events first prior to analysis for AND inputs. This will hold for the head event or any subsequent events that need further development.

The above rule is based on data collection completeness (think about all reasons for an occurrence of each event first) and probability considerations leading to an event. OR gate should be used just below the head event.

Rule 2: OR-Gate Event Rules

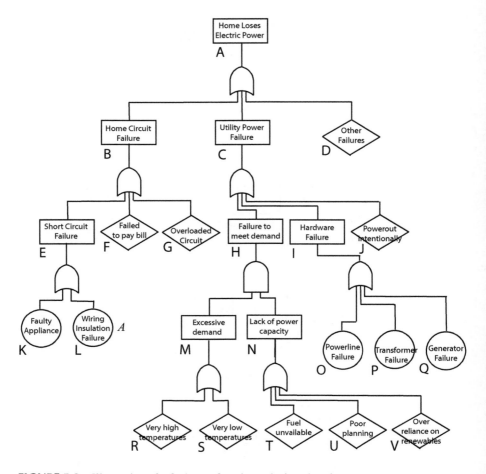

FIGURE 5.8 Illustration of a fault tree for a home losing electric power.

The input events to an OR gate must be defined such that together they constitute all possible ways that output event can occur. In addition, each event must include the occurrence of the output event.

This rule is based on inclusion of all possible causes for completeness. Otherwise, probability estimate will be underestimated. All events placed under an OR gate must collectively exhaust all possible events leading to the event above the OR gate. Thus, use an event (in a diamond shaped box; see Figure 5.8 second row of events from top) to include events occurring due to "all other" reasons (allows for future inclusions or expansions). (An example of this situation is in Figure 5.8 – see event D.) If all the events leading to an OR gate are mutually exclusive (i.e., the events do not overlap), then their probabilities can be simply added.

Rule 3: AND-gate Event Rule

The input events to an AND gate must be defined such that the second event is conditioned upon the first, the third event is conditioned upon the first and the second, and so on, and the last is conditioned on all others. In addition, at least one of the events must include the occurrence of the output event.

This rule is based on reduction of uncertainty related to independence of events under the AND gate and for accuracy in quantitative analysis.

FAULT TREE EXAMPLE: HOME LOSES POWER

Figure 5.8 illustrates a fault tree drawn for the following head event "Home Loses Power." The head event is labeled as "A." The head event can occur due to any of the three events B or C or D shown below the OR gate under the head event (see Rule 1). Note that event D is defined to satisfy Rule 2 above.

The events shown in the fault tree are briefly described below:

A = Home loses electrical power

B = Home circuit fails

C = Utility company fails to supply power

D = Electric power not available due to other (unknown) reasons

E = Short circuit in the home interrupts the power in the home

F = Utility company cut the power because the consumer failed to pay his bill

G = Overloaded circuit in the house trips the main circuit breaker in the house

H = The utility company fails to meet demand (i.e., could not provide the power).

I = Hardware of the utility company fails (Thus, the utility company cannot deliver power.)

J = Power turned off intentionally by the utility company (e.g., due to a forest fire near the home)

K = A faulty appliance in the home causes a short circuit

L = Wiring insulation failure in the home causes a short circuit

M = Excessive demand in the power grid causes the utility company to turn off the power to the home

N = Lack of electric power available from the power grid

O = Power line providing the electricity to the home fails (e.g., due to a windstorm)

P = Transformer failure at the substation providing electricity to the home

Q = Generator providing the power to the grid fails

R = Excessive electricity demand on the power grid due to very high outside temperatures requiring heavy use of air-conditioning systems by other customers

S = Excessive electricity demand on the power grid due to very low outside temperatures requiring heavy use of electric heating systems by other customers

T = Fuel not available for running the generators at the power plant supplying electricity to the power grid

U = Poor planning by the utility company causes power shutdown

V = Renewable energy power plants unable to supply the needed power to the power grid

Thus, using Boolean algebra, A can be defined as:

$$A = B + C + D$$

Similarly, other events can be defined as follows:

$$B = E + F + G$$
$$C = H + I + J$$
$$E = L + K$$
$$H = M \cdot N$$
$$I = O + P + Q$$
$$M = R + S$$
$$N = T + U + V$$

The probability of the occurrence of the head event can be computed by using the following equations.

$$P(A) = 1.0 - \left[P(\bar{B}) \cdot P(\bar{C}) \cdot P(\bar{D}) \right]$$

Where

$$P(B) = 1.0 - \left[P(\bar{E}) \times P(\bar{F}) \times P(\bar{G}) \right]$$
$$P(C) = 1.0 - \left[P(\bar{H}) \times P(\bar{I}) \times P(\bar{J}) \right]$$
$$P(E) = 1.0 - \left[P(\bar{L}) \times P(\bar{K}) \right]$$

$$P(H) = P(M) \times P(N) \left(\text{assuming } M \text{ and } N \text{ are independent events} \right)$$

$$P(I) = 1.0 - \left[P(\bar{O}) \times P(\bar{P}) \times P(\bar{Q}) \right]$$
$$P(M) = 1.0 - \left[P(\bar{R}) \times P(\bar{S}) \right]$$
$$P(N) = 1.0 - \left[P(\bar{T}) \times P(\bar{U}) \times P(\bar{V}) \right]$$

The probability of the head event can be computed by using the probabilities of the ending events D, F, G, J, K, L, O, P, Q, R, S, T, U, and V. It is assumed that the numeric values of the probabilities of each of the ending events are given (from estimates of experts or assumptions).

Advantages of Fault Tree Analysis

The advantages in using the FTA are as follows: (a) it allows better understanding of the system through causal relationships of events, (b) it helps in communicating issues among the design team members, (c) it provides quantitative estimates for better decision-making, and (d) it points out critical paths (branches in the fault tree with high probabilities of undesired events) and improvements in branches (or events in the branches) that can reduce probability of the occurrence of the undesired event.

ACCIDENT DATA ANALYSIS TOOLS

Purpose of Accident Data Collection

The purpose of the accident data collection is to get all the relevant facts about one or more accidents such as how, when, and what happened, who caused it, and so forth, to understand the factors and the causes associated with the accidents. The goal is to reduce frequency of such accidents by developing accident prevention programs and to monitor the effectiveness of implemented countermeasures for evaluating the state of safety. The collected accident data can be also compared with the accident data from past time intervals and the accident data available from other organizations or other situations.

Flow of Accident Data Collection

The accident data originates from the accident reports, which are generally prepared by the supervisor, the police officer, the medical staff, or the persons involved in the accident. The accident reports are collected and entered in summary sheets and databases by data entry personnel. A completed accident report form contains information such as persons involved, location of the accident, date and time of the accident, description of the injuries, equipment involved, description of the loss, the situation and the environment present at the time of the accident, accounts of how the accident occurred, diagram/map with the locations of various objects, movements of the equipment/vehicles, and so on. Some accidents are investigated at a greater detail by accident investigators or multidisciplinary teams (involving researchers, engineers, medical personnel, lawyers, and other experts), and the data are entered into databases. Their report is called the "Accident-Analysis Report."

The accident researchers and statisticians analyze the data by sorting, summarizing in tabular and other formats (e.g., pie charts, tree diagrams), and conducting statistical analyses to determine significant effects or differences due to certain variables. For example, the accidents can be sorted by conditions (e.g., day vs. night), situations (e.g., normal vs. emergency), before vs. after the implementation of the countermeasures, equipment used (e.g., make, model, brand of equipment/vehicle involved in the accidents), operator characteristics (e.g., age, gender, sober/intoxicated), and so on.

Accident Reporting Thresholds

All accidents have some type of reporting threshold, i.e., when an accident report must be filed. The thresholds are typically based on the level of severity of the injury or property damage over a certain value. Unfortunately, the thresholds vary between different organizations. For example, the police recordable accidents are defined based on occurrence of a personal injury or property damage over $100–$200. The workplace recordable accidents defined by the Occupational Safety and Health Administration (OSHA) are more severe than just the first-aid injuries and involve loss time or a doctor's judgment. Thus, what is reported can vary based on the accident location and the judgment of the person treating the injured or filing the report.

ACCIDENT INVESTIGATIONS

The primary reason for investigating an accident is not to identify a scapegoat, but to determine causes of the accident by gathering factual information about the details that led to the accident.

Generally, an accident report (supervisor or police reported) does not answer the "why" question. It answers who, what, where, and when related to the accident. Thus, the purpose of the accident investigation is to understand why the accident occurred, how it occurred so future accidents of the type can be prevented.

Ideally, the accidents should be investigated as soon as all emergency procedures have been accomplished. The accidents can be investigated by the supervisor, an investigative team, or an outside specialist to collect facts that led to causation of the accident. During the investigation, the accident scene is isolated to maintain the conditions that existed at the time of the accident and all the evidence (e.g., photographs and/or video of the accident scene, interviews of the witnesses) is gathered. Later, simulations may be created to understand and determine the series of events in the accident, and finally, a report is prepared.

ACCIDENT DATA: USERS AND SOURCES

The primary purposes of accident data analyses are (a) to estimate the magnitude of the accident problem, frequencies, and rates by accident types or categories and (b) to understand causes to develop prevention countermeasures to reduce accidents. The accident data are used by many decision-makers such as company management personnel (e.g., to determine state of safety in their products, operations, costs, and countermeasures), government officials (e.g., for standards development, recalls, and public notices), safety researchers (e.g., to analyze accident data, evaluate alternatives, and propose countermeasures) and lawyers and legal professionals (e.g., to establish evidence, link accidents to design or manufacturing defects of products, errors committed by persons involved in accident causation, and negligence of persons involved in development of the products/equipment related to the accident), and insurance company researchers (e.g., to determine accident causation and liability, incident rates, severity, costs, and insurance premiums).

The accident reports can be categorized by sources of the accident reports. Typical sources of accident reports are private organizations (e.g., companies, manufacturers, service providers, insurance companies, and health-care providers), public institutions (e.g., universities), law enforcement organizations (e.g., police in city, county, or state), state accident databases, and databases created by the federal agencies (see next section).

Accident Databases Maintained by Federal Agencies

The OSHA (under U.S. Department of Labor) established the annual OSHA Data Initiative (ODI) to collect data on injuries and acute illnesses attributable to work-related activities in private-sector industries from approximately 80,000 establishments in selected high-hazard industries. The Agency uses these data to calculate establishment-specific injury/illness rates, and in combination with other data sources, to target enforcement and compliance assistance activities. The OSHA data collection parallels aspects of the annual Bureau of Labor Statistics (BLS) Survey of Occupational Injuries and Illnesses in that both the ODI and the BLS Annual Survey collect summary information on occupational injuries and illnesses from private-sector establishments. BLS collects the data from a *sample* of all private-sector industry establishments. In addition, the BLS survey collects information on the demographics and circumstances of a sample of the injuries and illnesses that required recuperation away from work. The BLS survey is used to generate aggregate statistics on occupational injuries and illnesses at the state and national levels. However, the BLS Survey does not provide the establishment-specific data that OSHA needs.

The U.S. Consumer Products Safety Commission (CPSC) maintains a national-level accident database called the National Electronic Injury Surveillance System (NEISS). It represents a national probability sample of hospitals in the United States and its territories. Patient information is collected from each NEISS hospital for every emergency visit involving an injury associated with the consumer products (CPSC, 2020).

The National Highway Traffic Safety Administration (NHTSA, U.S. Department of Transportation) created the Fatality Analysis Reporting System (FARS) in 1975 (NHTSA, 2020). This accident data system was conceived, designed, and developed by the National Center for Statistics and Analysis (NCSA) to assist the traffic safety community in identifying traffic safety problems and evaluating both motor vehicle safety standards and highway safety initiatives.

SAFETY PERFORMANCE MONITORING, EVALUATION, AND CONTROL

INTERVIEW AND OBSERVATIONAL TECHNIQUES FOR NON-ACCIDENT MEASUREMENT OF SAFETY PERFORMANCE

In many situations requiring measurement of safety performance, non-accident events (e.g., unsafe acts, unsafe conditions, and near-accidents) can be measured instead of waiting for accidents to occur. Critical Incident Technique (CIT) and Behavioral Sampling are two commonly used techniques in the non-accident-based safety performance measurement area. CIT involves interviewing a preselected group of individuals to recall details about predefined critical incidents (e.g., potential or actual accidents). Behavioral sampling involves the traditional "Occurrence Sampling" or "Work Sampling" application by observing and recording of unsafe acts and unsafe conditions in workplaces. The collected information is used to monitor safety performance.

CRITICAL INCIDENT TECHNIQUE (CIT)

The CIT is a method of identifying errors and unsafe conditions that contribute to both potential and actual injurious accidents within a given population by means of a stratified random sample of participant-observers selected within the population.

Quantitative information (e.g., frequencies and proportion of unsafe behaviors) obtained from the questions asked during the interviews can be used to measure safety performance (Tarrants, 1980).

One of the oldest and famous applications of the CIT was by Fitts and Jones, who studied pilot errors to help design better aircraft controls and displays (Fitts and Jones, 1961a, 1961b). Soon after World War II, the Air Force launched a systematic study of errors made by pilots in situations where accidents and near-accidents occurred. The pilots were asked to recall incidents where they almost lost an airplane or witnessed a copilot make an error in reading aircraft displays or operating controls. From the analyses of the data gathered from these critical incidents, Fitts and Jones found that practically all the pilots, regardless of experience or skill, reported making errors in using cockpit controls and instruments. They also concluded that it should be possible to eliminate or reduce most of these pilot errors by designing equipment in accordance with human requirements. Similarly, human operator errors (e.g., errors committed by power plant operators, drivers) in using displays and controls can be reduced if they are designed in accord with the human engineering criteria.

Steps in conducting the CIT are as follows:

1. Define the objectives of the study. For example, to solve a safety problem by identifying causes of certain type of accidents and/or measuring level of safety (frequency or rates of unsafe occurrences) in a specified situation
2. Determine sample characteristics of participant-observers (their locations, representation)
3. Develop questions to ask and obtain (a) descriptions of situations related to the errors or unsafe conditions, (b) descriptions of the errors and unsafe conditions, (c) probe questions to assist in recall of incidents related to each particular type of error and unsafe condition
4. Carefully select a representative sample of participant-observers based on valid stratification criteria (e.g., stratification by shift, locations of product use)
5. Conduct preliminary interviews (to facilitate recall) with each participant-observer (about 24-hours before the data collection interview) to inform them about the purpose of the study, explain type of information desired, and explain the procedure
6. Conduct an interview with each participant-respondent separately for about 45 minutes. Record the interview for subsequent data retrieval
7. Replay audio-recording and extract details of each incident (when, where, how occurred, who, possible cause), classify the mentioned unsafe acts and conditions, and keep records on their counts (frequency) and data
8. Prepare a table of frequencies of unsafe acts and conditions
9. Conduct statistical analyses of the data for purposes such as (a) to determine distribution of unsafe acts and conditions (e.g., by types, occurrence time, and locations), (b) to compare with past data, and (c) to maintain control charts of frequency (C or U-charts), rates, or proportion of unsafe acts
10. Document and present conclusions to management

During the interviews (in step 6 above), the interviewer first starts with open-ended questions. Some examples of such open-ended questions are (a) Would you describe fully as you can the last time you saw an unsafe human error or unsafe condition in your department? (b) Would you please describe as fully as you can other unsafe human errors or conditions in your department during the past 2 years? Next, the interviewer probes systematically for other incidents that they may have forgotten. Examples of questions used here are (a) Have you ever seen anyone cleaning a machine or removing a part while the machine was in motion? (b) Have you ever seen anyone speeding or other improper handling of a power truck?, and (c) Have you ever seen anyone "beating" or "cheating" a machine guard?

BEHAVIORAL SAMPLING

Behavioral sampling is really the occurrence sampling (also known as work sampling) technique commonly used by the industrial engineers to obtain information on operational status (e.g., working or not working) of human operators in the workplaces (Konz and Johnson, 2004). Occurrence Sampling is a process of observing at discrete random points (times) to obtain estimate of occurrence proportions of certain preselected events. A trained observer observes a process at the preselected random observation times and classifies each observation into the preselected categories.

Steps in involved in conducting a behavior sampling study are as follows:

1. Define the objectives of the study. For example, to determine frequency and proportion or percentage of time spent by operators in working in unsafe manner and/or under unsafe conditions.
2. Set period of the study.
3. Define observable elements, e.g., safe and unsafe acts and conditions or situations. Train observers to identify and classify the observable elements.
4. Determine preliminary estimates of proportion of time spent in safe and unsafe elements (usually based on about 100–200 samples).
5. Determine required number of observations.

   The required number of observations (n) can be computed as follows:

$$n = z^2 p(1-p)/(A^2)$$

Where,

p = initial (or preliminary) estimate of proportion of time spent in unsafe conditions

(in decimal, e.g., $p = 0.05$ for 5%)

z = number of standard deviations for confidence level desired

(Note: $z = 1.96$ for 95% confidence level)

$A = sp$ = absolute accuracy desired on the value of proportion (p) of unsafe conditions

s = relative accuracy desired (in decimal, e.g., 0.1 for 10%)

The number of observations (n) obtained from the above formula represents the number of observations required to obtain estimate of p within the accuracy of $\pm A$ for the given confidence probability (defined by the input value of z).

6. Establish observation intervals and observation times to make n observations.
7. Design observation record sheet (e.g., tally sheets with predetermined randomized times and observation routes).
8. Conduct orientation of persons undergoing the study (e.g., keep people informed to get buy-in and train observers).
9. Conduct observations, classify elements, and record.
10. Evaluate the results (determine proportion of time spent in different types of observable elements).
11. Maintain control charts on proportion of time spent in unsafe conditions (P-chart) or number of unsafe acts and unsafe conditions (C-chart) observed in selected observation time interval.

To ensure that the data gathered during the observations are unbiased and representative of the situation in the observation area, many precautions must to exercised due to problems such as (a) the operators may change their behavior when they see an observer in their work area, (b) observer biases due to reasons such as lack of training and falsification, (c) inability of the observers to observe (e.g., due to obstructed or hidden workplaces or unlighted observation areas), (d) observation plan fails to include stratification based on influencing factors such as time of the day, type of work or department, and type of operations, and (e) certain behaviors may occur regularly and happen to take place (or not take place) in synch with the observations.

CONTROL CHARTS

Control charts can be maintained based on observations in the non-accident safety measurement studies or accident data gathered from available databases. The control charts provide visual as well as statistical information to monitor the observed processes over time. Some examples of types of controls charts that can be used are (a) C-charts for accident frequencies (or unsafe behavior), (b) P-charts for proportion of accidents of certain type (or type of observed unsafe behavior), (c) X-bar charts for lost workdays (accident severity). The control charts can be used when sufficient data are available. Generally, only large organizations can maintain and use charts.

BEFORE VS. AFTER STUDIES

The safety performance needs to be measured "before" implementation of a safety countermeasure and "after" the countermeasure is implemented to compare the data and determine if statistically significant difference exists between the "before" and the "after" situations. If the statistical analysis confirms that a significant difference in safety performance exists, then the magnitude of difference in safety performance can be also estimated. Based on the measured value of the difference, additional actions can be planned. The "before vs. after" study must be carefully planned to ensure that information gathered is valid and sufficient sample sizes are obtained for proper statistical inference.

SECURITY CONSIDERATIONS IN SYSTEMS DESIGN

Increased awareness of security issues with computers, databases, and terrorist attacks (e.g., cyber-attacks) has raised many issues with the security of large complex products and systems (e.g., commercial aircrafts, cruise ships, power plants, energy storage and distribution systems).

Some important security issues involve a) lockout of products and systems from unauthorized users, b) shielding products and systems from viruses and security threats, c) resilience from threats (how to make the product/system performance insensitive to threats), d) system shutdown or operating under high threat levels, and e) the abilities of the product/system to function during and after threat incidences.

Product and system designers, especially with embedded computerized control systems, need to include the security issues along with the safety issues discussed in this chapter.

CONCLUDING REMARKS

Product and systems safety are important areas, and the designers must make sure that their products and systems are safe. Safety requirements must be incorporated very early in the product (or system) design process. Safety reviews must be conducted during both the design and manufacturing phases to ensure that the products (or systems) do not have any design and manufacturing defects that will increase risks to the users as well as the manufacturer. It should be realized that during the product/system liability cases, the manufacturer is considered to be an expert and very knowledgeable about the product/system safety requirements, available safety devices, safety technologies, and safety-related design and manufacturing considerations. Further the manufacturer is also expected to provide warnings to the users about any potential hazard associated with the uses of the product. The design engineers should maintain proper records on safety analyses and safety-related decisions based on potential benefits of the product/system to the customers vs. costs incurred to make the product/system reasonably safe for defending their decisions if challenged during any future product reviews and liability cases.

REFERENCES

Bhise, V.D. 2012. *Ergonomics in the Automotive Design Process.* ISBN: 978-1-4398-4210-2. Boca Raton, FL: CRC Press.
Bhise, V.D. 2014. *Designing Complex Products with Systems Engineering Processes and Techniques.* ISBN: 978-1-4665-0703-6. Boca Raton, FL: CRC Press.
Brown, D.B. 1976. *Systems Analysis & Design for Safety- Safety Systems Engineering.* Englewood Cliffs, NJ: Prentice-Hall, Inc.
Colling, D.A. 1990. *Industrial Safety Management and Technology.* Englewood Cliffs, NJ: Prentice Hall.
Consumer Products Safety Commission (CPSC). 2020. National Electronic Injury Surveillance System (NEISS). Website: https://www.cpsc.gov/Research--Statistics/NEISS-Injury-Data (Accessed: October 2, 2020).

Energy Accidents. https://en.wikipedia.org/wiki/Energy_accidents.

Fitts, P.M. and R.E. Jones. 1947a. *Analysis of Factors Contributing to 460 "Pilot Errors" Experiences in Operating Aircraft Controls.* Memorandum Report TSEAA-694-12. Dayton, OH: Aero Medical Laboratory, Air Materiel Command, Wright-Patterson Air Force Base. In *Selected Papers on Human Factors in the Design and Use of Control Systems*, ed. H. Wallace. Sanaiko, NY: Dover Publications Inc.

Fitts, P.M. and R.E. Jones. 1947b. *Psychological Aspects of Instrument Display. I: Analysis of Factors Contributing to 270 "Pilot Errors" Experiences in Reading and Interpreting Aircraft Instruments.* Memorandum Report TSEAA-694-12A. Dayton, OH: Aero Medical Laboratory, Air Materiel Command, Wright-Patterson Air Force Base. In *Selected Papers on Human Factors in the Design and Use of Control Systems*, ed. H. Wallace. Sanaiko, NY: Dover Publications Inc.

Goetsch, D.L. 2007. *Occupational Safety and Health for Technologist, Engineers, and Managers.* 6th Edition. ISBN: 0132397609. Englewood Cliffs, NJ: Prentice Hall.

Hammer, W. 1980. *Product Safety Management and Engineering.* Englewood Cliffs, NJ: Prentice-Hall, Inc.

Hammer, W. 1989. *Occupational Safety Management and Engineering.* 4th edition. Englewood Cliffs, NJ: Prentice Hall.

Heinrich, H.W., Petersen, D. and N. Roos.1980. *Industrial Accident Prevention.* 5th edition. New York, NY: McGraw-Hill, Inc.

Konz, S. and S. Johnson. 2004. *Work Design: Occupational Ergonomics.* Holcomb Hathaway. ISBN 10: 1890871486 / ISBN 13: 9781890871482.

National Highway Traffic Safety Administration (NHTSA). 2020. Fatality Analysis Reporting System (FARS). Website: https://www.nhtsa.gov/research-data/fatality-analysis-reporting-system-fars (Accessed: October 2, 2020).

National Safety Council (NSC). 2012. *Injury Facts.* 2012 Edition. Itasca, IL: National Safety Council. Website: http://shop.nsc.org/Reference-Injury-Facts-2012-Book-P124.aspx (Accessed: October 25, 2012).

Petersen, D., and J. Goodale. 1980. *Readings in Industrial Accident Prevention.* New York, NY: McGraw-Hill, Inc.

Roland, H.E. and B. Moriarty. 1990. *System Safety Engineering and Management.* 2nd Edition New York, NY: John Wiley and Sons, Inc.

Tarrants, W.E. 1980. *The Measurement of Safety Performance.* New York, NY: Garland Publishing, Inc.

Section II

Measurements, Analysis, and Decision-Making

6 Decision-Making Approaches

INTRODUCTION

Decisions are made throughout the life cycle of every system or product program. Decisions are made whenever alternatives exist and the most desired alternative needs to be selected. The selected alternative should result in reducing risks and costs and increase in benefits. Many different principles can be used in selecting an alternative. Decisions are usually made based on values of one or more decision variables that can be achieved by considering each alternative. And the selected alternative is usually the one that has the highest values of the decision variables (assuming that higher values are desirable).

The early decisions are related to the type of system (or a product) (e.g., technology to be used in developing a new electricity generating plant) to be designed, requirements on the system (e.g., its maximum electricity generating capacity), and its characteristics (e.g., land space, operating temperatures, pressures voltages, polluting potential, and so forth). Later, the decisions are related to configuration and construction of the system, the number of subsystems (e.g., turbines, heat exchangers, and pumps) in the system, their functions, and how the subsystems should be configured, interfaced, and packaged within the overall system space.

Decisions are also made during each phase and at each milestone of a system/product development program where the management decides whether to proceed to the next phase, make changes to the system/product being designed, or even scrap the program because of problems such as technical and/or financial feasibility. Early decisions have a major impact on the overall costs and timings of the program – because the later decisions depend upon the design-specific parameters and their values selected in the earlier phases of the program. For example, the type of fuel to be used in the power plant will affect the type of turbines (e.g., steam or combustion) to be used in a power plant, and the later decisions will be related to the design of its systems, such as fuel handling system, cooling system, space available to package heat exchangers, and so on.

All the above described decisions involve risks. For example, adding more features (or capabilities) to the system than what the customer needs and overdesigning will waste resources. Conversely, failures in incorporating any of the customer-desired features and underdesigning the system will result in loss of sales or even degrade the reputation of the producer's brand in the marketplace.

Systems Engineering helps coordinate activities of the system/product design teams including decision-making, such as what needs to be done, when, how, how much, and determining trade-offs between different system attributes and their design requirements. This chapter covers various decision-making approaches and models and also provides an understanding into issues related to risks and methods to analyze the risks.

DOI: 10.1201/9781003107514-8

DECISION-MAKING PROBLEM FORMULATION

ALTERNATIVES, OUTCOMES, AND PAYOFFS

In a decision-making situation, the decision-maker (e.g., engineer, designer, or program manager) is faced with the task of deciding on an acceptable alternative among several possible alternatives. The decision-maker also needs to consider possible future outcomes (i.e., what events can happen in the future) and the costs or benefits (called the payoffs) associated with each combination of alternative and outcome. Further, each possible outcome may or may not occur in the future, but probabilities can be assigned for occurrences of the outcomes. There are many different decision models to determine a desired or an acceptable alternative (Blanchard and Fabrycky, 2011; Bhise, 2017; Floyd et al., 2016). A few of the models are described below.

DECISION MATRIX

Let us assume the following:

$A_i = i$th alternative, where $i = 1, 2,...., m$

$O_j = j$th outcome, where $j = 1, 2,..., n$

$P_j =$ the probability that jth outcome will occur, where $j = 1, 2,..., n$

$E_{ij} =$ evaluation measure = payoff value (positive for benefit [profit] and negative for cost [loss]) associated with ith alternative and jth outcome

The decision evaluation matrix associated for the above problem is presented in Table 6.1.

PRINCIPLES USED IN SELECTION OF AN ALTERNATIVE

Many principles can be used to select a desired alternative. The most commonly used principle is the maximum expected value principle. It is described here first and other principles are described below.

TABLE 6.1
Decision Evaluation Matrix

Alternative	Probabilities of Outcomes					
	P_1	P_2	P_3	.	.	P_n
	Outcomes					
	O_1	O_2	O_3	.	.	O_n
A_1	E_{11}	E_{12}	E_{13}	.	.	E_{1n}
A_2	E_{21}	E_{22}	E_{23}	.	.	E_{2n}
A_3	E_{31}	E_{32}	E_{33}	.	.	E_{3n}
.	.	.	.	.	.	.
.	.	.	.	.	.	.
A_m	E_{m1}	E_{m2}	E_{m3}	.	.	E_{mn}

Maximum Expected Value Principle

One commonly used principle to select an alternative is based on the maximum expected value of the payoff. The expected value of $A_i = \{E_i\}$ can be computed as:

$$\sum_j \left[P_j \times E_{ij} \right].$$

Thus, under this principle, the decision-maker will select the alternative with the maximum expected value of the payoff, which is defined as: max $\{E_i\}$ for $i = 1, 2, ..., m$.

The selection of the alternative and application of the above principle is illustrated in the following example.

Let us assume that an automotive manufacturer wants to select a powertrain for its new small vehicle. The manufacturer is considering the following five alternatives.

A_1 = Design a new small car using the current gasoline powertrain
A_2 = Do not design a new small car – continue with the present model
A_3 = Design a new small car with an electric powertrain
A_4 = Design a new small car with a diesel powertrain
A_5 = Design a small car with all the three (gasoline, diesel, and electric) powertrain options

Six possible outcomes assumed by the manufacturer are as follows:

O_1 = Economy does not change and the battery technology does not improve
O_2 = Economy improves by 5% and the battery technology does not improve
O_3 = Economy degrades by 5% and the battery technology does not improve
O_4 = Economy does not change and the battery technology improves by 50%
O_5 = Economy improves by 5% and the battery technology improves by 50%
O_6 = Economy degrades by 5% and the battery technology improves by 50%

The evaluation measures (payoff in dollars) associated with the combinations of above five alternatives and the six outcomes are provided in Table 6.2. The table also provides probabilities for each of the outcomes assumed by the manufacturer (see top two rows).

TABLE 6.2
Data for Powertrain Selection Decision Problem

	Probability of Outcome					
	0.15	0.3	0.15	0.15	0.2	0.05
	Outcomes					
Alternatives	O_1	O_2	O_3	O_4	O_5	O_6
A_1	$100,000	$120,000	$50,000	$80,000	$200,000	$100,000
A_2	-$200,000	$150,000	-$300,000	-$100,000	$100,000	$50,000
A_3	$50,000	$75,000	-$100,000	$100,000	$50,000	$150,000
A_4	$150,000	$50,000	$100,000	$75,000	$25,000	$25,000
A_5	$200,000	$100,000	$75,000	$100,000	$100,000	$75,000

The following computation illustrates the computation of the expected value of A_1 (i.e., expected value of payoff from alternative A_1).

$$\text{Expected value of } A_1 = \{E_1\} = (0.15 \times 100,000) + (0.3 \times 120,000)$$
$$+ (0.15 \times 50,000) + (0.15 \times 80,000)$$
$$+ (0.2 \times 200,000) + (0.05 \times 100,000)$$
$$= \$115,500$$

The expected values of $A_2, A_3, A_4,$ and A_5 are \$22,500, \$47,500, \$70,000, and \$110,000, respectively. Thus, the alternative A_1 has the maximum expected value of \$115,500 among the five alternatives, and it will be selected under the maximum expected value principle (see the column labeled as "Expected Value Principle" in Table 6.3).

OTHER PRINCIPLES

Six additional principles that can be used to select an alternative are described below.

1. *Aspiration level*: The principle of aspiration level assumes that the decision-maker needs to meet certain aspiration (or desired) level such as a minimum acceptable profit level or a maximum amount of tolerable loss. If we assume that the decision-maker in the above example (Table 6.3) wants to make at least \$200,000 profit, then he would consider alternatives A_1 and A_5 (because these two alternatives include the outcomes with payoff of \$200,000). On the other hand, if he does not want to incur any loss, he would not consider alternatives A_2 and A_3 (as these two alternatives can incur a loss in at least one outcome).

2. *Most probable future*: The decision-maker may decide based on the most likely (i.e., most probable) outcome, which has the highest probability of occurrence. In our example above (Table 6.3), the outcome O_2 has the highest probability, 0.3, of occurrence. Under this situation (outcome O_2), selection of alternative A_2 will ensure the maximum payoff (profit) of \$150,000.

3. *Laplace principle*: The Laplace principle assumes that the decision-maker does not have any information on the probability of occurrences of any of the outcomes, and thus, he assumes that all the outcomes are equally likely. In our example above under this principle, all the occurrence probabilities will be equal to 1/6. Thus, the decision-maker can simply take the average value of all E_{ij} for each alternative (i.e., over each i) and select the alternative with the maximum profit. In our example above, under this principle, the decision-maker would select alternative A_1 or A_5 with the maximum average payoff (profit) of \$108,333 (see the column labeled as "Laplace Principle" in Table 6.3).

4. *Maximin principle*: This principle is based on the "Extremely Pessimistic View" of the decision-maker (i.e., the nature will do its worst under every alternative). Therefore, the decision-maker will select the alternative that maximizes the value of the payoff among the minimum values of payoff

TABLE 6.3
Alternatives Selected by Five Principles

	Probability of Outcome										
	0.15	0.3	0.15	0.15	0.2	0.05					
	Outcomes						Expected Value Principle	Laplace Principle (Average Value)	Maximin Principle (Min Value)	Maxmax Principle	Hurwicz Principle (with $\alpha = 0.5$)
Alternatives	O_1	O_2	O_3	O_4	O_5	O_6					
A_1	$100,000	$120,000	$50,000	$80,000	$2000,000	$100,000	$115,500	$108,333	$50,000	$200,000	$125,000
A_2	-$200,000	$150,000	-$300,000	-$100,000	$100,000	$50,000	-$22,500	-$50,000	-$300,000	$150,000	-$75,000
A_3	$50,000	$75,000	-$100,000	$100,000	$50,000	$150,000	$47,500	$54,167	-$100,000	$150,000	$25,000
A_4	$150,000	$50,000	$100,000	$75,000	$25,000	$25,000	$70,000	$70,833	$25,000	$150,000	$87,500
A_5	$200,000	$100,000	$75,000	$100,000	$100,000	$75,000	$110,000	$108,333	$75,000	$200,000	$137,500

Note: The selected alternatives are shown in the shaded cells of the last five columns of this table.

of all alternatives (i.e., the decision-maker will reduce his loss by selecting the alternative with the least loss [or select the alternative with the highest profit among the minimum values]). The profit (P_i) in ith alternative can be defined as follows:

$$P_i = \max_i \left\{ \min_i E_{ij} \right\}$$

Table 6.3 shows that under this principle, the decision-maker will select alternative A_5, which has the highest value ($75,000) among the lowest possible values of the evaluation measure among all the alternatives (see the column labeled as the "Maximin Principle" in Table 6.3).

5. *Maxmax principle*: This principle is based on the "Extremely Optimistic View" (i.e., think about the best possible) of the decision-maker. The decision-maker will select the alternative that maximizes the maximum payoff values in each alternative, i.e., to take the maximum of the maximum values in each alternative. The profit (P_i) in i^{th} alternative can be defined as follows:

$$P_i = \max_i \left\{ \max_j E_{ij} \right\}$$

Table 6.3 shows that under this principle, the decision-maker will select alternative A_1 orA_5, which has the highest value of $200,000 among the highest possible values of the evaluation measure among all the alternatives (see the column labeled as the "Maximax Principle" in Table 6.3).

6. *Hurwicz Principle*: This principle is based on a compromise between the optimism (Maxmax principle) and pessimism (Maxmin principle). The profit (P_i) in ith alternative is computed based on selection of a value of index of optimism (α) as follows:

$$P_i = \alpha \left[\max_i \left(\max_j E_{ij} \right) \right] + (1-\alpha) \left[\max_i \left(\min_j E_{ij} \right) \right]$$

Where α = index of optimism

And α can vary as follows: $0 \le \alpha \le 1$

Note: $\alpha = 1$ indicates that the decision-maker is extremely optimistic

$\alpha = 0$ indicates that the decision-maker is extremely pessimistic

The value of P_i should be computed for each alternative using the above formula and the alternative with the maximum value of P_i should be selected.

The last column of Table 6.3 illustrates that for $\alpha = 0.5$, alternative A_5 will be selected because it has the highest value of $137,500 in the last column when the values were computed using the above expression for P_i.

SUBJECTIVE VS. OBJECTIVE METHODS

A decision can be made based on subjective judgments of the decision-maker or by use an objective decision variable. The subjective judgments are measured (or provided) by a human decision-maker (or a subject matter expert), and the objective measures are measured by use of physical instruments (e.g., kWh of electricity sold as measured by an electric meter).

For subjective measurements (or judgments), the human decision-maker should be carefully selected to ensure that the person is unbiased, very knowledgeable, or an expert in the subject matter related to the problem. Similarly, care must be exercised to ensure that the values of objective measures are obtained from physical measurement instruments (or by use of analytic models developed from analyses of relevant objective data) that are unbiased and error-free.

Several subjective and objective methods that can be used in problem-solving are presented in subsequent chapters of this book. In subjective methods, the experts (or representative users of systems or products) are asked to provide judgments or ratings based on one or more scales. In objective methods, the collected data on performance variables are usually converted into monetary units to measure profits (i.e., revenues generated minus costs). The revenues are generated from selling the output (e.g., generated electricity) plus any additional income that organization may have from sources such as investments, incentives, rebates, or credits received from other organizations (e.g., local, state, and federal tax credits). The revenues can be called as "benefits or rewards." The risks that an organization faces can be expressed as potential costs. Thus, cost–benefit analysis is commonly used in business decision-making activities (see Chapter 8).

The decision to select an alternative is based on values of a decision variable obtained from all the alternatives considered in problem-solving. The decision variable is generally selected by the decision-maker (or higher authority such as senior management personnel). For most organizations, the decision variable usually involves profits.

The subjective and objective methods used in problem-solving and described in this book are as follows:

SUBJECTIVE METHODS

1. Subjective rating methods based on a single evaluation attribute
   a. Ordering alternatives based on an evaluation attribute
   b. Rating on a scale
   c. Thurstone's method of paired comparisons (Bhise, 2012; Thurstone, 1927)
   d. Analytical hierarchical method (Satty, 1980)
2. Subjective rating methods based on multiple evaluation attributes
   a. Sum of ratings on multiple attributes
   b. Weighted sum of ratings on multiple attributes
   c. Pugh diagram based on unweighted attributes
   d. Pugh diagram based on weighted attributes
   e. Three-step analytical hierarchical method
3. Hazard analysis
4. Fault tree analysis
5. Event tree analysis
6. Failure modes and effects analysis

Objective Methods

1. Methods based on a single or multiple evaluation attributes
   a. Test methods using physical measurement methods
      Laboratory and bench tests
      Field tests
   b. Application of computer models to predict output value of an evaluation attribute
      Computational models
      Simulation models
2. Cost–benefit models to evaluate alternatives
   a. Using a single outcome
   b. Using multiple outcomes with probabilities of occurrence of outcomes

Important and useful methods from the above list are described and illustrated with examples in Chapters 5, 8, 9, 16, and 17.

INFORMATIONAL NEEDS IN DECISION-MAKING

The key to making good decisions is to have sufficient information and good understanding of issues related to the alternatives, outcomes, trade-offs, and payoffs associated with the decision situation. Therefore, it is important to select a decision-maker carefully and make sure that the person is familiar with the system (or product) and its operation (or uses). In some situations, customers who have used similar systems (or products) are asked to provide their ratings on each system/product (or alternative) used in the evaluation. On the other hand, experts who are very familiar and very knowledgeable about the alternative can be very discriminating (or even more discriminating than the most familiar customers) and can provide unbiased evaluations.

In addition, the experts can obtain additional information through other methods such as (a) benchmarking other products or systems, (b) literature surveys, (c) exercising available models (e.g., models to predict performance of systems/products under different situations) and using the information obtained from the model results, and (d) conducting experiments involving tests and evaluations.

Exercising available models under various "what if" scenarios (e.g., conducting sensitivity analyses) can provide more insights into the variability (or robustness) in the performance of a product or a system and thus, can prepare the decision-maker to make more informed decisions (see Chapter 17). Design reviews with different groups, disciplines, and experts can also generate information on strengths and weaknesses of the product/system (or product/system concept) being reviewed.

IMPORTANCE OF EARLY DECISIONS DURING SYSTEM OR PRODUCT DEVELOPMENT

"Designing right the first time" is very important because reworking any system or product design in later phases is always very time-consuming and costly. Early in the system/product development, key decisions are generally made on what technologies to use and how the system/product should be configured. Any changes to these early assumptions in the later stages of system/product development can increase costs substantially. Because such changes may require throwing away much of the early design work (and even some hardware development work) and redoing all the analyses again with a different set of assumptions and requirements.

Involvement of specialists from all key technical areas (i.e., use of multidisciplinary teams) is very important aspect of the systems engineering process as it ensures that all possible technologies and design configurations are considered as possible alternatives before converging on one or a few alternatives. The subsequent decisions are dependent upon the selected technologies and design configurations. For example, during the development of Boeing 777, the management decided on a two-engine airplane as compared with the four-engine approach used in the past long-distance commercial aircrafts. The two-engine approach required substantially more work in designing bigger engines, improving reliability of the engines, and performing additional flight tests to prove to the Federal Aviation Administration that the two-engine Boeing 777 aircrafts were as safe or safer than the four-engine aircrafts in the long trans-oceanic flights (PBS, 1995). Another example in a new material-related technology is as follows. Early during the program planning, Boeing decided to produce the Boeing 787 Dreamliner using the carbon-fiber materials as compared with using aluminum for its exterior panels and many internal components. The development of large parts (e.g., airplane wings, tail, and fuselage) with the carbon fiber materials involved many developmental challenges related to understanding and implementing the carbon-fiber technology). Modern wind turbines also use carbon fiber for its blades to reduce weight, increase strength, and in turn reduce blade breakage.

CONCLUDING REMARKS

This chapter covered the basic decision matrix-based model and issues in decision-making. Decision-making in the real world involves consideration of many issues (both internal and external to the organization), many variables and their effects, likelihoods of outcomes and associated costs that cannot be well-quantified due to reasons such as missing facts, uncertainties in the readiness of new technologies, unknown future developments, global economy, and so on.

Many models involving varied levels of complexity using many independent variables can be created to analyze effects of many risk-related variables. The models can be exercised under different assumptions (conducting sensitivity analysis) to get a good understanding of underlying variables and their effect on the decisions. However, a good decision-maker will also inject some subjectivity based on his/her intuition or judgment to make the final decisions. The decisions are never final and can be revisited after new and more reliable information is available. Delaying decisions also involves risks as late changes in decisions may involve a lot of rework.

REFERENCES

Bhise, V.D. 2012. *Ergonomics in the Automotive Design Process.* Boca Raton, FL: CRC Press.
Bhise, V.D. 2017. *Automotive Product Development: A Systems Engineering Implementation.* Boca Raton, FL: CRC Press.

Blanchard, B.S. and W.J. Fabrycky. 2011. *Systems Engineering and Analysis*. Fifth Edition. Upper Saddle River, NJ: Prentice Hall PTR.

Floyd, P., Nwaogu, T. A., Salado, R. and C. George, 2006. *Establishing a Comparative Inventory of Approaches and Methods Used by Enforcement Authorities for the Assessment of the Safety of Consumer Products Covered by Directive 2001/95/EC on General Product Safety and Identification of Best Practices*. Final Report dated February 2006 prepared for DG SANCO. Norwich: European Commission by Risk & Policy Analysts Limited.

Public Broadcasting Service (PBS). 1995. 21st Century Jet – The Building of the 777. Producers: Karl Sabbagh, David Davis and Peggy Case. PBS Home Video (5 hours). Produced by Skyscraper Products for KCTS Seattle and Channel 4 London.

Satty, T.L. 1980. *The Analytic Hierarchy Process*. New York, NY: McGraw Hill.

Thurstone, L.L. 1927. The Method of Paired Comparisons for Social Values. *Journal of Abnormal and Social Psychology*, 21: 384–400.

7 Costs, Revenues, and Time Considerations

INTRODUCTION

A cost–benefit analysis is a useful tool as an aid in decision-making. To conduct a cost–benefit analysis, costs and benefits associated with different alternatives involved in solving a problem need to be determined. Therefore, the objective of this chapter is to provide information on different types of costs and benefits that need to be considered in solving energy systems problems. We will also study concepts and review background information on variables used in computation of values of different costs and benefits. The steps in conducting cost–benefit analysis and some examples of its application will be covered in the next chapter.

TYPES OF COSTS

FIXED COSTS

These costs do not vary as a function of output (e.g., operating level of the power plant). Thus, they are called fixed costs. These costs generally incur before a plant begins to supply the output sold to the customers. The fixed costs include capital costs, which have the following categories of costs (EIA, 2016):

1. *Project equipment and financing costs*: These include costs to purchase plant equipment (e.g., boilers, turbines, generators, pumps, pipes, cables, and so forth) and to secure required project funding. i.e., costs of borrowing the required capital. These costs generally include interest, fees, and licenses. It should be noted that overnight costs do not include financing costs.
2. *Design costs*: These costs include all the funds spent in paying for resources such as manpower and equipment needed to design, analyze, and test to verify that the plant design would meet its stated requirements.
3. *Land acquisition costs*: These costs include costs to purchase or lease the land and costs for necessary permissions, permits, negotiation with selling or leasing parties to acquire land needed for the operation of the power plant.
4. *Civil and construction costs*: These costs include all construction-related costs that include manpower, equipment, and materials (used in building and constructing). It includes costs related to transportation, installation of power generating equipment, and testing tasks related to bringing the project to its operational state.

DOI: 10.1201/9781003107514-9

The capital costs thus include all the above four costs. Table 2.3 (in Chapter 2) provides capital costs (expressed in dollars per kWh produced) associated with electricity generating plants using different technologies.

Variable Costs

These costs include all costs that vary proportional to the amount of output (kWh of electrical energy) generated by the plant. It includes plant operating and maintenance costs, manpower costs, cost to purchase, transport, and fuels and process materials required to operate the plant.

The plant operating and maintenance costs generally have both the fixed and variable components. Some of the maintenance costs that do not vary with the level of plant output are generally included as fixed maintenance costs. Similarly, the costs of manpower that do not vary with plant output level are included as fixed costs. Or depending upon the organization's accounting practices, these costs may be lumped into the overhead costs (described below). Some plants may also include fixed costs such as lighting, insurance, safety inspections (that are required and do not vary proportional to the plant output level) into the fixed plant operating costs category.

Table 2.3 (in Chapter 2) provides fixed and variable operating and maintenance costs associated with electricity generating plants using different technologies.

Overheads

These costs are generally fixed costs associated with management and support staff and services provided for indirect activities such as maintenance of certain equipment, office, warehouses, parking lots, and so forth.

Safety and Accident Costs

These are additional costs the companies incur to improve safety (i.e., to reduce accidents by investing into accident prevention programs and equipment) and accident-related costs. These costs include safety training and providing safety equipment such as fire-fighting equipment, steps and ladders, hand holds, lockouts, protective suits, and hard hats. Costs due to accidents such as property damage and medical costs for injuries are included here. Insurance premiums to cover accident costs will also be a part of the safety costs. The safety and accident costs are organized into the following categories.

1. *Safety costs related to design*: These costs are associated in incorporating safety in the design of products (equipment), plant, and processes and ensuring that the products and/or systems operate safely during their operational life. It should be noted that safety is a product/system attribute. Thus, the costs related to managing the safety attribute related to the development of the system/product are considered in this category. These costs include:

a. Costs associated in gathering data on safety regulations, past accidents, and litigations

b. Costs associated in developing safety requirements, cascading safety requirements to various systems, subsystems, and components of the product/system

c. Costs to design and implement safety features in the product/system (e.g., costs associated in conducting safety analyses, product safety design reviews, meetings with experts, management, government agency experts, lawyers). These costs also include verification and validation costs related to safety systems (e.g., safety testing costs).

2. *Safety costs incurred during product/system uses/operations*: These costs incur after the product or equipment is purchased, installed, and used. These costs include:

a. Costs associated in following safety practices and precautions (e.g., training, checks, inspections) during operation of the system or product (e.g., operating electricity generating plant, maintain transmission lines or oil/gas pipelines, and storage systems)

b. Costs associated in gathering data on safety-related incidences (e.g., safety audits, meeting with customers, users, line operators, repair shop personnel, government agency personnel, and lawyers investigating product failures and accidents)

c. Conducting ongoing safety analyses and tests

d. Providing technical and legal support on product litigations, recalls, repairs, fines, customer relations campaigns, and so forth.

e. Costs associated with fixing safety-related defects (i.e., product recalls, repairs, retests)

3. *Safety costs related to system/product discontinuation and disposal*: Safety-related costs are also incurred after the product uses or plant operations are discontinued. These costs include:

a. Disposal or recycling of retired products and/or systems

b. Disposal of plant equipment and hazardous/toxic substances (e.g., toxicity tests after disposal)

4. *Accident costs*: Accident costs generally include (a) medical costs for the injured persons, (b) cost of lost work hours due to accident, (c) insurance premiums and administration costs, (d) costs to repair property damage, (e) fire-related losses, and (f) other indirect costs related to the accident.

Accident costs are typically underestimated due to many unreported and unaccounted components of the costs. Incidental costs of accidents have been estimated to be four times as great as the actual costs. Many safety researchers have explained the accident costs using the analogy of an iceberg. The visible portion of the iceberg is usually very small as compared with the submerged part of the iceberg in the water. Similarly, the reported accident costs are like the visible portion of the iceberg; and the unreported accident costs are like the submerged portion of the iceberg.

Examples of some costs that may be overlooked (or unreported) are as follows:

a. paid time to the injured employee on the day of the accident
b. paid time for any emergency-responder personnel involved in the accident case
c. paid time for all employees who were interviewed as a part of the accident investigation
d. paid time for the safety personnel who conducted the accident investigation
e. paid time for the human resources personnel who handled the worker's compensation and medical aspects of the accident
f. paid time of the supervisor involved in the accident investigation and accident response
g. paid time of the employees working near the accident who slowed down temporarily because of the accident
h. paid time of the employees who spent time talking about the accident as news of it spread through the company's grapevine

Further, comprehensive costs of accidents include not only the economic cost components, but also a measure of the value of lost quality of life associated with the deaths and injuries, i.e., what society is willing to pay to prevent them. For example, the comprehensive cost of accidents in 2013 was estimated by the National Safety Council as follows: (a) $2,600 for noninjury accident, (b) $28,600 for a possible injury accident, (c) $60,000 non-capacitating evident injury accident, (d) $235,400 for an incapacitating injury accident, and (e) $4,628,000 for an accident involving a death. Worker's compensation claims data show that average total cost of an injury claim was about $36,551 (NSC, 2013).

Estimation of Accident Costs: Accident costs are generally estimated by sum of products of number of accidents of each accident type multiplied by costs of accidents of each accident type, both obtained (or predicted) from historic data and number of hours worked by all employees in the organization within a given time period. The number of accidents is estimated from incident rates for each type of accident (i.e., number of accidents per 200,000 hours worked) multiplied by number of hours worked, and the costs of accidents of different types of accidents based on accident severity (e.g., first aid accidents, non-disabling [temporary work stoppage] injuries, disabling injuries with lost workdays, permanent partial injuries, permanent total disabling injuries, and fatalities).

Table 7.1 provides an illustration of how the accident costs for 2021 calendar year can be estimated from historic incident rates (accidents per 100 full-time workers) data collected in 2020 calendar year for a plant. Each worker is assumed to work 50 weeks/year×40 hours/week=2,000 hours/year. Total 2021 accident cost for 1,200 full-time workers was estimated to be $123.516 million.

ENVIRONMENTAL COSTS

These environmental costs generally include costs incurred to comply with regulations (e.g., the Clean Air act) by changing or modifying plants (or processing

TABLE 7.1

Illustration of Estimation of Accident Costs

Accident Type	Historic Data from 2020		2021 Data Estimated from 1200 Full-time Employees	
	Average Cost of Accident ($)	Incident Rate (Accidents/200,000 hours)	Estimated Number of Accidents	Est. Cost of Accidents
First aid	$250	30	360	$90,000
Non-disabling [temporary work stoppage] injuries	$1,500	7	84	$126,000
Disabling injuries with lost work days	$15,000	5	60	$900,000
Permanent partial disabling injuries	$3,000,000	2	24	$72,000,000
Permanent total disabling injuries	$6,000,000	0.5	6	$36,000,000
Fatalities	$6,000,000	0.2	2.4	$14,400,000
			Total est. cost of accidents→	$123,516,000

equipment) that generate polluting materials and costs associated with their cleanup. For example, changing a coal-fired plant to a natural-gas-fired plant will require substantial costs in plant modifications, equipment installation, and environment monitoring. Adding carbon capture and sequestering capability to a coal or natural gas power plant will also require substantial and costly modifications (see Chapter 2).

The social costs of carbon, methane, and nitrous oxide (described below) are concepts developed and used by federal agencies to provide comprehensive cost estimates of climate-change-related damages. These estimates have been used in cost–benefit analyses (e.g., in estimating impact of fuel economy and emissions requirements proposed jointly by NHTSA and EPA in 2012) (EPA, 2020a).

Social Cost of Carbon

The social cost of carbon (SC-CO_2) is a measure, in dollars, of the long-term damage done by one metric ton of carbon dioxide (CO_2) emissions in a year. EPA and other federal agencies use estimates of the social cost of carbon (SC-CO_2) to value the climate impacts of rulemakings (EPA, 2020a). This dollar figure also represents the value of damages avoided for a small emission reduction (i.e., the benefit of a CO_2 reduction).

The SC-CO_2 is meant to be a comprehensive estimate of climate change damages and includes changes in net agricultural productivity, human health, property damages from increased flood risk, and changes in energy system costs, such as reduced costs for heating and increased costs for air-conditioning. However, given current modeling and data limitations, it does not include all important damages. The IPCC (Intergovernmental Panel on Climate Change) Fifth Assessment report observed that SC-CO_2 estimates omit various impacts that would likely increase damages.

One of the most important factors influencing SC-CO_2 estimates is the discount rate.

A large portion of climate change damages are expected to occur many decades into the future, and the present value of those damages (the value at present of damages that occur in the future) is highly dependent on the discount rate.

Present value effect: To understand the effect that the discount rate has on present value calculations, consider the following example. Let us say that you have been promised that after 50 years, you will receive $1 billion. In "present value" terms, that sum of money is worth $291 million today with a 2.5% discount rate. In other words, if you invested $291 million today at 2.5% and let it compound, it would be worth $1 billion in 50 years. A higher discount rate of 3% would decrease the present value today to $228 million, and the present value would be even lower, i.e., 87 million with 5% discount rate. This effect is even more pronounced when looking at the present value of damages further out in time. (Also see "Effect of Time on Costs and Revenues" section provided later in this Chapter).

Table 7.2 presents the values of social cost of CO_2 for 2015–2050 for each metric ton of CO_2 increased (benefit if one metric ton of CO_2 is reduced) by assuming different discount rates (5% average, 3% average, 2.5% average discount rate, and 95th percentile value [as a worse case at 3% discount rate]).

The timing of the emission release (or reduction) is key to estimation of the SC-CO_2, which is based on a present value calculation. The integrated assessment models first estimate damages occurring after the emission release and into the future, often as far out as the year 2300. The models then discount the value of those damages over the entire time span back to present value to arrive at the SC-CO_2.

TABLE 7.2
Social Cost of CO_2, 2015–2050 (in 2007 Dollars per Metric Ton CO_2)

Year	5% Average	3% Average	2.5% Average	High Impact (95th pct at 3%)
2015	$11	$36	$56	$105
2020	$12	$42	$62	$123
2025	$14	$46	$68	$138
2030	$16	$50	$73	$152
2035	$18	$55	$78	$168
2040	$21	$60	$84	$183
2045	$23	$64	$89	$197
2050	$26	$69	$95	$212

Note: The SC-CO_2 values are dollar-year and emissions-year specific.

For example, the SC-CO$_2$ for the year 2020 represents the present value of climate change damages that occur between the years 2020 and 2300 (assuming 2300 is the final year of the model run); these damages are associated with the release of one metric ton of carbon dioxide in the year 2020.

Calculations Related to Social Cost of Carbon

The following illustration provides steps involved to computing the social cost of carbon for a 500 MW natural-gas-fueled combined cycle (NGCC) power plant.

1. Natural gas-fueled combined cycle (NGCC) power plant power plant emits 117 lb of CO$_2$/MMBtu (see Table 2.3).
2. The social cost of carbon (SS-CO$_2$) is \$42/Metric Ton of CO$_2$ for 2020 at 3% Discount Rate (see Table 7.2).
3. 1 kwh of energy is equal to 3,412 Btu.
4. Therefore, the output of a 500 MW NGCC Plant $= 500,000$ kW $\times 3,412/10^6$
   $= 1,706$ MMBtu/h
5. The amount of carbon emitted by the 500 MW NGCC plant $= 117 \times 1,706/2,204.62$
   $= 90.54$ Metric Tons of CO$_2$/h. (Note: 1 metric ton $= 2,204.62$ lb)
6. The social cost of carbon (SSC) of the plant $= 90.54 \times \$42 = 3,802.60$ \$/h
   $= 3,802.60 \times 365 \times 24$ \$/y $= \$33.31$ million/year

Discussions – Who Pays for the SSC?

As stated earlier, many government agencies have used the SS-CO$_2$ concept to compute costs of carbon (if CO$_2$ pollution is increased) or benefits from reduction of carbon in conducting cost–benefit analysis of proposed changes in government actions (e.g., new regulations on pollutions). These costs or benefits are generally incurred or gained, respectively, by the society (i.e., neighborhoods in the vicinity of the proposed changes). The additional costs due to increased pollution are generally borne by a combination of number of entities such as individuals/patients, families, insurance companies, employers, and/or shareholders. To reduce carbon output from fossil-fuel-powered plants, the utility companies need to invest in costly carbon capture systems, and it is also costly to operate the carbon capture and sequestration (CCS) equipment. Thus, the CCS systems eventually end up reducing their profits and earnings per share. Higher taxes or fines on the polluters will also increase their operating costs. On the other hand, the renewable energy sources such as wind turbine or solar have no SC-CO$_2$.

Social Cost of Methane (SC-CH$_4$) and the Social Cost of Nitrous Oxide (SC-N$_2$O)

EPA and other federal agencies also use estimates of the social cost of methane (SC-CH$_4$) and the social cost of nitrous oxide (SC-N$_2$O) in analyses of regulatory actions that are projected to influence CH$_4$ or N$_2$O emissions in a manner consistent with how CO$_2$ emission changes are valued. The SC-CH$_4$ and SC-N$_2$O estimates are taken from a paper by Marten and Newbold (2011), which provided the first set of published SC-CH$_4$ and SC-N$_2$O estimates that are consistent with the modeling assumptions underlying the SC-CO$_2$ estimates. These costs are provided in

Tables 7.3 and 7.4. Comparing the cost data provided in Table 7.2 to corresponding data in Tables 7.3 and 7.4, the social cost values for methane and nitrous oxide are much higher than those of CO_2. This is because the severity of toxicity effects of methane and nitrous oxide is much higher than that of CO_2. For example, the EPA's equivalency calculator shows that 1 metric ton of methane is equivalent to 25 metric tons of CO_2, and 1 metric ton of N_2O is equivalent to 298 metric tons of CO_2 (EPA, 2020b).

LEVELIZED COST OF TECHNOLOGIES

EIA calculates two following measures, levelized cost of electricity (LCOE) and levelized cost of avoided electricity (LACE) that, when used together, largely explain the economic competitiveness of electricity generating technologies (EIA, 2020).

TABLE 7.3
Social Cost of CH_4, 2015–2050 (in 2007 Dollars/Metric Ton CH_4)

Year	5% Average	3% Average	2.5% Average	High Impact (95th pct at 3%)
2015	$450	$1,000	$1,400	$2,800
2020	$540	$1,200	$1,600	$3,200
2025	$650	$1,400	$1,800	$3,700
2030	$760	$1,600	$2,000	$4,200
2035	$900	$1,800	$2,300	$4,900
2040	$1,000	$2,000	$2,600	$5,500
2045	$1,200	$2,300	$2,800	$6,100
2050	$1,300	$2,500	$3,100	$6,700

Note: The SC-CH_4 values are dollar-year and emissions-year specific.

TABLE 7.4
Social Cost of N_2O, 2015–2050 (in 2007 Dollars/Metric Ton N_2O)

Year	5% Average	3% Average	2.5% Average	High Impact (95th pct at 3%)
2015	$4,000	$13,000	$20,000	$35,000
2020	$4,700	$15,000	$22,000	$39,000
2025	$5,500	$17,000	$24,000	$44,000
2030	$6,300	$19,000	$27,000	$49,000
2035	$7,400	$21,000	$29,000	$55,000
2040	$8,400	$23,000	$32,000	$60,000
2045	$9,500	$25,000	$34,000	$66,000
2050	$11,000	$27,000	$37,000	$72,000

Note: The SC-N_2O values are dollar-year and emissions-year specific.

Levelized Cost of Electricity

The LCOE represents the installed capital costs and ongoing operating costs of a power plant, converted to a level stream of payments over the plant's assumed financial lifetime. Installed capital costs include construction costs, financing costs, tax credits, and other plant-related subsidies or taxes. Ongoing costs include the cost of the generating fuel (for power plants that consume fuel), expected maintenance costs, and other related taxes or subsidies based on the operation of the plant.

The levelized cost of energy (LCOE), or levelized cost of electricity, is a measure of the average net present cost of electricity generation for a generating plant over its lifetime. The LCOE is calculated as the ratio between all the discounted costs over the lifetime of an electricity generating plant divided by a discounted sum of the actual energy amounts delivered. It is measured in $/MWh. The LCOE is used to compare different methods of electricity generation on a consistent basis.

LCOE represents the average revenue per unit of electricity generated that would be required to recover the costs of building and operating a generating plant during an assumed financial life and duty cycle. LCOE is often cited as a convenient summary measure of the overall competitiveness of different generating technologies.

Key inputs to calculating LCOE include capital costs, fuel costs, fixed and variable operations and maintenance (O&M) costs, financing costs, and an assumed utilization rate for each plant type. The importance of each of these factors varies across technologies.

The LCOE is calculated as:

$$LCOE = (\text{Sum of costs over lifetime}) /$$
$$(\text{Sum of electrical energy produced over lifetime})$$

Where,

$$\text{Sum of costs over lifetime} = \Sigma (I_t + M_t + F_t) / (1+r)^t$$

$$\text{Sum of electrical energy produced over lifetime} = \Sigma\, E_t / (1+r)^t$$

I_t = investment expenditures in the year t
M_t = operations and maintenance expenditures in the year t
F_t = fuel expenditures in the year t
E_t = electrical energy generated in the year t
r = discount rate
n = expected lifetime of system or power station
Σ = Sum over from $t=1$ to n years

Figure 7.1 presents levelized cost of electricity for different electricity generating technologies. (EIA, 2020).

Levelized Avoided Cost of Electricity

The levelized avoided cost of electricity (LACE) represents a power plant's value to the grid. A generator's avoided cost reflects the costs that would be incurred to provide the electricity displaced by a new generation project (e.g., a solar generator

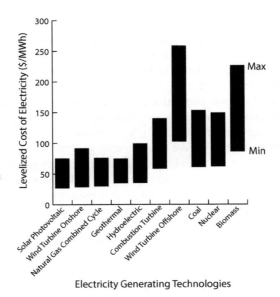

FIGURE 7.1   Levelized cost of electricity for different electricity generating technologies. (Data projected for 2025 from: Annual Energy Outlook, EIA, 2020.)

produced power sold back to the utility company) as an estimate of the revenue available to the plant. As with LCOE, these revenues are converted to a level stream of payments over the plant's assumed financial lifetime.

BENEFITS

Benefits include the amount of cash gained (income) by an organization (e.g., a utility company) by (a) selling electricity to its customers (i.e., the revenue generated from selling electricity), (b) avoided cost of producing energy (within utility company's own plants) or purchasing additional energy from other high cost suppliers, (c) income from investments, and (d) credits or incentives gained by the organization in the form of tax credits.

REVENUES

The revenues are generally considered as the amount collected by selling products or services. The revenue is computed by multiplying number of units sold times the selling price per unit. Thus, the revenue generated by selling electricity is number of kWh of electricity sold times rate of electricity sold at dollars per kWh. (The utility companies also add other costs such as transmission costs, improvement charges, surcharges, meter, and equipment charges, etc., to the bill to recover charges in addition to the electricity generating charges. These are cost recoveries and not revenues.)

POWER PURCHASE AGREEMENTS

A power purchase agreement (PPA) is a legal contract between an electricity generator (provider, typically an owner of renewable electric generating facility, e.g., a solar, wind turbine, or a biomass plant) and a power purchaser (buyer, typically a utility or large power buyer/trader). Contractual terms may last anywhere between 5 and 20 years, during which time the power purchaser buys energy, and sometimes also capacity and/or ancillary services, from the electricity generator. Such agreements play a key role in the financing of independently owned (i.e., not owned by a utility) electricity generating assets. The seller under the PPA is typically an independent power producer (IPP).

AVOIDED COSTS

In the context of electricity generation, the costs a utility would otherwise incur to generate electricity if it did not purchase the electricity from another source, such as a qualifying facility (QF) is defined as the avoided cost. Thus, it is a benefit to the utility company.

To promote energy conservation and the rational pricing of electricity, Congress enacted the Public Utilities Regulatory Policies Act of 1978 (PURPA), which required utilities to buy power from nonutility producers if the cost of doing so was less than the utility's avoided cost rate to the consumer. This was intended to force utilities to buy electricity from small power producers at a price that would keep them in business and provide cost savings to consumers. However, following the deregulation of the electricity industry, the avoided costs calculation is no longer as relevant. This is because (a) many utilities no longer own generation facilities and buy a lot of their power in the wholesale markets from power marketers and other suppliers of electricity, and (b) The Energy Policy Act of 2005 amended the avoided cost requirement to provide that no utility will be required to buy electricity from a QF if certain conditions (including the QF having nondiscriminatory access to the grid), as determined by the Federal Energy Regulatory Commission, are met. Where the avoided cost calculation applies, it is determined by the applicable state utilities commission.

AVOIDED COSTS OF ELECTRICITY

For the customer of a utility company, it is the amount of cost of electricity that the customer does not have to pay (i.e., avoided paying for it) to the utility company because the customer has electricity generating capability (e.g., a solar plant or a wind turbine) that is used to supply his own electricity needs. Thus, it is a benefit the customer.

EFFECT OF TIME ON COSTS AND REVENUES

As the costs are incurred over time, in determining all the above costs, the effect of time due to factors such as interest rate (or discount rate), inflation rate, and fluctuations in currency exchange rates (if applicable) must be taken into account. Similarly, since the revenues are generated over the selling periods of the products and payments are received over time, the effects of changes in interest rates, inflation, and currency exchange rates should be also considered.

Most complex product (or power plant programs) programs extend over many years. Therefore, the cost computations need to consider the effects of interest and inflation. The computations can be made by using the following variables and the formulas.

Let P = present value (or value at a time assumed to be the present)

i = combined annual interest and inflation rate

$$= i_r + i_f$$

i_r = annual interest rate

i_f = annual inflation rate

n = number of annual interest periods

F = future value after n periods

With the annual compounding of the combined interest and inflation, the relationship between P and F is as follows (Blanchard and Fabrycky, 2011):

$$F = P(1+i)^n \text{ or } P = F\left[1/(1+i)^n\right]$$

Using the above formula, the value of $100 today will be $128 in 5 years at 5% combined annual interest and inflation rate [Note: $128 = 100 (1+0.05)^5$]. This means that $128 spent 5 years from now will be equivalent to $100 today assuming 5% rate of combined interest and inflation.

For a program extending over many periods, the present value of revenues minus the present value of costs can be computed for each period. The present values for each of the periods (assumed to be monthly, quarterly, or annually) can be summed over the entire duration of the program to obtain present value of the cumulative cash flow. The present value is generally computed at the beginning of the program to provide the management the estimate of cash flow over the life of the program.

EFFECT OF TIME

During the entire life cycle of a program (from its conception till its discontinuation and disposal), revenues are generated, and costs are incurred. Because of the time value of money is different due to effect of discount rate (interest rate) and inflation, dollars gained or spent at different time periods are different. Thus, to account for the differences, revenues earned, and costs incurred are converted into their value at the present time, called the "present value" (PV). Since spending $110 a year from now at 10% discount rate will be equivalent to spending $100 today, the present value of $110 spent next year will be worth $100. The present value approach thus allows to estimate all revenues and costs spent in different time periods in their respective present values, which can be summed to get a more accurate financial picture.

The formula for computation of present value (P) of a future value (F) spent at nth annual time period at a discount rate (i) is given below.

$$P = F/(1+i)^n$$

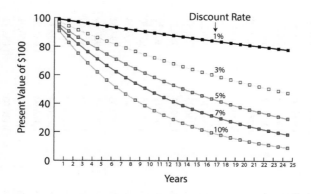

FIGURE 7.2 Present value of $100 at the end of each future year over the next 25 years at discount rates of 1%, 3%, 5%, 7%, and 10%.

Using the above formula, Figure 7.2 illustrates how the present value of 100 dollars paid in a future year decreases as a function of future year and discount rate. For example, the figure shows that $100 paid after 17 years (in future) at 3% discount rate will be worth about $60 at the present time.

Utility Company Example: Present Value Calculations

Let us look the following example illustrated in Table 7.5. A utility company wants to evaluate its business of providing electric power for 25 years from 2020 to 2044. The energy demand in 2020 is assumed to be 20,000 kWh at the rate of $0.12/kWh. The power required assumed to increase by 2% each year and the electricity rate is assumed to increase at 2.5% each year. The revenue generated from selling the electricity is thus calculated to be the multiplication of energy sold by price per kWh for each year. The revenue is shown in fifth column of the table. The utility company is assumed to obtain their power from the electric grid at a fixed rate of $0.10/kWh on a long 25-year contract with another utility company that is connected to the same grid. The present value of revenue and cost for each year was computed by using discount rate of 5% each year. The last column of the table shows the net present value computed by subtracting present value of the cost from the present value of the revenue for each year. The sums over 25 years for each of the last three columns are shown at the bottom of the table. The net gain in the present value over the 25 years is $ 20,926.61.

Sum of net present value over the 25 years is $20,926.61. This value is positive indicating that the energy business was profitable. The ratio of net present value of revenue to net present value of costs was 1.578 (i.e., 57,013.33/36,086.72). Since the value of 1.578 is larger than 1.0, it also indicates that the business was profitable.

From the data presented in Table 7.5, the following two observations can be made:

a. The revenue generated (fifth column) increased every year due to annual increases in both, the energy used (third column) and price per kWh (fourth column). However, the present value of the revenue (seventh column) decreased every year because of the 5% discount rate over time.

TABLE 7.5

Illustration of Net Present Values for a 25-Year Electric Power Supply Situation

Sr. No.	Year	Energy Used (kWh)	Price per kWh ($)	Revenue Generated ($)	Cost ($)	Present Value of Revenue ($)	Present Value of Costs ($)	Net Present Value ($)
1	2020	20,000.00	0.1200	2,400.00	2,000.00	2,400.00	2,000.00	400.00
2	2021	20,400.00	0.1230	2,509.20	2,040.00	2,389.71	1,942.86	446.86
3	2022	20,808.00	0.1261	2,623.37	2,080.80	2,379.47	1,887.35	492.13
4	2023	21,224.16	0.1292	2,742.73	2,122.42	2,369.27	1,833.42	535.85
5	2024	21,648.64	0.1325	2,867.53	2,164.86	2,359.12	1,781.04	578.08
6	2025	22,081.62	0.1358	2,998.00	2,208.16	2,349.01	1,730.15	618.86
7	2026	22,523.25	0.1392	3,134.41	2,252.32	2,338.94	1,680.72	658.22
8	2027	22,973.71	0.1426	3,277.02	2,297.37	2,328.92	1,632.70	696.22
9	2028	23,433.19	0.1462	3,426.13	2,343.32	2,318.94	1,586.05	732.89
10	2029	23,901.85	0.1499	3,582.02	2,390.19	2,309.00	1,540.73	768.27
11	2030	24,379.89	0.1536	3,745.00	2,437.99	2,299.10	1,496.71	802.39
12	2031	24,867.49	0.1575	3,915.40	2,486.75	2,289.25	1,453.95	835.30
13	2032	25,364.84	0.1614	4,093.55	2,536.48	2,279.44	1,412.41	867.03
14	2033	25,872.13	0.1654	4,279.80	2,587.21	2,269.67	1,372.05	897.62
15	2034	26,389.58	0.1696	4,474.53	2,638.96	2,259.94	1,332.85	927.09
16	2035	26,917.37	0.1738	4,678.12	2,691.74	2,250.26	1,294.77	955.49
17	2036	27,455.71	0.1781	4,890.98	2,745.57	2,240.61	1,257.78	982.84
18	2037	28,004.83	0.1826	5,113.52	2,800.48	2,231.01	1,221.84	1,009.17
19	2038	28,564.92	0.1872	5,346.18	2,856.49	2,221.45	1,186.93	1,034.52
20	2039	29,136.22	0.1918	5,589.44	2,913.62	2,211.93	1,153.02	1,058.91
21	2040	29,718.95	0.1966	5,843.75	2,971.89	2,202.45	1,120.08	1,082.37
22	2041	30,313.33	0.2015	6,109.65	3,031.33	2,193.01	1,088.07	1,104.94
23	2042	30,919.59	0.2066	6,387.63	3,091.96	2,183.61	1,056.99	1,126.63
24	2043	31,537.99	0.2118	6,678.27	3,153.80	2,174.25	1,026.79	1,147.47
25	2044	32,168.74	0.2170	6,982.13	3,216.87	2,164.94	997.45	1,167.49
				Sum of present values→		5,7013.33	36,086.72	20,926.61

b. Net present value (last column, which represents the net present value of profits – which is obtained by subtracting the net present value of costs from the net present value of revenue) increased with years from 2020 to 2044.

CONCLUDING REMARKS

To stay in business, the energy companies must make profit. Therefore, they must keep track of their costs and make sure that the revenue exceeds their costs. Since the costs and revenues generated over many years are affected by time, discount rate, and inflation rate, conversion of all costs and revenues at a predefined present time is a useful method to determine economic success or failure of any business. The next chapter shows that the present values can be computed for each alternative and used to determine the most desired alternative.

REFERENCES

Blanchard, B.S. and W.J. Fabrycky. 2011. *Systems Engineering and Analysis*. 5th Edition. Upper Saddle River, NJ: Prentice Hall PTR.

EIA. 2016, November. *Capital Cost Estimates for Utility Scale Electricity Generating Plants*.

EIA. 2020. EIA Uses Two Simplified Metrics to Show Future Power Plants' Relative Economics. Website: https://www.eia.gov/todayinenergy/detail.php?id=35552 (Accessed: October 3, 2020).

EPA. 2020a. Social Cost of Carbon. Website: https://19january2017snapshot.epa.gov/climatechange/social-cost-carbon_.html (Accessed: October 3, 2020).

EPA. 2020b. Greenhouse Gas Equivalencies Calculator. Website: https://www.epa.gov/energy/greenhouse-gas-equivalencies-calculator (Accessed July 21, 2020).

Environmental Protection Agency and National Highway Traffic Safety Administration. 2012. 2017 and Later Model Year Light-Duty Vehicle Greenhouse Gas Emissions and Corporate Average Fuel Economy Standards. Federal Register, Vol. 77, No. 199, October 15, 2012, Pages 62623–63200. Environmental Protection Agency, 40 CFR Parts 85, 86, and 600. Department of Transportation National Highway Traffic Safety Administration, 49 CFR Parts 523, 531, 533, 536, and 537. [EPA-HQ-OAR-2010-0799; FRL-9706-5; NHTSA-2010-0131]. RIN 2060-AQ54; RIN 2127-AK79.

Marten, A.L. and S.C. Newbold. 2011. Estimating the Social Cost of Non-CO_2 GHG Emissions: Methane and Nitrous Oxide. Working Paper # 11-01. Washington, DC: U.S. Environmental Protection Agency National Center for Environmental Economics. Website: http://www.epa.gov/economics. January, 2011. Website: https://www.epa.gov/sites/production/files/2014-12/documents/estimating_the_social_cost_of_non-co2_ghg_emissions_0.pdf (Accessed: June 10, 2021).

National Safety Council (NSC). 2013. *Injury Facts*. 2013 Edition. Itasca, IL: NSC.

8 Cost–Benefit Analysis

INTRODUCTION

Cost–benefit analysis is a useful method for decision-making. The analysis requires understanding and use of costs and benefits associated with each alternative involved in the problem being solved. The benefits generated from the use of a system should be higher than the costs associated in operating and maintaining the system. This chapter provides information on how to create a cost–benefit analysis and presents an application of the method to determine the acceptability of a solar photovoltaic power source for residential purposes. The basics of costs and benefits of an energy system are described in Chapter 7.

COST–BENEFIT ANALYSIS: WHAT IS IT?

A cost–benefit analysis is a process by which a decision-maker can analyze available alternatives and make decisions related to selection of an alternative involving a system or a project. To conduct a cost–benefit analysis, a decision model is first developed by identifying alternatives and outcomes and by determining benefits and costs of each combination of alternatives and outcomes. The decision is made based on one or more decision principles and maximum values of evaluation measures obtained by (a) subtracting the costs from benefits and/or (b) computing ratios of benefits-to-costs (see decision matrix covered in Chapter 6).

WHY USE COST–BENEFIT ANALYSIS?

Cost–benefit analysis is an objective method of decision-making. However, to determine the alternatives, outcomes, costs, and benefits (associated with each combination of alternatives and outcomes), the cost–benefit analysis relies on the abilities of the analyst and/or the team involved in its formulation and estimation of values of the variables included in the analysis.

Organizations rely on cost–benefit analysis to support decision-making because it provides an objective and evidence-based view of the issues being evaluated – without the influences of opinions, politics, or biases of decision-makers and other individuals within and outside the organization who may directly or indirectly be affected by the decision. By providing an unclouded view of the consequences of a decision, the cost–benefit analysis is an invaluable tool in developing business strategy, evaluating a proposal, or making resource allocation or purchase decisions. In business, government, finance, and even the nonprofit world, the cost–benefit analysis offers unique and valuable insights. Some examples of its applications are as follows:

a. Comparing project proposals
b. Deciding whether to pursue a proposed project

DOI: 10.1201/9781003107514-10

c. Evaluating alternate locations for a new power plant or an oil well
d. Weighing different investment opportunities
e. Measuring social benefits of proposed changes in regulations
f. Appraising the desirability of suggested policy alternatives
g. Assessing change initiatives
h. Determining effects of economic or political change on stakeholders and participants

STEPS INVOLVED IN COST–BENEFIT ANALYSIS

While there is no "standard" format for performing a cost–benefit analysis, there are certain core elements that will be present across almost all such analyses. The five basic steps to performing a cost–benefit analysis include the following:

1. Identify alternatives, outcomes, and probabilities of the outcomes
2. Identify costs and benefits so they can be categorized by type (see Chapter 7)
3. Calculate costs and benefits for each combination of alternative and outcome over the assumed life of the project or initiative at the beginning of the project planning time using the "present value" method (see Chapter 7)
4. Select principles to be used in evaluating the alternatives (see Chapter 6)
5. Calculate evaluation measures for all alternatives and make recommendations. The most used evaluation measures are (a) the net present value (i.e., sum of the present value of benefits over the life of the project minus the sum of the present value of costs over the life of the project), and (b) the benefit-to-cost ratio (i.e., ratio of sum of the present value of benefits over the life of the project to the sum of the present value of costs over the life of the project). The evaluation measures are computed for each combination of alternatives and outcomes, and the alternative that has the highest value of the evaluation measure is usually selected (or considered further for company management concurrence).

As with any process, it is important to work through all the steps thoroughly and not give in to the temptation to cut corners or base assumptions on opinion or guesses. It is important to ensure that the analysis is as comprehensive as possible (i.e., it covers all costs and benefits incurred over the entire life cycle of the project). Cost–benefit analysis does not require any specific tool to display and calculate costs, benefits, and evaluation measures. However, tabular formats are generally used to display data and spreadsheets are commonly used to perform calculations.

SOME EXAMPLES OF PROBLEMS FOR APPLICATION OF COST–BENEFIT ANALYSIS

Any decision-making problem in the energy systems area can be analyzed by applying the cost–benefit analysis. Some examples of problems in the energy systems area that involve computations of costs and benefits associated with different alternatives considered in problem solving are described below:

1. *Select a technology for building a new electric power plant*: Alternatives considered here are different technologies (e.g., natural-gas-fired plant, nuclear power plant, geothermal power plant, wind turbines, and so forth) used to produce electricity. The costs involved are plant construction and equipment installation costs (capital costs), operating and maintenance costs, insurance costs, safety, and accident costs. The benefits are typically revenue generated, rebates and tax credits, and economic impact on the community where the plant will be located (see Chapter 16 for a detailed example covering this problem).

2. *Select a technology for carbon capture and sequestration for fossil fuel power plants*: Alternatives here will be different methods used to capture and sequestration of the carbon and the percentage of emitted carbon that is captured. The costs will include construction and installation of the carbon capturing and sequestration equipment, operating and maintenance of the specialized equipment, safety, and accident costs. The benefits will include reduction in pollution-related costs (e.g., social cost of carbon based on reduction of metric tons of carbon dioxide over a period and discount rate).

3. *Develop a plan for improvement in existing electric grid*: Here the alternatives will be different approaches and specific levels of improvements that will be considered in improving the grid (e.g., adding more energy sources and power lines, incorporating energy demand measurement, and electricity distribution control center, storage of excess energy). The costs will include installation costs of additional equipment and additional operating and maintenance and safety and accident costs. The benefits will be gains due to improved grid efficiency, reduced power outages, increased power usage, and so forth.

4. *Develop future requirements to control pollution from various industries*: The alternatives that are considered here are different methods that can be used in controlling pollution and level of pollution or percent reduction in pollution that can be achieved by each alternative (e.g., pollution controls on factories producing paints and plastics, steel producing plants, and retiring coal-fueled power plants). The costs will be those associated with various changes that need to be made in different industries affected by the new requirements. The gains will be based on reduction in social and health-related costs and additional economic benefits gained (e.g., more people moving into the area with reduced pollution levels and additional jobs created) due to the implementation of the proposed requirements.

5. *Develop future requirements on transportation and distribution of fuel and/or energy*: The alternatives to be considered here are type of requirements and their applicability to different fuel and power distribution systems. The costs will include capital costs in incorporating the changes, operating and maintenance costs associated with the changes, and changes in other costs (e.g., safety costs). The benefits will include revenues from additional

sale of fuel/energy and avoided costs in fuel and power distribution due to the changes and economic impact in the local community due to the changes.

6. *Develop future requirements on fuel economy and emissions from motor vehicles*: The alternatives will be type of changes to be made to future motor vehicles (e.g., weight reduction, changes in powertrains, and aerodynamic improvements). The costs will include costs to develop and incorporate new changes in future vehicles; and benefits will include reduction in cost of fuel and reduction in pollution-related costs (e.g., medical costs, climate-change-related costs).

COST–BENEFIT ANALYSIS OF RESIDENTIAL SOLAR PANELS: AN EXAMPLE

PROBLEM

The problem considered in this section is to conduct two cost–benefit analyses for installing 12 kW residential photovoltaic (PV) solar energy systems to generate electric energy in two average size homes, one in Detroit, MI and other in Phoenix, AZ. The analysis procedure used here was similar to one described by the Solar Foundation (2012).

Two spread sheets (one for each city) were prepared by using assumptions and data from the websites given below. The data were used to compute benefit-to-cost ratios for homes in the two cities.

Assumptions
1. Life of Equipment: 25 years
2. Financing Period: 25 years
3. Loan Interest (or discount rate): 5% per year

Websites
1. Solar Photovoltaic System Costs:
   U.S. Solar Photovoltaic System Cost Benchmark: Q1 2017 (Fu et al., 2017; Barbose and Dargh, 2019).
2. Energy Consumption:
   http://www.eia.gov/consumption/residential/reports/2009/state_briefs/
   http://www.eia.gov/consumption/residential/reports/2009/state_briefs/pdf/mi.pdf
   http://www.eia.gov/consumption/residential/reports/2009/state_briefs/pdf/az.pdf
3. PV Watts:
   http://rredc.nrel.gov/solar/calculators/PVWATTS/version1/
4. Financial Incentives for Solar PV:
   http://programs.dsireusa.org/system/program?state=MI
   http://programs.dsireusa.org/system/program?state=AZ

Costs of Going Solar
1. The average weighted installed cost of solar for a residential nonutility solar energy system was assumed to be $3.30/Watt (Barbose and Dargh, 2019).

2. Equipment costs are those associated with purchasing the hardware necessary for installing a solar energy system. For a rooftop PV system,
hardware components include the PV modules, solar power inverters,
mounting and racking hardware, meters, disconnect devices, and system
wiring.
3. Securing interconnection approvals and system inspection costs
4. Project financing costs
5. Sales taxes on purchase of equipment and services
6. Operation and maintenance costs (O&M costs) (about $6–27/kW): System
cleaning, replacement of broken panels, and inverter replacement.
7. Cost of supplemental insurance (0.25%–0.50% of installed costs)
8. Power Backup System Costs: Solar plant can only provide power during
daytime and the power output will depend upon intensity of sun illumination. Thus, a backup power supply (e.g., from the utility company) or a
battery would be needed to provide power when the solar system cannot
provide the required level of electric power.

Benefits of Going Solar
1. Avoided energy costs: These are long-term energy cost savings (avoid paying retail costs of energy consumed and other fixed costs per month for
service) due to use of electricity generated by the solar system.
2. Value of excess solar generated energy sold to the utility company
3. Benefits from reduced pollution: These are reduced costs due to less pollution generated by utility company supplied power. These can be in the form
of rebates provided by the utility company to the solar electricity producers
(e.g., solar renewable energy credits [SREC]).
4. Incentives provided by federal, state, and local governments: Cash rebates
and grants, federal investment tax credit, and low interest loans.
5. Utility company provided incentives

Solar Energy Output
1. Capacity Rating is a measure of the size of a solar energy system, typically
measured in watts (W) or kilowatts (kW).
2. The output of the solar panels is reduced due to losses in DC-to-AC conversion. Derate factor is used to account for energy losses in conversion.
3. Peak Sun Hours is a measure of the number of hours per day that solar irra-
diance (the amount of solar radiation falling on a particular area) is at it is
maximum (i.e., $1,000\,W/m^2$).
4. The annual energy output of the solar plant can be calculated by using the
following formula:

$$\text{Annual output} = DCR \times DF \times PSH \times 365 \ kWh/y \qquad (8.1)$$

Where, $DCR = DC$ Rating of the solar plant (kW)
$DF =$ Derate factor
$PSH =$ Peak sun hours per day ($kWh/m^2/day$)

Cost–Benefit Analysis and Calculations

In this study two cost–benefit analyses were conducted for installing 12 kW residential PV solar energy systems to generate energy in two average size homes, one in Detroit, MI, and the other in Phoenix, AZ (Schwager et al., 2020). The spreadsheets for the two analyses are presented in Tables 8.1 and 8.2. The life of the equipment and finance period evaluated was over a 25-year period, and the loan interest (or discount rate) assumed was 5% per year. This study compared the present value of the costs, present value of the benefits, the net present value of the benefits (benefits minus the costs), and the ratio of present value of benefits to present value of costs over 25 years for each PV solar system.

The costs that were considered included the PV solar system installation cost, the O&M cost, and the cost of insurance for a PV system. The benefits considered included the avoided electrical utility cost, SREC benefits, tax credits, and net metering.

Installed Costs

The cost of installing a 12 kW solar panel system was determined for Michigan and for Arizona. While the cost of installing the solar panel system has decreased over time, according to the 2019 annual "Tracking the Sun" report from Berkeley National Labs (Barbose and Dargh, 2019), the median cost to install a solar panel system in Arizona was $3.30/W in 2018. Prices in the state of Michigan were not included in the report. Therefore, the national median cost for installation, $3.30/W (Barbose and Dargh, 2019) was used for Michigan.

The installed cost assumed the customer financed the entirety of the system over 25 years, with a 5% interest loan, compounded annually, where the interest was spread out over the life of the loan and included in payments. The equal payment capital recovery equation below was used.

$$A = P\left[\left(i(1+i)^n\right)/\left((1+i)^n - 1\right)\right] \tag{8.2}$$

Where A is the yearly payment, P is the principal (total amount financed), i is the annual interest rate, and n is the number of annual payments

For both systems (installed in Arizona and Michigan), the total cost of installation, including interest, was $70,242.93 at a yearly payment of $2,809.72.

Operation and Maintenance Cost

According to the U.S Department of Energy (Walker, 2020), the yearly O&M cost of a residential PV system was about $20/kW/year. These costs include replacing solar power inverters, cleaning the panels, and other potential maintenance costs. This is one of the costs that increases with larger PV systems.

Insurance

In both Arizona and Michigan, there is not a separate insurance to cover solar panels for your home. However, you can still have your solar panels covered by your homeowner's insurance by increasing the coverage for your home (Nationwide, 2020).

The insurance cost in Phoenix, Arizona is 0.0068 of the value of the home. Thus, multiplication of 0.0068 by the total cost of the solar system of $ 39,699 (i.e., 12 kW×3.3 $/kW) provides an estimate of $ 269/year cost to insure the solar system. Michigan charges an insurance cost of 0.017 times $ 39,600 (the added value from the solar system). Thus, the cost to insure the solar system in Michigan will be $ 672 per year.

Present Value of Cost

For each of the 25 years, the sum of the yearly installment cost, O&M cost, and insurance cost was computed. Then the present value of the cost was calculated for each year with the equation below.

$$P = F/(1+i)^n \tag{8.3}$$

where P is the present value, F is the future value of the payment made in nth annual payment with discount rate (i).

Using the above equation and summing the present values of cost for 25 years for each PV system, it was found that the present value of cost for the PV system in Detroit was $55,081 and the present value of the cost for the PV system in Phoenix was $49,116.65 (see Tables 8.1 and 8.2, respectively).

Avoided Electric Utility Cost

The average electrical utility usage for the average residential house was taken at the state level for both Detroit and Phoenix. The average residential yearly consumption for Michigan was 8,000 kWh/year and for Arizona, 14,000 kWh/year (EIA, 2020a, b, c). The lower electrical consumption in Michigan can largely be explained by the colder climate and the resultant lower air-conditioning usage.

Using the equation of annual output (see Equation 8.1 above), a derate factor of 0.77, and the peak sun hours average from PV Watts (NREL, 2020), 4.6 kWh/m²/day for Michigan and 6.52 kWh/m²/day for Phoenix, the energy generated by PV system in Detroit was 15,514 kWh/y and the energy generated by PV system in Phoenix was 21,989 kWh/y. PV Watts gave similar numbers of 15,742 kWh/year for Detroit and 20,482 kWh/y for Phoenix with inputs of a 12 kW system that was roof-mounted with a 20 degree tilt and 14.08% efficiency loss. PV Watts (NREL, 2020) values were within 2% for Detroit and 7% for Phoenix of the above estimates. A 1% efficiency loss each following year was included in the calculations to account for equipment degradation.

The avoided yearly electricity cost was computed by determining the energy supplied by the PV generation up to the yearly electrical consumption. The 12 kWh systems in both states generated more than the consumption, so the avoided electrical usage generation was less than the PV generation for each year. Then the avoided electrical usage was multiplied by the electrical cost to obtain the avoided electrical utility cost. The average electricity cost used for Detroit was $0.15/kWh and for Phoenix was $0.12/kWh. It was also assumed that the electrical usage and electricity costs would increase by 1% a year.

SREC, Net Metering, and Tax Credit Revenue

Revenue generated by the solar power system is highly dependent on the state and even local utility regulations. Revenue generation opportunities made up the difference in why the Michigan system is a negative value proposition and the Arizona system is a positive one. For this analysis, three sources of revenue from the solar power system were considered, namely SREC revenue, Net Metering Credits, and Tax Credits.

SREC are credits that the owners of the solar energy generators can sell to utilities in states where there is a requirement that power companies produce a certain portion of their energy output using solar power. If a utility company does not own the solar infrastructure themselves, they can pay solar energy generators (owners of the solar generators) for credit for producing that energy. While neither Michigan nor Arizona has such requirements for utilities, Michigan solar producers could sell credits to Ohio, which did have a solar carve-out, until 2019. The passage of HB6 eliminated the solar energy requirement and with it, the market for Ohio SRECs (Callender et al., 2019). For the analysis, the amount for SREC credits was zero for all 25 years (DSIRE, 2020).

A federal tax credit of 26% of the total cost to install the solar power system is available to homeowners for systems installed in 2020 (IRS, 2020). Taxpayers have the option of claiming the credit all in 1 year or splitting the amount between 2 years. As it is more financially advantageous for this analysis, the tax credit was claimed in the first year for both the Michigan and Arizona analyses. In addition to the federal tax credit, Arizona residents are eligible for a tax credit of $1,000, or 25% of the cost of installation, whichever is less (Arizona State Government, 2020). For this system, $1,000 is the lesser value, and this value was used. This $1,000 was also claimed entirely in the first year of installation, for a total value of tax credits in the first year of $11,296.00 for Arizonans and $10,296.00 for Michiganders.

Net metering, which is the ability to sell excess energy generated back to utilities, is available in Michigan and Arizona. In Arizona, residents receive $0.0964 per kWh sent to the grid (Wichner, 2018). This credit is disbursed monthly in the form of a bill credit and at the end of the year can be cashed out (TEP, 2018). Since the system produces more energy than the house consumes, net metering credits produced a total of $11,816 over 25 years, assuming the homeowner elected to cash out bill credits at the end of each billing year. While net metering is available in Michigan, there is no provision to allow a homeowner to cash out excess solar bill credits (DTE, 2020). For this reason, net metering revenue was not included in the calculation of Michigan benefits, since a bill credit is not a realizable gain. Ideally, to take advantage of net metering in Michigan, the solar power system would be smaller than required to always meet all energy demand to allow a homeowner to use the bill credits to offset the cost of energy from the utility.

Net Present Value

The present value of the benefits was calculated using the same equation (Equation 8.3) used for the present value of costs. Then the net present value for each year was calculated by subtracting the present value costs from the present value of benefits.

The net cumulative present value of a PV system in Detroit was –$8,668, and the net cumulative present value of a PV system in Phoenix was $8,024 (see Tables 8.1 and 8.2).

The benefits-to-costs ratios obtained for the two analyses show that the value of the ratio for Detroit, MI, was 0.843 (less than 1.0); whereas the value of the ratio was 1.163 for Phoenix, AZ. The curves for cumulative present value shown in Figure 8.1 also show that implementation of solar PV system for an average home in Phoenix will be profitable, but it will not be profitable for an average home in Detroit.

Conclusions of the Cost–Benefit Analyses

Based on the data presented in Tables 8.1 and 8.2, the net cumulative present value over 25 years for the Detroit PV system was a loss of $8,668 and for the Phoenix PV system was net benefit of $8,024.

Many different factors impact the overall cost–benefit analysis. In particular, the environment at the location affects the consumption and the amount of energy that can be generated. A home in cooler city uses less air-conditioning and thus less electricity; and a home located in higher peak sun hours city allows for more electricity generation by the PV system. In general, more electrical usage with more peak sun hours allows for more electricity utility cost avoidance, and this was one of the factors that lead to the Phoenix systems showing more profitability than the Detroit system.

It was also found that a 12 kW system was too large for the average residential house. PV systems are more expensive with the larger DC rating, and the excess electrical generation does not give that much advantage. Additional calculations run for an 8 kW system showed that with scaling the system down to meet the consumption of electricity, the loss over 25 years for the Detroit system decreased by $2,800 and the profit by the Phoenix system increased by $8,500. Phoenix gives out a net metering credit for excess electrical generation, but it was still financially better over the 25 years to downsize the PV system.

The analyses also showed that factors such as cost of insurance, government incentives, and net metering affect the evaluation measures. Michigan had higher insurance costs, while Arizona had larger government incentives. And Arizona's net metering credit made generating excess electricity profitable.

On the overall, it is important to consider all the factors when looking into if a PV system will have a positive net cumulative present value. Adding more factors in such analysis will make the analysis more comprehensive and representative of the actual situation. Further, the assumptions based on the past data may not apply well over future years because the financial incentives and costs of the solar system will change in the future. Such analysis should also be iterated by changing the kW power output capability of the PV system to determine the optimum capability of the PV system to achieve maximum long-term profitability. These calculations showed that 8 kW systems fit the average household better than a 12 kW system for both the locations analyzed. Finally, not all PV systems will necessarily be profitable, so homeowners must consider how much they are willing to pay to lower emissions.

TABLE 8.1

Spreadsheet for PV Solar Residential System in Detroit, MI

	Costs					Benefits										Net present Value=Present Value of Benefits – Present Value of Total Costs($)	
Year	Installed Costs ($)	O&M Costs ($)	Insurance ($)	Total Costs ($)	Present Value of costs ($)	Total Electrical Energy Usage (kWh)	Energy Generated by PV (kWh)	Energy Usage Avoided (kWh)	Average Electrical Costs ($/kWh)	Avoided Electricity Costs ($)	Excess Generated (kWh)	SREC Revenue($)	Tax Credit ($)	Total Benefits ($)	Present Value of Benefits ($)		Net Cumulative Present Value
1	2,810	240	672	3,722	3,722	8,000	15,514	8,000	0.12	2,327	7,514	0	10,296	12,623	12,623	8,901	8,901
2	2,810	240	672	3,722	3,545	8,080	15,436	8,080	0.12	2,339	7,356	0	0	2,339	2,227	–1,318	7,583
3	2,810	240	672	3,722	3,376	8,161	15,359	8,161	0.12	2,350	7,198	0	0	2,350	2,132	–1,244	6,339
4	2,810	240	672	3,722	3,215	8,242	15,282	8,242	0.12	2,362	7,040	0	0	2,362	2,040	–1,175	5,164
5	2,810	240	672	3,722	3,062	8,325	15,206	8,325	0.12	2,374	6,881	0	0	2,374	1,953	–1,109	4,055
6	2,810	240	672	3,722	2,916	8,408	15,130	8,408	0.13	2,385	6,722	0	0	2,385	1,869	–1,047	3,007
7	2,810	240	672	3,722	2,777	8,492	15,054	8,492	0.13	2,397	6,562	0	0	2,397	1,789	–989	2,018
8	2,810	240	672	3,722	2,645	8,577	14,979	8,577	0.13	2,409	6,402	0	0	2,409	1,712	–933	1,085
9	2,810	240	672	3,722	2,519	8,663	14,904	8,663	0.13	2,421	6,241	0	0	2,421	1,639	–881	205
10	2,810	240	672	3,722	2,399	8,749	14,830	8,749	0.13	2,433	6,080	0	0	2,433	1,568	–831	–627
11	2,810	240	672	3,722	2,285	8,837	14,755	8,837	0.13	2,445	5,919	0	0	2,445	1,501	–784	–1,411
12	2,810	240	672	3,722	2,176	8,925	14,682	8,925	0.13	2,457	5,756	0	0	2,457	1,497	–740	–2,150
13	2,810	240	672	3,722	2,073	9,015	14,608	9,015	0.14	2,469	5,594	0	0	2,469	1,375	–698	–2,848
14	2,810	240	672	3,722	1,974	9,105	14,535	9,105	0.14	2,481	5,431	0	0	2,481	1,316	–658	–3,506
15	2,810	240	672	3,722	1,880	9,196	14,463	9,196	0.14	2,494	5,267	0	0	2,494	1,259	–620	–4,126
16	2,810	240	672	3,722	1,790	9,288	14,390	9,288	0.14	2,506	5,103	0	0	2,506	1,205	–585	–4,711
17	2,810	240	672	3,722	1,705	9,381	14,318	9,381	0.14	2,518	4,938	0	0	2,518	1,154	–551	–5,263
18	2,810	240	672	3,722	1,624	9,474	14,247	9,474	0.14	2,531	4,772	0	0	2,531	1,104	–520	–5,782
19	2,810	240	672	3,722	1,547	9,569	14,175	9,569	0.14	2,543	4,606	0	0	2,543	1,057	–490	–6,272
20	2,810	240	672	3,722	1,473	9,665	14,105	9,665	0.14	2,556	4,440	0	0	2,556	1,011	–461	–6,734
21	2,810	240	672	3,722	1,403	9,762	14,034	9,762	0.15	2,569	4,273	0	0	2,569	968	–435	–7,168
22	2,810	240	672	3,722	1,336	9,859	13,964	9,859	0.15	2,581	4,105	0	0	2,581	927	–409	–7,578
23	2,810	240	672	3,722	1,272	9,958	13,894	9,958	0.15	2,594	3,936	0	0	2,594	887	–386	–7,963
24	2,810	240	672	3,722	1,212	10,057	13,825	10,057	0.15	2,607	3,767	0	0	2,607	849	–363	–8,326
25	2,810	240	672	3,722	1,154	10,158	13,756	10,158	0.15	2,620	3,598	0	0	2,620	812	–342	–8,668
Total→	70,243	6,000	16,809	93,052	55,082	225,946	365,446	225,946		61,768		0	0	72,064	46,413	–8,668	

Benefits-to-Costs Ratio = 0.843.

TABLE 8.2

Spreadsheet for PV Solar Residential System in Phoenix, AZ

	Costs					Benefits											Net present Value=Present value of benefits minus present value of Total Costs ($)	Net Cumulative Present Value
Year	Installed Costs ($)	O&M Costs ($)	Insurance ($)	Total Costs ($)	Present Value of costs ($)	Total Electrical Energy Usage (kWh)	Electrical Energy Generated by PV (kWh)	Energy Usage Avoided (kWh)	Average Electrical Costs ($/kWh)	Avoided Electricity Costs ($)	Excess Generated (kWh)	SREC Revenue($)	Net Metering Credit ($)	Tax Credit ($)	Total Benefits ($)	Present Value of Benefits ($)		
1	2,810	240	269	3,319	3,319	14,000	21,989	14,000	0.15	2,100	7,989	0	770	10,296	14,166	14,166	10,847	10,847
2	2,810	240	269	3,319	3,161	14,140	21,879	14,140	0.15	2,142	7,739	0	746	0	2,751	2,751	–410	10,437
3	2,810	240	269	3,319	3,010	14,281	21,770	14,281	0.15	2,185	7,489	0	722	0	2,637	2,637	–374	10,063
4	2,810	240	269	3,319	2,867	14,424	21,661	14,424	0.15	2,229	7,237	0	698	0	2,528	2,528	–339	9,725
5	2,810	240	269	3,319	2,731	14,568	21,553	14,568	0.16	2,274	6,984	0	673	0	2,425	2,425	–306	9,419
6	2,810	240	269	3,319	2,601	14,714	21,445	14,714	0.16	2,320	6,731	0	649	0	2,326	2,326	–275	9,144
7	2,810	240	269	3,319	2,477	14,861	21,338	14,861	0.16	2,366	6,477	0	624	0	2,232	2,232	–245	8,899
8	2,810	240	269	3,319	2,359	15,010	21,231	15,010	0.16	2,414	6,221	0	600	0	2,142	2,142	–217	8,682
9	2,810	240	269	3,319	2,246	15,160	21,125	15,160	0.16	2,462	5,965	0	575	0	2,056	2,056	–191	8,492
10	2,810	240	269	3,319	2,139	15,312	21,019	15,312	0.16	2,512	5,708	0	550	0	1,974	1,974	–166	8,326
11	2,810	240	269	3,319	2,038	15,465	20,914	15,465	0.17	2,562	5,450	0	525	0	1,896	1,896	–142	8,184
12	2,810	240	269	3,319	1,941	15,619	20,810	15,619	0.17	2,614	5,190	0	500	0	1,821	1,821	–120	8,064
13	2,810	240	269	3,319	1,848	15,776	20,706	15,776	0.17	2,666	4,930	0	475	0	1,749	1,749	–99	7,966
14	2,810	240	269	3,319	1,760	15,933	20,602	15,933	0.17	2,720	4,669	0	450	0	1,681	1,681	–79	7,887
15	2,810	240	269	3,319	1,676	16,093	20,499	16,093	0.17	2,775	4,406	0	425	0	1,616	1,616	–60	7,826
16	2,810	240	269	3,319	1,596	16,254	20,397	16,254	0.18	2,830	4,143	0	399	0	1,554	1,554	–43	7,784
17	2,810	240	269	3,319	1,520	16,416	20,295	16,416	0.18	2,887	3,879	0	374	0	1,494	1,494	–26	7,757
18	2,810	240	269	3,319	1,448	16,580	20,193	16,580	0.18	2,945	3,613	0	348	0	1,437	1,437	–11	7,746
19	2,810	240	269	3,319	1,379	16,746	20,092	16,746	0.18	3,005	3,346	0	323	0	1,383	1,383	3	7,749
20	2,810	240	269	3,319	1,313	16,914	19,992	16,914	0.18	3,065	3,078	0	297	0	1,330	1,330	17	7,766
21	2,810	240	269	3,319	1,251	17,083	19,892	17,083	0.18	3,127	2,809	0	271	0	1,280	1,280	30	7,796
22	2,810	240	269	3,319	1,191	17,253	19,792	17,253	0.18	3,189	2,539	0	245	0	1,233	1,233	41	7,837
23	2,810	240	269	3,319	1,135	17,426	19,693	17,426	0.19	3,254	2,267	0	219	0	1,187	1,187	52	7,890
24	2,810	240	269	3,319	1,081	17,600	19,595	17,600	0.19	3,319	1,995	0	192	0	1,143	1,143	63	7,952
25	2,810	240	269	3,319	1,029	17,776	19,479	17,776	0.19	3,386	1,721	0	166	0	1,101	1,101	72	8,024
Total→	70,243	6,000	6,732	82,975	49,117					67,350		0	11,816	0	90,462	57,141	8,024	

Benefits-to-Costs Ratio= 1.163.

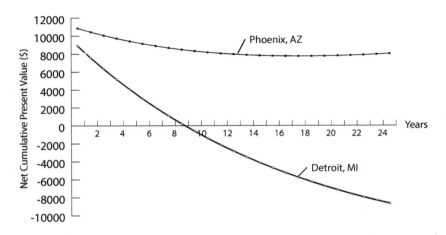

FIGURE 8.1 Net cumulative present value of solar PV systems in Detroit, MI and Phoenix, AZ Homes.

EXERCISING COST–BENEFIT MODEL FOR SENSITIVITY ANALYSIS

After creation of the cost–benefit model (i.e., the spreadsheet) such as those shown in Tables 8.1 and 8.2, they can be exercised (i.e., iterated) by using different values of input variables to determine the sensitivity of changes in the input variables to overall conclusions. Sensitivity analyses thus conducted can provide a better understanding into the risks associated in determining the best course of action. Chapter 17 provides examples of sensitivity analysis and applications of Monte Carlo simulation to aid in decision-making.

RISKS AND UNCERTAINTIES IN COST–BENEFIT ANALYSIS

Despite its usefulness, cost–benefit analysis has several associated risks and uncertainties that are important to note. These risks and uncertainties can result from incomplete and erroneous formulation of the analysis, inappropriate use of evaluation measures and/or criteria to reach conclusions. Much of the risk involved with cost–benefit analysis can be related to deliberate manipulations of the data leading to selection of a wrong or inappropriate alternative. Stakeholders or interested parties may try to influence results by over- or understating costs and benefits. In some cases, supporters of a project may insert a personal or organizational bias into the analysis.

On the data side, there can be a tendency to rely too much on data compiled from previous projects. This may inadvertently yield results that do not directly apply to the situation being considered. Since data leveraged from an earlier analysis may not directly apply to the circumstances at hand, this may yield results that are not consistent with the requirements of the situation being considered. Using heuristics to assess the dollar value of intangibles may provide quick, "ballpark-type" information, but it can also result in errors that produce an inaccurate picture of costs that can invalidate findings.

In addressing risk, it is helpful to utilize probability theory (i.e., assign probabilities to cost and benefit estimates and to outcomes) to identify and examine key patterns that can influence the selection of an alternative.

UNCERTAINTIES

There are several considerations that can influence the results of any cost–benefit analysis, and while they would not apply in every situation, it is important to keep them in mind as these analyses are conducted and exercised. These considerations are as follows:

1. Inaccuracies in cost and benefit information can diminish credibility of findings.
2. Revenue and costs are moving targets, and thus, selecting values of their parameters can be very challenging.
3. Income level can influence a customer's or decision-maker's ability or willingness to make decisions.
4. Some benefits cannot be directly reflected in dollar amounts because of the subjective considerations and judgments associated with values of many variables.
5. Many cost and benefits are not just linear functions of input or output variables. Thus, nonlinearities must be carefully studied before simply linear-scaling the values of evaluation measures.

CONTROVERSIAL ASPECTS

The most controversial aspects of cost–benefit analysis are costs and benefits that are associated with subjective nature of their values that lead to intangibles. Many variables related to costs and benefits are difficult to quantify, such as human life, pain, and sufferings due to adverse health effects of environment or injuries due to accidents, and other considerations such as level of comfort and convenience, brand equity, customer loyalty, and so forth.

With respect to intangibles, using the cost–benefit analysis process to drive more critical thinking around all aspects of value can provide beneficial outcomes. Cost–benefit analysis assumes that a monetary value can be placed on all the costs and benefits of a program, including tangible and intangible returns. A major advantage of cost–benefit analysis lies in forcing people to explicitly and systematically consider the various factors, which should influence strategic choice. Next chapter covers several useful decision-making techniques that are based on subjective judgments of people, especially of those who are experts and very familiar with the issues and problems being analyzed.

CONCLUDING REMARKS

During the development of any cost–benefit analysis, care must be exercised to ensure that uncertainties in (a) the prediction of scenarios, (b) model development, and (c) data must be considered. Uncertainty in scenarios can occur due to errors in selecting alternatives, outcomes, and their probabilities. The model could be conceptually imperfect due to omission or inappropriate consideration of many variables. And the data used to estimate costs and benefits may not be valid and contain errors. Model validation is very important and necessary in many cases and involves comparison of model predictions with other existing or future observations. Confidence in model results can be significantly increased by model calibration, verification, and validation exercises.

REFERENCES

Arizona State Government. 2019. Arizona Form 310: Credit for Solar Energy Devices. Website: https://azdor.gov/forms/tax-credits-forms/credit-solar-energy-credit (Accessed: January 22, 2020).

Barbose, G. and N. Darghouth. 2019. *Tracking the Sun: Pricing and Design Trends for Distributed Photovoltaic Systems in the United States.* Lawrence Berkeley National Laboratory, Oct. Website: https://emp.lbl.gov/tracking-the-sun (Accessed: July 27, 2020).

Callender, W., et al. 2019. As Passed by Senate: Am. Sub.H. B. No. 6. *The Ohio Legislature, 133 General Assembly,* 2019. Website: https://www.legislature.ohio.gov/legislation/legislation-documents?id=GA133-HB-6 (Accessed: January 22, 2020).

DSIRE. 2020. *Database of State Incentives for Renewables & Efficiency.* NC State University. Websites: http://programs.dsireusa.org/system/program?state=AZ and http://programs.dsireusa.org/system/program?state=MI (Accessed: January 22, 2020).

DTE Energy (DTE). 2020. Distributed Generation FAQs. Website: https://www.newlook.dteenergy.com/wps/wcm/connect/8dc7d28e-5621-41ba-a874-5dddc8c56703/DistributedGenerationFAQ.pdf?MOD=AJPERES (Accessed: January 20, 2020).

EIA. 2020a. Household Energy Use in Arizona. Website: http://www.eia.gov/consumption/residential/reports/2009/state_briefs/pdf/az.pdf (Accessed January 20, 2020).

EIA. 2020b. Household Energy Use in Michigan. Website: http://www.eia.gov/consumption/residential/reports/2009/state_briefs/pdf/mi.pdf (Accessed January 20, 2020).

EIA. 2020c. Residential Energy consumption Survey (RECS) -Analysis & Projections. Website: http://www.eia.gov/consumption/residential/reports/2009/state_briefs/ (Accessed Jan 20, 2020).

Fu, R., Feldman, D., Margolis, R., Woodhouse, M. and Ardani, K. 2017. *U.S. Solar Photovoltaic System Cost Benchmark: Q1 2017.* National Renewable Energy, Technical Report NREL/TP-6A20-68925, September 2017.

Internal Revenue Service (IRS). 2020. *Instructions for IRS form 3468.* Department of the Treasury, Internal Revenue Service, 2019. Website: https://www.irs.gov/pub/irs-pdf/i3468.pdf (Accessed: January 22, 2020).

Nationwide Mutual Insurance Company (Nationwide). 2020. Are solar panels covered by home insurance? Website: https://www.nationwide.com/lc/resources/home/articles/solar-panel-insurance (Accessed: January 20, 2020).

National Renewable Energy laboratory (NREL). 2020. PV Watts Calculator. Website: https://pvwatts.nrel.gov/ (Accessed Jan 20, 2020).

Schwager, M. Tate, G. and Kim, J. 2020. *P-4 Project Report*. Prepared for ESE 504 Course, University of Michigan-Dearborn, April 2020.

The Solar Foundation. 2012. Solar Accounting: Measuring the Costs and Benefits of Going Solar. Sponsored by the U.S. Department of Energy under Award Number DE-EE0003525. August 23, 2012. Website: https://farm-energy.extension.org/wp-content/uploads/2019/04/DOE-Solar-Foundation.pdf (Accessed: July 26, 2020).

Tucson Electric Power Company (TEP). 2018. *Rider 4: Net Metering for Certain Partial Requirements Service*. Tucson Electric Power Company, 21 Sep 2018. Website: https://www.tep.com/wp-content/uploads/2018/10/704__tep_rider.pdf (Accessed: January 22, 2020).

Walker, A. 2020. *PV O&M Cost Model and Cost Reduction*. NREL, US Department of Energy. NREL/PR-7A40-68023 217AD. Website: www.nrel.gov/docs/fy17osti/68023.pdf (Accessed: October 4, 2020).

Wichner, D. 2018. New Tucson Electric Power Customers to Get Lower Credits for Excess Electricity. *Arizona Daily Star*, 12 Sep 2018. Website: https://tucson.com/news/local/new-tucson-electric-power-solar-customers-to-get-lower-credits/article_c1c16729-8333-57e1-9c39-3a03cdd4b3c2.html (Accessed: January 22, 2020).

9 Subjective Methods for Risk Assessment

INTRODUCTION

Subjective methods to gather data and conduct analyses to solve problems play an important role in decision-making in energy systems area. Subjects (or participants) need to be carefully selected to ensure that they possess the necessary technical background and expertise to make judgments on many characteristics and issues related to the energy systems problems. The judgments made by these subjects are used as data to evaluate alternatives selected for problem-solving. In some cases, the subjects are asked also to participate in selection of the alternatives to formulate a given problem.

The researcher or administrator of the subjective evaluation process must be also very knowledgeable about the energy systems problems to ensure that the process used to select the subjects (or experts), the variables used to evaluate the problem, the methods used to collect the data, and procedures used to analyze data are valid (i.e., they represent the characteristics associated with the evaluation measures) and are free from biases and errors.

The subjective methods are used by the engineers because in many situations (a) suitable objective measures do not exist, (b) the subjects are better able to perceive characteristics and issues with the product (or system), and thus, the subjects can be used as the measurement instruments, and (c) the data on subjective measures are easier to obtain than objective measures, which generally require physical measurement instruments.

Four important types of subjective methods, namely (a) rating on a scale, (b) Pugh analysis, (c) paired comparison methods, and (d) failure modes and effects analysis, are presented in this chapter along with their applications in solving problems in energy systems areas. It should be noted that Chapter 6 provided an overview of the problem-solving approaches, the decision matrix, and subjective and objective methods.

RATING ON A SCALE

The use of rating scales to obtain judgments of subjects (of experts) on an issue or characteristic of a system or a product is a very common. "Rating on a scale" is a relatively simple method to apply and use to collect data on opinions or judgments of customers and/or experts. However, care must be taken to ensure that the subjects (also called respondents) understand the rating scale and are very familiar with the system (or product) and its characteristic being evaluated.

DOI: 10.1201/9781003107514-11

In this method of rating, the subject is first given instructions on the procedure involved in evaluating a given system (or product) including explanations on one or more of the product attributes and the rating scales to be used for scaling each attribute. Interval scales are used most commonly. Many different variations are possible in defining the rating scales. The interval scales can differ due to (a) how the end points of the scales are defined, (b) number of intervals used (Note: odd number of intervals allow use of a mid-point), and (c) how the scale points are specified (e.g., without descriptors vs. with word descriptors or numerals). Bhise (2012, 2014) presents a number of examples of interval scales.

A 10-point scale, with 10 = excellent and 1 = poor, is commonly used to evaluate a number of characteristics of systems such as durability, quality, acceptance, functionality, and so forth. Many adjectives can be used to describe the scale values to assist the respondents to assess the levels of characteristics (e.g., 10 = excellent, 8 = very good, 7 = good, 5 = fair, 3 = poor, 1 = very poor). Other specialized adjectives can be also used to better describe the scale values for specialized characteristic of a system. For example, noise level can be rated by using the following adjectives: 10 = Very pleasant, 8 = pleasant, 6 = acceptable, 4 = somewhat unpleasant, 2 = unpleasant, and 1 = very unpleasant.

Five-point scales with following adjectives are also commonly used: 5 = very acceptable, 4 = somewhat acceptable, 3 = neither acceptable nor unacceptable, 2 = somewhat unacceptable, and 1 = very unacceptable.

The advantages of ratings on a scale method are that the scales can be easily and quickly created and used to collect data of a large number of subjects (customers, users, experts – depending upon the objective of the study) without costly expenditure of creating intricate instruments to make objective measurements of the system characteristics. Further, some system of product characteristics can be only measured by use of human subjects based on their perception and use of the system (e.g., measurement of acceptance, preference, quality, comfort, satisfaction, appearance, and smoothness/roughness). The disadvantage is that all subjects may not understand the characteristic being evaluated and may have difficulty in scaling the levels despite the use of the adjectives – which can be also interpreted differently by different subjects. Systems or products that have characteristics that fall near the extreme (i.e., top or bottom ends of the scales) are also difficult to judge unless the subjects are given opportunity to familiarize with the systems (or products) involving the extreme levels of characteristics. Designing of instructions to be provided to the subjects prior to rating sessions is also very important to ensure that the subjects understand the scales and their tasks.

AN EXAMPLE OF RATING CHARACTERISTICS OF ELECTRICITY GENERATING POWER PLANT TECHNOLOGIES

During early planning phase of selecting a power plant technology, the team involved in making the recommendation needed to evaluate many attributes (characteristics) of different power plant technologies. Table 9.1 provides the ratings on nine attributes (listed in the first column of the table) on a five-point preference scale (from "Very Preferred" to "Very Non-preferred"; provided in the top two rows of the table) for

TABLE 9.1
Illustration of Ratings on Attributes of Power Plants with Different Technologies

Attribute ↓	Five-Point Scale				
	Very Non-Preferred	Somewhat Non-Preferred	Neither Preferred nor Non-Preferred	Somewhat Preferred	Very Preferred
Construction cost	N, S	CCS, G	H	B	C, NGCC, W, B
Operating and maintenance Costs			CCS, N, S	O, NGCC, B	H, G, W
Land/Space requirement	NGCC, H, S	C, CCS, N, B		W	G
Water requirements	C, CCS, NGCC, N, B	G	H		W, S
CO_2 emissions	C	NGCC, B		CCS	H, N, 5, W, S
Non-CO_2 emissions	C	CCS, B	NGCC	N , G	H, W, S
Wildlife danger		CCS	C, B	N, G, W, S	NGCC, H
Flexibility-- ability to quickly respond to changes in demand	G, W, S	C, CCS, B		NGCC, H, N	
Safety during malfunctions	N	C, CCS, H, B		NGCC, G, W	S

Notation: C, Coal-fired plant; CCS, Coal-fired plant with carbon capture and sequestration; NGCC, Natural gas combined cycle plant; N, Nuclear power plant; H, Hydroelectric power plant; W, Wind turbines; S, Solar photovoltaic plant; B, Biomass plant; G, Geothermal plant.

nine different technologies used to construct electricity generating power plants. The nine technologies and their notations are provided at the bottom of Table 9.1. The ratings of each of the technologies for each attribute (in each row) are provided by the placing the alphabetic identification notation) of each technology within the scale shown in last five columns of the table.

The preference ratings for the attributes provided in Table 9.1 show the following: (a) Coal-fired power plants [C] are cheaper to construct, operate, and maintain but is not preferred by considering emissions. (b) Geothermal plants [G] are expensive to construct, but cheaper to operate and maintain, safer, and do not produce harmful emissions. (c) Solar photovoltaic (PV) power plants [S] are expensive to construct, require a lot of land space, and are safe during malfunctions. (d) The coal-fired power plants with carbon capture and sequestration [CCS] are expensive to construct but are flexible and safer to wildlife and produce lesser emissions than coal-fired plants, and (e) Observing the distribution of ratings for the nine attributes, no one technology has ratings on the preferred side of all the attribute scales. Wind turbines ratings fall on the preferred side of most of the attributes except the flexibility attribute (as the output of wind turbines is totally dependent on wind velocity).

The ratings data thus provides a good understanding into preferred and non-preferred attributes of different power plant technologies. The ratings of each technology can be summed, or a weighed sum can be obtained to help in deciding the overall preferred plant technology. The weights for each attribute can be obtained by use of subjective methods such as ratings on a scale or ratings obtained by paired comparison methods covered in later sections of this chapter.

PUGH ANALYSIS

Pugh diagram is a tabular formatted tool consisting of a matrix of system (or product) attributes (or characteristics) and alternate system (or product) concepts along with a benchmark (reference) system (or product) called the "*datum.*" The diagram helps to undertake a structured concept selection process and is generally created by a multidisciplinary team of experts to converge on a superior system (or product) concept. The process involves creation of the matrix by inputs from all the team members. The rows of the matrix consist of system (or product) attributes based on the customer needs, and the columns represent different alternate system concepts (or products).

The evaluations of each system (or product) concept on each attribute are made with respect the *datum*. (Usually, the datum is selected as the manufacturer's existing system or [product]).The process uses classification metrics of "same as the datum" (S), "better than the datum" (+), or "worse than the datum" (−). The scores for each system (product) concept are obtained by simply adding the number of plus and minus signs in each column. The system (or product) concept with the highest net score (the "sum of pluses" minus the "sum of minuses") is the preferred system (or product) concept. Several iterations are employed to improve system (or product) superiority by combining the best features of highly ranked concepts until a superior concept emerges and becomes the new benchmark.

An Example of Pugh Diagram Application

An automotive powertrain engineer wanted to determine if the performance of a transient turbo-charged gasoline engine can be improved over the gasoline turbo direct injection (GTDI) methodology by employing the following three concepts: (a) Concept #1: an electric turbo-boost (e-Turbo), (b) Concept #2: a hybrid turbo using an electric motor assist in parallel with the turbo operated by the exhaust gases, or (c) Concept #3: use of electrical compressor only (Black, 2011). The engineer created a Pugh diagram to compare the above three technologies with the GDTI as the *datum*. Table 9.2 presents the Pugh diagram. The product attributes used to compare the above three technologies are presented in the second column from the left.

TABLE 9.2

Pugh Diagram for Product Concept Selection

Attribute No.	Customer-Based Product Attribute	Product Concept #1	Product Concept #2	Product Concept #3	Datum
1	Life-cycle durability	–	–	–	
2	Cost	–	–		
3	Package and ergonomics	–	+	+	
4	Performance	+	+	+	
5	Fuel economy	$	+	$	
6	Safety/Security	–	–	–	
7	Vehicle dynamics	+	+	+	
8	Emissions	+	+	+	
9	Electrical and electronics	–	–	–	
10	Weight	–	–	–	
11	Noise, vibrations, and harshness	–	–		
	Sum of pluses	3	5	4	
	Sum of minuses	6	6	6	
	Net	–3	–1	–2	

The four right-hand columns represent product concepts #1, #2, #3, and the datum as the last column.

All the three product concepts improve the "performance" attribute (attribute #4) as compared with the datum by eliminating the turbo-lag (a transient condition during quick accelerations). This is shown by the "+" signs in all the three product concepts (columns) of the row corresponding to attribute #4. However, they introduce additional negatives into the system due to additional cost (attribute #2), weight (attribute #10), noise (attribute #11), electrical and electronics (attribute #9), and life-cycle durability (attribute #1). The bottom row of "Net" score ("sum of pluses" minus "sum of minuses") shows that none of the three product concepts were better than the datum – since the net scores of all were negative. The concept #2 (Hybrid Turbo) is the least negative based on the net score. Life-cycle durability, cost, safety/security, electrical and electronics, weight, and NVH (noise, vibrations, and harshness) are all additional problem issues with concept #2 compared with the datum (traditional turbo [GTDI]). In the above analysis, all the product attributes were considered to have equal weight, and the number of plus signs minus the number of minus signs was used as the net score.

The example illustrated above assumes that each attribute is weighted equally in the evaluation process. The method can be modified by assigning importance weight for each attribute, and then a weighted score is computed for each product concept. The modified method called the "Weighted Pugh Analysis" is described below.

WEIGHTED PUGH ANALYSIS

This modified Pugh analysis method allows use of different weights for each of the attributes. This method can also be used in Pugh selection process. Table 9.3 presents an additional analysis on the above turbo-boost problem using importance weighting for each of the product attributes. The importance of each product attribute was rated by using a 10-point scale, where 10 = most important and 1 = least important (see column with heading "Importance Rating" in Table 9.3). The importance scores were converted to "Importance Weight" (by dividing importance rating of each attribute by the sum of the all the importance ratings). The Importance Weights are shown in the column to the right side of the Importance Rating column. Each product concept was evaluated with respect to the datum (the current product GTDI system) for each attribute by using another 10-point scale ranging from −5 to +5. Here, +5 score indicates that a given product concept is very much better than the datum; and −5 score indicates that the product concept is very much inferior to the datum. The sum of the weighted scores of each product concept was obtained by summing of the multiplied values of importance weight and product rating over the entire set of product attributes. The

TABLE 9.3
Modified Pugh Analysis with Importance Ratings for Attributes

Customer-Based Product Attribute	Importance Rating	Importance Weight	Preference Rating Using −5 to +5 Scale Compared with the Datum			
			Product Concept#1	Product Concept #2	Product Concept #3	Datum
Customer life-cycle durability	5	0.06	−3	−3	−3	
Cost	10	0.13	−3	−5	−5	
Package and ergonomics	3	0.04	−3	3	3	
Performance	10	0.13	5	5	5	
Fuel economy	10	0.13	0	5	0	
Safety/security	10	0.13	−3	−3	−3	
Vehicle dynamics	B	0.10	5	5	5	
Emissions	3	0.04	5	5	5	
Electrical and electronics	5	0.06	−3	−5	−5	
Weight	5	0.06	−3	−5	−5	
Noise, vibrations, and harshness	9	0.12	−3	−5	−5	
Sum or weighted sum	78	1.00	−0.46	−0.33	−0.97	

weighted sums of the three product concepts are -0.46, -0.33, and -0.97 (see last line of Table 9.2). Concept #2 had the largest value of the weight sum (-0.33). Thus, product concept #2 (hybrid turbo) emerged as the winner among the three concepts. However, it is still worse than the datum. If fuel economy becomes more important in the future, then this concept has the potential to be implemented. The adoption of a 42-V electrical system in the future could aid in the implementation of the concept.

The benefit of the hybrid turbo is that it enables a completely independent intake and exhaust and enable many modes of operation including additional fuel savings. This also helps eliminate some of the air intake system routing and packaging issues. The penalties in the trade-offs are due to (a) higher electrical load to drive the electric motor driven compressor, (b) poor reliability and durability of the electrically driven compressor, (c) added complexity due to extra parts, and (d) additional costs of extra hardware.

PUGH ANALYSIS TO SELECT ENERGY SOURCE FOR ELECTRICITY GENERATING PLANTS: AN EXAMPLE

This section provides another example of a Pugh Analysis conducted during the planning phase of an electricity generation plant by considering six alternate energy sources and a coal-fired source (used as the datum) for the selection of the plant.

Alternate Energy Sources

The following six energy sources were used as the alternatives in the Pugh analysis:

Alternative 1: Natural Gas Combined Cycle

Natural gas is a popular fuel source alternative. The design of a natural gas power plant is different from coal-fired plant in terms of its combustion chamber. It uses a combustion turbine instead of a steam turbine in a coal-fired plant (see Figures 2.8 and 2.9). For this analysis, it is assumed that the natural gas power plant is a combined cycle, meaning that there are two different turbine stages (see Figure 2.10). The first turbine is an internal combustion type using natural gas as the fuel. The exhaust from the combustion turbine is then used to heat water and convert into steam for running a steam turbine in the second stage. The combined cycle is used to improve overall efficiency of the power plant system.

Alternative 2: Nuclear Power Plant

Nuclear power plant uses a similar process to coal power plants, essentially driving a steam turbine to run the electric generator. The difference being that the source of heat to generate steam comes from recurring fission reaction as compared with burning coal. The nuclear power plant requires management of radiation and very strict safety practices.

Alternative 3: Biomass Plant

The biomass plant uses a similar process as the coal-fired power plant, but it uses plant-based materials (e.g., wood, saw dust, corn oil) as the source of heat to produce steam and to run a steam turbine that drives a generator to produce electricity.

Alternative 4: Wind Turbines

A wind turbine uses wind velocity as a power source to rotate its turbine, which drives an electric generator. For this analysis, a wind farm with multiple turbines can be considered as the major power source.

Alternative 5: Solar Photovoltaic Power Plant

This alternative uses PV solar panels. The solar panels generate DC voltage directly from the incident sunlight. The major disadvantage of the solar plant is that it does not produce electricity during nighttime and thus a backup electricity source is needed to provide power during nighttime hours and on cloudy days.

Alternative 6: Hydroelectric Power Plant

In a hydroelectric plant, the potential energy of stored water in a reservoir is converted to kinetic energy by releasing water flow at a lower elevation into turbines. The water-driven turbines run generators that produce the electricity.

Datum: Coal

The datum energy source considered here is a coal-fired power plant. It is assumed that pulverized coal is used as the fuel source to heat water and to produce steam in a boiler. The pressurized steam is used to run a steam turbine, which drives an electric generator and produces electricity.

Attributes Used for Pugh Analysis:

The following attributes were used to compare power plant technologies:

1. *Capital cost*: The capital cost is an initial fixed cost covering the design of the plant, purchase of equipment, construction of the plant, installation of the equipment, testing, and inspection to get the plant to its operational state. The capital cost is often stated in terms of dollars per kilowatt of energy generated.
2. *Construction time*: The time required from groundbreaking to build a fully functional power plant of a specified technology and electricity generating capacity.
3. *Fuel cost*: This is the dollar cost of fuel used in a power plant, typically specified in $/MWh. For energy sources that do not require a purchased fuel source (wind, solar, hydro), this value is assumed to be zero.
4. *Operating and maintenance cost*: The recurring cost to operate and maintain a power plant. This can include standard maintenance, cleaning, employee salary, and any other recurring cost that is expected over the life of the system. The operating and maintenance cost is often stated as a variable cost in terms of dollars per MWh of electricity generated.
5. *Energy generation capacity*: It is the amount of electricity a generator (or a power plant) can produce when it is running at maximum (peak demand) level. This maximum amount of power is typically measured in megawatts (MW) or kilowatts (kW) and helps utilities determine the maximum electricity load a generator (or power plant) can handle.

6. *Environmental impact due to emissions*: The environmental impact related to emissions (primarily greenhouse gases) from use of a specified fuel consumed to generate electricity. The environmental impact is generally measured in terms dollars spent on medical/health-related costs and costs due to climate change. It can be also estimated by concepts such as social cost of carbon, methane, and nitrous oxides (see Chapter 7).

7. *Environmental impact on wildlife*: The environmental impact of power plant and its operation specifically related to the changes in wildlife (e.g., effect on wild animals, birds, aquatic animals, and fish, and so forth such as wildlife displaced by reservoir created by building a dam, birds killed by wind turbines, animals displaced by solar farms). It is a concern for some consumers and activists more than the energy companies producing the power plants.

8. *Land/space requirements*: The land required to build and operate a power plant is generally expressed in terms of acres or square miles per MW of plant capacity. The land use depends upon fuel source, e.g., a typical 1,000-MW nuclear facility in the United States needs a little more than 1 square mile to operate. The wind farms require 360 times more land area to produce the same amount of electricity as a nuclear power plant; and solar PV plants require 75 times more space. The focus of this attribute is more toward alternatives that require large amounts of land to provide a similar generation capacity (e.g., wind, solar, hydro).

9. *Safety during failure*: The safety of a power plant during a failure. Some of the effects of fuel sources are more harmful than others (e.g., nuclear radiation). Historically we can look at different accidents that have occurred with the use of different fuel sources and compare the severity ratings of accidents.

Pugh Analysis

Historical data for the above attributes (capital cost, fuel cost, O&M cost) was collected from the U.S. Energy Information Administration (see Table 2.3 in Chapter 2) and used in conducting the Pugh analysis presented in Table 9.4. The Pugh diagram was created by students as a part of a project in the author's course to understand the use of Pugh analysis.

The scores based on the sum of pluses (+) minus the sum of minuses (−) show that the solar PV is the most desirable technology alternative given that all attributes being used for analysis are equally weighted. The next comparable system would have been biomass or coal, which both had a zero score.

PAIRED COMPARISON-BASED METHODS

The method of paired comparison involves evaluating items presented in pairs. The items can be alternatives, systems, products, concepts, or attributes that need to be compared and evaluated with respect to each other. In this evaluation method, each subject is essentially asked to compare two items in each pair using a predefined procedure and is asked to simply identify the better item in the pair based on a given

TABLE 9.4

Pugh Diagram Comparing Different Power Plant Alternatives by Attributes of the Power Plants

Attributes	Datum: Coal-Fired	Alternatives					
		Natural Gas Combined Cycle	Nuclear	Biomass	Wind Turbines	Solar PV	Hydro-Electric
Capital Cost		+	-	-	-	-	-
Construction Time		S	-	S	+	+	-
Fuel Cost		-	+	+	+	+	+
Operating and Maintenance Cost		-	S	S	+	+	+
Energy Generation Capacity		S	+	S	-	-	+
Environmental Impact - Emissions		+	-	S	+	+	+
Environmental Impact - Wildlife		S	-	S	-	+	-
Space Requirements		S	S	S	-	-	-
Safety During Failure		-	-	S	-	+	-
Sum of (+)	0	2	2	1	4	6	4
Sum of (-)	0	3	5	1	5	3	5
Sum of (S)	0	4	2	7	0	0	0
Net Score = Sum of (+) - Sum of (-)	0	-1	-3	0	-1	3	-1

criterion (e.g., pollution potential of a power plant). (If the respondent states that there is no difference between the two items, the instruction would be to randomly pick one of the items in the pair. The idea is that, if there truly is no difference in that pair among the respondents, the result will average out to 50:50.) The evaluation task of the subject is, thus, easier as compared with rating on a scale. However, if n items need to be evaluated, then the subject is required to go through each of the $n(n-1)/2$ possible number of pairs and identify the better item in each pair. Thus, if five items need to be evaluated, then the total number of possible pairs would be $5(5-1)/2 = 10$.

The major advantage of the paired comparison approach is that it makes the subject's task simple and more accurate as the subject is asked to only compare the two items in each trial and only identify the better item in the pair based on the given criterion. The disadvantage of the paired comparison approach is that as the number of alternatives (n) to be evaluated increases, the number of possible paired comparison judgments that each subject is asked to make increases rapidly (proportional to the square of n), and the entire evaluation process becomes very time-consuming.

We will review two commonly used methods based on the paired comparison approach, namely (a) Thurstone's Method of Paired Comparisons and (b) Analytical Hierarchy Method. Thurstone's method allows us to develop scale values for each of the n items on a z-scale of desirability (z is a normally distributed variable with mean equal to 0 and standard deviation equal to 1) (Thurstone, 1927); whereas the Analytical Hierarchy Method allows us to obtain relative importance weights of each of the n items (Satty, 1980). Both the methods are simple and quick to administer and have the potential of providing more reliable evaluation results as compared with other subjective methods where a subject is asked to evaluate one item at a time using a given criterion and evaluate all items.

THURSTONE'S METHOD OF PAIRED COMPARISONS

Let us assume that we have nine attributes (included in the above example of Pugh Diagram for Selection of Energy Source for a Power Plant) that need to be evaluated in terms of their importance to the power generating company. The nine attributes are: (a) capital cost, (b) construction time, (c) fuel cost, (d) operation and maintenance costs, (e) energy generation capacity, (f) environmental impact due to emissions, (g) environmental impact on wildlife, (h) space requirements, and (i) safety when failure. These nine attributes can be compared in possible $9 \times 8/2 = 36$ pairs. The 36 pairs are located above the diagonal shown in the 9×9 matrix in Table 9.5. (Note that i and j are indices used here to define pairs of the above nine attributes. The indices i and j range from 1 to 9.) The steps to be used in the procedure are presented below.

Step 1: Select a Criterion for Evaluation of the Attributes

The purpose of the evaluation is to order the nine attributes along an interval scale based on their importance in power plant technology selection from the business viewpoint. Thus, importance of the attributes was considered as the criterion used for obtaining desirability (or preference) of the attributes. (Note: The scale values of the attributes obtained by use of this technique shown in Table 9.10 can be used to judge relative importance.)

Step 2: Prepare Description of the Nine Attributes to Be Evaluated

The descriptions of attributes (presented in the above section) were given on a sheet of paper to the evaluators (subjects) to use as a reference during their evaluations.

Step 3. Obtain Responses of Each Subject on All Pairs

It is also assumed that subjects (who are experts, i.e., specialized in the power plant technologies) were asked to serve as evaluators.

Each subject was brought in a test area separately by an experimenter. The experimenter provided instructions to the subject and asked the subject to read the description of each the nine attributes of the power plant technologies. Each subject was asked to make 36 paired comparisons in a random order. In each pair the subject was asked to determine the attribute that was more important from the business viewpoint (i.e., to improve profitability of power plant operation) of the power company.

The responses of an individual subject are illustrated in Table 9.5. Each cell of the table presents "Yes" or "No" depending upon if the attribute shown in the column (represented by the cell) was more important than the other attribute in the pair (i.e., the attribute shown in the row represented by the cell). It should be noted that only the 36 cells above the diagonal (marked by x) needed evaluations.

Note: A "Yes" response indicates that the attribute shown in the column is more important (or preferred) than the attribute in the row. A "No" response indicates that the attribute shown in the row was more important than the attribute shown in the column.

TABLE 9.5

Responses of an Individual Subject for the 36 Possible Attribute Pairs

		*i*th Attribute								
		1	**2**	**3**	**4**	**5**	**6**	**7**	**8**	**9**
*j*th Attribute	1	X	No	No	Yes	Yes	Yes	Yes	Yes	Yes
	2		X	Yes	Yes	Yes	Yes	Yes	No	Yes
	3			X	Yes	Yes	No	No	No	Yes
	4				X	No	Yes	No	No	Yes
	5					X	Yes	No	No	Yes
	6						X	Yes	No	No
	7							X	No	No
	8								X	Yes
	9									X

TABLE 9.6

Ratio of Subjects Preferring Attribute in the Column over the Attribute in the Row Divided by Number of Subjects

		*i*th Attribute								
		1	**2**	**3**	**4**	**5**	**6**	**7**	**8**	**9**
*j*th Attribute	1	X	4/9	2/3	2/3	8/9	8,/9	8/9	5/9	1
	2		X	5/9	8/9	1	8/9	8/9	5/9	8/9
	3			X	2/3	7/9	2/3	2/3	1/3	8/9
	4				X	2/3	7/9	2/3	1/9	7/9
	3					X	7/9	2/3	2/9	7/9
	6						X	5/9	1/3	2/3
	7							X	2/9	2/3
	S								X	7/9
	9									X

Step 4. Summarize Responses of All Subjects in Terms of Proportion of Subjects Who Rated the Attribute in the Column More Important than the Attribute in the Row

Nine subjects were used for the evaluations. After all the subjects provided responses, the responses were summarized as shown in Table 9.6 by assigning a "1" to a "Yes" response and a "0" to a "No" response. Thus, the cell corresponding to fifth column and first row indicates that only eight out the nine subjects judged the attribute in the fifth column (energy generating capacity) more important than the attribute in the first row (capital cost).

TABLE 9.7

Response Ratio Matrix with Lower Half of the Matrix Filled with Complementary Ratios

		*i*th Attribute								
		1	2	3	4	5	6	7	8	9
*j*th Attribute	1	X	4/9	2/3	2/3	8.9	8/9	8/9.	5/9	1
	2	5/9	X	5/9	8/9	1	8/9	8/9	5/9	8/9
	3	1/3	4/9	X	2/3	7/9	2/3	2/3	1/3	8/9
	4	1/3	1/9	1/3	X	2/3	7/9	2/3	1/9	7/9
	5	1/9	0/9	2/9	1/3	X	7/9	2/3	2/9	7/9
	6	1/9	1/9	1/3	2/9	2/9	X	5/9	1/3	2/3
	7	1/9	1/9	1/3	1/3	1/3	4/9	X	2/9	2/3
	8	4/9	4/9	2/3	8/9	7/9	2/3	7/9	X	7/9
	9	0/9	1/9	1/9	2/9	2/9	1/3	1/3	2/9	X

TABLE 9.8
Proportion of More Important (Preferred) Responses (p_{ij})

		*i*th Attribute								
		1	2	3	4	5	6	7	8	9
*j*th Attribute	1	X	0.444	0.667	0.667	0.889	0.889	0.889	0.556	1.000
	2	0.556	X	0.556	0.889	1.000	0.889	0.889	0.556	0.889
	3	0.333	0.444	X	0.667	0.778	0.667	0.667	0.333	0.889
	4	0.333	0.111	0.333	X	0.667	0.778	0.667	0.111	0.778
	5	0.111	0/9	0.222	0.333	X	0.778	0.667	0.222	0.778
	6	0.111	0.111	0.333	0.222	0.222	X	0.556	0.333	0.667
	7	0.111	0.111	0.333	0.333	0.333	0.444	X	0.222	0.667
	8	0.444	0.444	0.667	0.889	0.778	0.667	0.778	X	0.778
	9	0/9	0.111	0.111	0.222	0.222	0.333	0.333	0.222	X

Step 5: Summary of Response Ratios and Proportions

The complements of the summarized ratios in Table 9.6 were then entered in the cells below the diagonal as shown in Table 9.7. For example, the complement of "4/9 responses of attribute 2 more important than attribute 1" is "5/9 responses of attribute 1 more important than attribute 2."

The proportions in Table 9.7 are expressed in decimals in Table 9.8. Each cell in the matrix presented in Table 9.8 thus represents proportion (p_{ij}) indicating the proportion of responses in which the attribute in the *i*th column was more important (or preferred) over the attribute in the *j*th row.

Step 6: Adjusting p_{ij} Values

To avoid the problem of distorting the scale values (computed in Step 7) of the attributes (when p_{ij} values are very small [close to 0.00], or very large [close to 1.00]), the

TABLE 9.9

Adjusted Table of p_{ij} (If $p_{ij} > 0.977$, then set $p_{ij} = 0.977$; and if $p_{ij} < 0.023$, then set $p_{ij} = 0.023$)

		ith Attribute								
		1	2	3	4	5	6	7	8	9
jth Attribute	1	X	0.444	0.667	0.667	0.889	0.889	0.889	0.556	0.977
	2	0.556	X	0.556	0.889	0.977	0.889	0.889	0.556	0.889
	3	0.333	0.444	X	0.667	0.778	0.667	0.667	0.333	0.889
	4	0.333	0.111	0.333	X	0.667	0.778	0.667	0.111	0.778
	5	0.111	0.023	0.222	0.333	X	0.778	0.667	0.222	0.778
	6	0.111	0.111	0.333	0.222	0.222	X	0.556	0.333	0.667
	7	0.111	0.111	0.333	0.333	0.333	0.444	X	0.222	0.667
	8	0.444	0.444	0.667	0.889	0.778	0.667	0.778	X	0.778
	9	0.023	0.111	0.111	0.222	0.222	0.333	0.333	0.222	X

p_{ij} values (in Table 9.8) above 0.977 are set to 0.977 and the p_{ij} values below 0.023 are set to 0.023 as shown in Table 9.9.

Step 7: Computation of Z-values and Scale Values for the Attributes

In this step, the values of the proportions (p_{ij}) in each cell are converted into Z-values by using the table of standardized normal distribution found in any standard statistics textbook. For example, the value of $p_{19} = 0.023$ is obtained by integrating the area under the standardized normal distribution curve (with mean equal to 0 and standard deviation equal to 1.0) from minus infinity to -1.995. Thus, Z-value of -1.995 provides p-value of 0.023. The Z-values can also be obtained by using a function called NORMINV by setting its parameters as (p_{ij}, 0,1) in Microsoft Excel. The Z-values (Z_{ij}) obtained by converting all the proportion (p_{ij}) values in Table 9.9 by using the above conversion procedure are shown in the 9x9 matrix on the top part (i.e., Z matrix) of Table 9.10.

The Z-values obtained in each column are summed (i.e., summed over all j's) and the scale values for each attribute (S_i) are obtained by using the following formula (see last two rows of Table 9.10):

$$S_i = (\sqrt{2}/n)Z_{ij}$$

Where $n =$ number of attributes used in paired comparisons

The bottom row of Table 9.10 presents the scale values (S_i) for each attribute. (Note: using $n=9$ in the above formula). It should be noted that the sum of the scale values (i.e., summing over all i's) computed from the above formula is equal to 0.0 (i.e., $\sum S_i = 0.0$).

Figure 9.1 presents a bar chart of the scale values (S_i) of the nine attributes shown in Table 9.10. Thus, the above procedure shows that by using the Thurstone's method of paired comparisons, scale values of the nine attributes are obtained. The scale

TABLE 9.10

Values of Z_{ij} Corresponding to Each p_{ij} and Computation of Scale Values (S_i)

		\textit{i}th Attribute								
		1	**2**	**3**	**4**	**5**	**6**	**7**	**8**	**9**
\textit{j}th	1	X	−0.140	0.431	0.431	1.221	1.221	1.221	0.140	1.995
Attribute	2	0.140	X	0.140	1.221	1.995	1.221	1.221	0.140	1.221
	3	−0.431	−0.140	X	0.431	0.765	0.431	0.431	−0.431	1.221
	4	−0.431	−1.221	−0.431	X	0.431	0.765	0.431	−1.221	0.765
	5	−1.221	−1.995	−0.765	−0.431	X	0.765	0.431	−0.765	0.765
	6	−1.221	−1.221	−0.431	−0.765	−0.765	X	0.140	−0.431	0.431
	7	−1.221	−1.221	−0.431	−0.431	−0.431	−0.140	X	−0.765	0.431
	8	−0.140	−0.140	0.431	1.221	0.765	0.431	0.765	X	0.765
	9	−1.995	−1.221	−1.221	−0.765	−0.765	−0.431	−0.431	−0.765	X
Σ Z-values		−6.519	−7.297	−2.276	0.912	3.216	4.262	4.207	−4.097	7.592
Scale value (Si)		−1.024	−1.147	−0.358	0.143	0.505	0.670	0.661	−0.644	1.193

Note: Z_i = Value of NORMIN V(p_{ij},0,1) function from the Microsoft Excel.

values indicate the strength of the relative importance of each of the attributes in the set of the n attributes. The unit of the scale values is in number of standard deviations, and the "zero" value on the scale corresponds to the point of indifference (i.e., the attribute with the zero scale value is neither important [preferred] nor unimportant [not preferred]). Thus, in this example, attribute 9 (safety when failure) is the most important (most preferred) among the nine attributes products, and attribute 2 (construction time) is least important.

ANALYTICAL HIERARCHICAL METHOD

In the analytical hierarchical method (AHM), the items (e.g., products, systems, alternatives) are also compared in pairs based on a selected criterion (or an attribute). However, the better item in each pair is also rated in terms of the strength of the attribute it possesses in relation to the strength of the same attribute in the other item in the pair. The strength of the attribute is expressed using a ratio scale. The scale (or the weight) value of 1 is used to denote equal strength of the attribute in both the items in the pair. And the scale value of 9 is used to indicate extreme or absolute strength of the attribute in the better item as compared with the other item in the pair. And the item with the weaker strength is assigned the inverse of the scale (weight) value of the better item.

The AHM is a simple technique to determine relative importance of different items (e.g., alternatives). It is based on subjective judgments made by one or more decision-makers. Each decision-maker is assumed to be an expert in the problem area and is free from any biases. The decision-maker's task is further simplified by paired comparisons of items. For example, if there are n possible items, then there

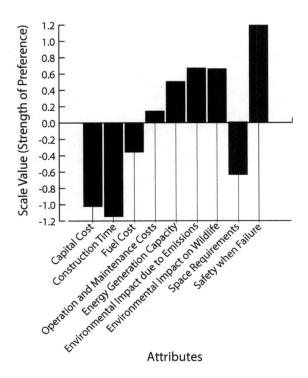

FIGURE 9.1   Scale values of the nine attributes of power plant technologies.

will be $n(n-1)/2$ number of possible pairs of items. The decision-maker is given each pair separately and is asked to select the better of the two items and assign a relative importance rating (weight) to the selected item based on a preselected criterion (or an attribute). The importance ratings are then used to compute relative weights of each of the items. The item with the highest weight is selected as the most preferable item. The method is described by Satty (1980) and Bhise (2012). The following example illustrates the procedure.

Let us assume that there are two products, W and S in a pair; and the attribute to compare the products is "ease of use." The scale values assigned to the products using the ratio scale would be as follows:

1. If product W is "extremely or absolutely easy" to use as compared with product S, then, the weight of W preferred over S will be 9, and the weight of S preferred over W will be 1/9.
2. If product W is "very easy" to use as compared with product S, then, the weight of W preferred over S will be 7, and the weight of S preferred over W will be 1/7.
3. If product W is "easy" to use as compared with product S, then, the weight of W preferred over S will be 5, and the weight of S preferred over W will be 1/5.

4. If product W is "moderately easy" to use as compared with product S, then, the weight of W preferred over S will be 3, and the weight of S preferred over W will be 1/3.
5. If product W is "equally easy" to use as compared with product S, then, the weight of W preferred over S will be 1, and the weight of S preferred over W will be also 1.

Satty (1980) described the nine-point scale with the following adjectives to indicate the level of preference (or importance) for comparing the two items in each pair.

1 = Equal preference
2 = Weak preference
3 = Moderate preference
4 = Moderate plus preference
5 = Strong preference
6 = Strong plus preference
7 = Very strong or demonstrated preference
8 = Very, very strong preference
9 = Extreme or absolute preference

From the viewpoint of making the scales more understandable and easier to apply, usually only the odd numbered scale values (shown in bold case above) are described and presented to the subjects. To allow the subjects to decide on the weight, the author found that the scale presented in Figure 9.2 works very well. Here, the subject will be asked to put a "X" mark on the scale on the left side if product W is preferable over product S. The higher numbers on the scale indicate higher preference. If both products are equally preferred, then the subject will be asked to place the "X" mark at the mid-point of scale with value equal to 1. If product S is preferred over product W, then the subject will use the right side of the scale.

Let us assume that we have to compare six products, namely W, S, Q, M, P, and L, by using the AHM. A subject will be asked to compare the products in pairs. The 15 possible pairs of the six products will be presented to the subject in a random order. (Note: for $n = 6$, $n(n-1)/2 = 15$.) The subject will be given a preselected attribute (e.g., ease of use) and asked to provide strength of preference ratings for the preferred product in each of the 15 pairs by using scales such as the one presented in Figure 9.2. The data obtained from the 15 pairs will then be converted into a matrix of paired comparison responses as shown in Table 9.11. Each cell of the matrix indicates the ratio of preference weight of the product in the row over the product in the column. Thus, the ratio 9/1 in the first row and second column indicates that the product in the row (W) was "absolutely preferred" (i.e., considered extremely or absolutely preferred = rating weight of 9) over the product in the column (S).

To compute the relative weights of preference of the products, the fractional values in Table 9.11 are first converted into decimal numbers as shown in the left side matrix in Table 9.12. All the six values in each row are then multiplied together

FIGURE 9.2 Scale used to indicate strength of the preference when comparing two products (W and S).

TABLE 9.11

Matrix of Paired Comparison Responses for One Evaluator

	W	S	Q	M	P	L
W	1	9./1	1/1	5/1	1/3	1/1
S	1/9	1	1.2	5/1	1/1	3/1
Q	1/1	2/1	1	3/1	5/1	3/1
M	1/5	1/5	113	1	1/1	1/3
P	3/1	1/1	1/5	1/1	1	1/3
L	1/1	1/3	1/3	3/1	3/1	1

Note: The value in a cell (of Table 9.11 above) indicates the preference ratio for comparing the product in a row with the product in a column corresponding to the cell.

---

and entered in the column labeled as "Row Product" in Table 9.12. The geometric mean of each row product is computed. It should be noted that the geometric mean of the product of n numbers is the $(1/n)^{th}$ root of the product (e.g., $1/6^{th}$ root of 90.00 is 2.1169. Note: $2.1169^6 = 90$). All the six geometric means in the column labeled as "Geometric Mean" are then summed. The sum, as shown in Table 9.12, is 6.9796. Each of the geometric means is then divided by their sum (6.9796) to obtain the normalized weight of the products (see last column of Table 9.12). It should be noted that due to the normalization, the sum of the normalized weights over all the products is 1.0.

The normalized weights (called the normalized preference values) are plotted in Figure 9.3. The figure, thus, shows that the most preferred product (based on the ease of use) was Q (with its normalized weight of 0.3033) and the least preferred product was M (with its normalized weight of 0.0581).

The above example was based on data obtained from one subject. If more subjects are available, then the normalized weights for each subject can be obtained by using the above procedure and then average weight of each product can obtained by averaging over the normalized weights of all the subjects for each product.

TABLE 9.12

Computation of Normalized Weights of the Products

	W	S	Q	M	P	L	Row Product	Geometric Mean	Normalized Weight
W	1.00	9.00	1.00	5.00	0.33	1.00	15.0000	1.5704	0.2250
S	0.11	1.00	0.50	5.00	1.00	3.00	0.83333	0.9701	0.1390
Q	1.00	2.00	1.00	3.00	5.00	3.00	90.0000	2.1169	0.3033
M	0.20	0.20	0.33	1.00	1.00	0.33	0.0044	0.4055	0.0581
P	3.00	1.00	0.20	1.00	1.00	1.00	0.60000	0.9184	0.1316
L	1.00	0.33	0.33	3.00	3.00	1.00	0.9900	0.9983	0.1430
							Sum→	6.9796	1.000

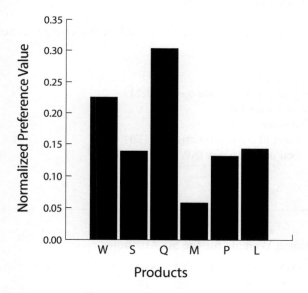

FIGURE 9.3 Normalized preference values of the six products.

APPLICATION OF ANALYTICAL HIERARCHY METHOD TO A MULTI-ATTRIBUTE PROBLEM

The application of AHM for evaluating a set of items (or alternatives) using multiple attributes involves a three-step procedure. In step 1, the weights of attributes are determined by applying the AHM described in the preceding section. Then in step 2, weights of the alternatives are determined for each attribute separately using the AHM. And in step 3, the weights obtained for the attributes in step 1 and weights of alternatives obtained for each of the attributes in step 2 are used to obtain the final weights of the alternatives.

The three steps are illustrated in Tables 9.13–9.15 for steps 1, 2, and 3, respectively. The problem illustrated here involves determination of weights of eight alternatives (i.e., eight technologies used in developing power plants) based on six attributes of the power plants.

The eight alternatives are as follows:

1. Coal-fired plant
2. Natural-gas-fired powered plant
3. Nuclear power plant
4. Biomass-fired power plant
5. Geothermal power plant
6. Hydroelectric power plant
7. Wind turbines
8. Solar photovoltaic power plant

The six attributes of the power plants considered for evaluating the above eight alternatives are:

1. Capital costs (CC)
2. Operating and maintenance costs (O&M)
3. Pollution potential (PP)
4. Wildlife danger (WD)
5. Displacement of other infrastructure (DI)
6. Safety when malfunctions (SWM)

TABLE 9.13

Step 1: Obtaining Weights for the Six Attributes

	CC	O&M	PP	WD	DI	SWM
CC	1	1/5	1/9	5/1	1/7	1/9
O&M	5/1	1	1/5	5/1	1/7	1/9
PP	9/1	5/1	1	7/1	3/1	1/1
WD	1/5	1/5	1/7	1	1/5	1/9
DI	7/1	7/1	1/3	5/1	1	1/5
SWM	9/1	9/1	1/1	9/1	5/1	1

	CC	O&M	PP	WD	DI	SWM	Row Product	Geometric Mean	Normalized Weight
CC	1.00	0.20	0.11	5.00	0.14	0.11	0.002	0.348	0.035
O&M	5.00	1.00	0.20	5.00	0.14	0.11	0.079	0.656	0.066
PP	9.00	5.00	1.00	7.00	3.00	1.00	945.000	3.133	0.317
WD	0.20	0.20	0.14	1.00	0.20	0.11	0.000	0.224	0.023
DI	7.00	7.00	0.33	5.00	1.00	0.20	16.333	1.593	0.161
SWM	9.00	9.00	1.00	9.00	5.00	1.00	3,645.000	3.923	0.397

TABLE 9.14

Step 2: Obtaining Weights for Alternatives for Each Attribute

A. Capital Costs (CC) Preferred = Low Capital Costs

	Importance of Criterion in Row Over Criterion in Column								Row Product	Geometric Mean	Normalizes Weight
	Coal	Nat Gas	Nuclear	Biomass	G-thermal	Hyd-elec	Wind	Solar			
Coal	1.00	1.00	5.00	0.33	0.33	0.20	7.00	9.00	6.861	1.272	0.121
Nat gas	1.00	1.00	5.00	0.33	0.33	0.20	7.00	9.00	6.861	1.272	0.121
Nuclear	0.20	0.20	1.00	0.11	0.11	3.00	3.00	3.00	0.013	0.581	0.055
Biomass	3.03	3.03	9.09	1.00	0.33	0.20	3.00	3.00	49.587	1.629	0.155
G-thermal	3.03	3.03	9.09	3.03	1.00	0.33	3.00	5.00	1,252.191	2.439	0.233
Hyd-elec	5.00	5.00	0.33	3.03	3.03	1.00	5.00	7.00	2,678.298	2.682	0.256
Wind	0.14	0.14	0.33	0.33	0.33	0.20	1.00	5.00	0.001	0.407	0.039
Solar	0.11	0.11	0.11	0.33	0.20	0.14	0.20	1.00	0.000	0.201	0.019
									Sum→	10.484	1.000

B. Operating Cost and Maintenance Cost (O&M) Preferred = Low O&M Costs

	Coal	Nat Gas	Nuclear	Biomass	G-thermal	Hyd-elec	Wind	Solar	Row Product	Geometric Mean	Normalizes Weight
Coal	1.00	0.33	3.00	1.00	0.20	0.14	0.20	0.20	0.001	0.427	0.035
Nat Gas	3.03	1.00	0.14	3.00	0.20	0.11	0.20	0.20	0.001	0.428	0.035
Nuclear	0.33	7.14	1.00	0.11	0.14	0.14	0.20	0.20	0.000	0.346	0.029
Biomass	1.00	0.33	9.09	1.00	0.33	0.20	0.20	0.20	0.008	0.547	0.045
G-thermal	5.00	5.00	7.14	3.03	1.00	0.33	0.33	0.33	19.623	1.451	0.120
Hyd-elec	7.14	9.09	7.14	5.00	3.00	1.00	5.00	7.00	243,750.244	4.714	0.390
Wind	5.00	5.00	5.00	5.00	3.03	0.20	1.00	3.00	1,136.364	2.410	0.200
Solar	5.00	5.00	5.00	5.00	3.03	0.14	0.33	1.00	90.188	1.755	0.145
									Sum→	12.077	1.000

(Continued)

TABLE 9.14 (Continued)
Step 2: Obtaining Weights for Alternatives for Each Attribute

A. Capital Costs (CC)　　　　Preferred = Low Capital Costs

	Coal	Nat Gas	Nuclear	Biomass	G-thermal	Hyd-elec	Wind	Solar	Row Product	Geometric Mean	Normalizes Weight
	Importance of Criterion in Row Over Criterion in Column										

C. Polluting Potential(PP)　　Preferred: Low Polluting Potential

	Coal	Nat Gas	Nuclear	Biomass	G-thermal	Hyd-elec	Wind	Solar	Row Product	Geometric Mean	Normalizes Weight
Coal	1.00	0.33	0.14	0.50	0.11	0.11	0.11	0.11	0.000	0.207	0.016
Nat gas	3.03	1.00	0.14	1.00	0.11	0.11	0.11	0.11	0.000	0.298	0.022
Nuclear	7.14	7.14	1.00	3.00	0.33	0.11	0.11	0.11	0.067	0.714	0.054
Biomass	2.00	1.00	0.33	1.00	0.20	0.11	0.11	0.11	0.000	0.340	0.026
G-thermal	9.09	9.09	3.03	5.00	1.00	1.00	0.50	0.50	313.048	2.051	0.154
Hyd-elec	9.09	9.09	9.09	9.09	1.00	1.00	0.50	1.00	1,707.534	2.535	0.190
Wind	9.09	9.09	9.09	9.09	2.00	2.00	1.00	1.00	27,320.538	3.586	0.269
Solar	9.09	9.09	9.09	9.09	2.00	2.00	1.00	1.00	27,320.538	3.586	0.269
								Sum→		13.316	1.000

D. Wildlife Danger(WD)　　Preferred: Low or No Danger to Wildlife

	Coal	Nat Gas	Nuclear	Biomass	G-thermal	Hyd-elec	Wind	Solar	Row Product	Geometric Mean	Normalizes Weight
Coal	1.00	1.00	0.14	0.50	0.11	0.14	0.20	0.11	0.000	0.264	0.026
Nat gas	1.00	1.00	1.00	1.00	0.33	0.20	0.20	0.20	0.003	0.476	0.047
Nuclear	7.14	1.00	1.00	2.00	0.50	0.20	1.00	0.20	0.286	0.855	0.085
Biomass	2.00	1.00	0.50	1.00	0.50	0.50	2.00	9.00	4.500	1.207	0.120
G-thermal	9.09	3.03	2.00	2.00	1.00	1.00	3.00	1.00	330.579	2.065	0.205
Hyd-elec	7.14	5.00	5.00	1.00	1.00	1.00	3.00	1.00	535.714	2.193	0.218
Wind	5.00	5.00	1.00	0.33	0.33	0.33	1.00	0.14	0.130	0.775	0.077
Solar	9.09	9.09	9.09	0.11	1.00	1.00	7.14	1.00	596.282	2.223	0.221
								Sum→		10.058	1.000

(Continued)

TABLE 9.14 (Continued)
Step 2: Obtaining Weights for Alternatives for Each Attribute

A. Capital Costs (CC) Preferred = Low Capital Costs

	Coal	Nat Gas	Nuclear	Biomass	G-thermal	Hyd-elec	Wind	Solar	Row Product	Geometric Mean	Normalizes Weight
			Importance of Criterion in Row Over Criterion in Column								

E. Displacement of Other Infrastructure (DI) Preferred: No or Low Displacement of People or Properties

	Coal	Nat Gas	Nuclear	Biomass	G-thermal	Hyd-elec	Wind	Solar	Row Product	Geometric Mean	Normalizes Weight
Coal	1.00	0.50	1.00	0.33	2.00	1.00	1.00	0.20	0.066	0.712	0.068
Nat gas	2.00	1.00	3.00	2.00	0.20	1.00	3.00	2.00	14.400	1.396	0.133
Nuclear	1.00	0.33	1.00	0.33	0.14	1.00	0.33	0.11	0.001	0.392	0.037
Biomass	3.03	0.50	3.03	1.00	0.20	0.33	2.00	0.11	0.067	0.714	0.068
G-thermal	0.50	5.00	7.14	5.00	1.00	1.00	5.00	0.33	148.661	1.869	0.178
Hyd-elec	1.00	1.00	1.00	1.00	1.00	1.00	3.00	0.33	0.999	1.000	0.095
Wind	1.00	1.00	3.03	0.20	0.20	0.33	1.00	0.20	0.008	0.548	0.052
Solar	5.00	5.00	5.00	9.09	3.00	3.00	5.00	1.00	51,238.790	3.879	0.369
									Sum→	10.508	1.000

F. Safety When Malfunctions (SWM) Preferred: Safe during Accidents

	Coal	Nat Gas	Nuclear	Biomass	G-thermal	Hyd-elec	Wind	Solar	Row Product	Geometric Mean	Normalizes Weight
Coal	1.00	3.00	5.00	1.00	0.20	0.20	0.20	0.11	0.013	0.582	0.051
Nat gas	0.33	1.00	0.11	1.00	0.14	0.33	0.20	0.11	0.000	0.280	0.024
Nuclear	0.20	9.09	1.00	0.11	0.20	0.11	0.20	0.11	0.000	0.315	0.027
Biomass	1.00	1.00	9.09	1.00	0.33	0.20	0.33	0.14	0.028	0.640	0.056
G-thermal	5.00	7.14	5.00	3.00	1.00	1.00	1.00	0.20	107.250	1.794	0.156
Hyd-elec	5.00	3.03	9.09	1.00	1.00	1.00	1.00	1.00	137.741	1.851	0.161
Wind	5.00	5.00	5.00	1.00	1.00	1.00	1.00	0.14	17.500	1.430	0.125
Solar	9.09	9.09	9.09	7.14	5.00	1.00	7.14	1.00	191,661.939	4.574	0.399
									Sum→	11.466	1.000

TABLE 9.15

Step 3: Obtaining Final Weights for Alternatives

Final Weighting of Scores

Attribute→		CC	O&M	PP	WD	DI	SWM	Final
Normalized Weights→		0.035	0.066	0.317	0.023	0.161	0.397	Weights
Normalized	Coal	0.121	0.035	0.016	0.026	0.068	0.051	0.043
weights	Nat gas	0.121	0.035	0.022	0.047	0.133	0.024	0.046
	Nuclear	0.055	0.029	0.054	0.085	0.037	0.027	0.040
	Biomass	0.155	0.045	0.026	0.120	0.068	0.056	0.052
	G-thermal	0.233	0.120	0.154	0.205	0.178	0.156	0.161
	Hyd-elec	0.256	0.390	0.190	0.218	0.095	0.161	0.180
	Wind	0.039	0.200	0.269	0.077	0.052	0.125	0.160
	Solar	0.019	0.145	0.269	0.221	0.369	0.399	0.319
Sum→		1.000	1..000	1.000	1.000	1.000	1.000	1.000

The final weight of the coal-fired power plant is computed as follows (see Table 9.15):

$$(0.035 \times 0.121) + (0.066 \times 0.035) + (0.317 \times 0.016) + (0.023 \times 0.026)$$
$$+ (0.161 \times 0.068) + (0.397 \times 0.051) = 0.043.$$

Similarly, the final weights of other power plants are computed by obtaining sum of multiplications over all attributes. Each multiplication (for each attribute) is computed by multiplying the normalized weight of each attribute by normalized weight of each plant for the same attribute, as shown above for the coal-fired plant.

Figure 9.4 presents a plot of the final weights computed in step 3.

When more than one subjects (or experts) are available, the above described three-step process should be applied by using each subject (or expert) separately. The final weights of alternatives obtained for each alternative in step 3 for each subject can be aggregated by computing arithmetic (or geometric) means of the weights of the alternatives obtained for each subject.

FAILURE MODES AND EFFECTS ANALYSIS (FMEA)

The FMEA is a proactive and qualitative tool used by Quality, Safety, and Product/ Process engineers to improve reliability (i.e., to eliminate failures – thus, improve quality and customer satisfaction). It is based on knowledge and subjective judgment of engineers and experts involved in the design and development of the product or process being evaluated. It is very effective when performed early in the product or process development and conducted by experienced multifunctional team members as a team exercise.

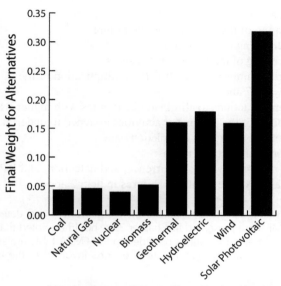

Alternatives -- Power Plant Technologies

FIGURE 9.4 Weights of the eight alternatives obtained in step 3.

FMEA method was initially developed in the 1960s as a Systems Safety Analysis tool. It was used in the early days in the defense and aerospace systems design to ensure that the product (e.g., an aircraft, spaceship, or a missile) was designed to minimize probabilities of all major failures by brainstorming and evaluating all possible failures that could occur and acting on the resulting prioritized list of the corrective actions.

For over the past 20 years, the method is routinely used by the product design and process design engineers to reduce the risk of failures in the designs of products and processes used in production, operation, and maintenance of various systems used in the many industries (e.g., automotive, aviation, utilities, and construction). The FMEA conducted by product design engineers is typically referred as DFMEA (The first letter "D" stands for design.). And the FMEA conducted by the process designers is referred as PFMEA (The first letter "P" stands for process.).

The method involves creating a table with each row representing a possible failure mode of the given product (or a process) and providing information about the failure mode in the following columns of the FMEA table:

1. Description of a system, subsystem, or component
2. Description of a potential failure mode of the system, subsystem, or component
3. Description of potential effect(s) of the failure on the product/systems, its subsystems, components, or other systems

4. Potential causes of the failure
5. Severity rating of the effect due to the failure
6. Occurrence rating of the failure
7. Detection rating of the failure or its causes
8. Risk priority number (RPN). It is the multiplication of the three ratings in items 5, 6, and 7 above.
9. Recommended actions to eliminate the failures with higher RPNs
10. Responsibility of the persons or activities assigned to undertake the recommended actions and target completion date
11. Description of the actions taken
12. Resulting ratings (severity, occurrence, and detection) and RPNs (after the action is taken) of the identified failures in item 2 above

Examples of rating scales used for severity, occurrence, and detection and an example of FMEA are presented in Chapter 3. It should be noted that the severity, occurrence, and detection ratings provided in the FMEA table are subjective and are based on the collective judgments of experts involved in the rating of each failure mode.

FAILURE MODES AND EFFECTS AND CRITICALITY ANALYSIS (FMECA)

The Failure Modes and Effects and Criticality Analysis (FMECA) is very similar in format and content to the FMEA described above. It contains an additional column of criticality. The criticality column provides a rating illustrating level of criticality of the failure (in each row) in accomplishing the major goal (or mission) of the product. The criticality ratings are generally based on subjective judgment of experts involved in performing the analysis. The technique is also called Failure Modes and Criticality Analysis (Hammer, 1980).

Criticality can be rated by using different scales for different products. The criticality ratings typically cover range from low criticality involving stoppage of equipment (requiring minor maintenance) to high criticality levels involving failures resulting in potential loss of life.

CONCLUDING REMARKS

Subjective methods covered in this chapter are very useful in decision-making where a complex problem with many variables cannot be easily solved because quantitative data relating the variables to the decision evaluation measures are not available. Models of such complex problems are generally not available for decision-making exercises. Development of models of such situations is also very costly. On the other hand, experienced subject matter experts who have a good understanding of many variables that affect certain outcomes can provide useful data for subjective analyses that can in turn aid decision-makers in recommending solutions for complex problems.

REFERENCES

Bhise, V.D. 2012. *Ergonomics in the Automotive Design Process*. ISBN 978-1-4398-4210-2. Boca Raton, FL: CRC Press.

Bhise, V.D. 2014. *Designing Complex Products with Systems Engineering Processes and Techniques*. ISBN: 978-1-4665-0703-6. Boca Raton, FL: CRC Press, Taylor and Francis Group.

Satty, T.L. 1980. *The Analytic Hierarchy Process*. New York, NY: McGraw Hill.

Thurstone, L.L. 1927. The Method of Paired Comparisons for Social Values. *Journal of Abnormal and Social Psychology*, 21: 384–400.

Section III

Customers, Governments, and Future Changes

10 Energy Users, Societal Needs, and Energy Providers

INTRODUCTION

The purpose of this chapter is to provide the reader basic data on characteristics of U.S. energy generation and consumption from the users' viewpoints and considerations. Many different sources of energy are used in the United States. They can be grouped into general categories such as primary and secondary, renewable and non-renewable, and fossil and nonfossil fuels. Primary energy sources include fossil fuels (petroleum, natural gas, and coal), nuclear energy, and renewable sources of energy. Electricity is a secondary energy source that is generated (produced) from primary energy sources.

Energy Information Agency (EIA) has developed comprehensive databases on customers, energy consumption, energy generation, energy sources and their locations, capacities of different energy-related facilities such as transmission and distribution power lines and pipelines, prices of different energy from different sources to different customers (EIA, 2020a). These databases are useful in understanding the overall energy picture from the viewpoint of energy generating and distribution capabilities and needs of its varied customers.

To satisfy the customers and their varying demands, the energy providers face many challenges. Some of the major challenges are all also discussed in this chapter.

CUSTOMER AND SOCIETAL NEEDS

The present and future needs of the customers of the energy systems and hence the societal needs can include the following:

a. providing reliable uninterrupted energy to meet the level of customer demand any time of the day or night
b. providing energy at affordable prices
c. providing healthier and safer environment, e.g., reduced pollution potential (clean air and water) and accidents, reduced danger to people and wildlife during normal operating and failure situations
d. increased use of renewable energy sources, and allowing customers to sell excess energy generated by their own renewable sources (e.g., wind turbines, solar photovoltaic electric generators)
e. providing improved convenience and comfort during usage, e.g., easy to use controls and for providing and monitoring energy system access (with smart

DOI: 10.1201/9781003107514-13

    grid features, wireless and remote control interfaces, ability to interface and operate autonomous equipment, appliances, and vehicles)

    f. supporting development and use of more energy efficient equipment, appliances, and vehicles for all customers

    g. providing quick response to customer service requests related to loss of power, power fluctuations, brownouts, and so forth

    h. providing increased customer satisfaction with energy services

ENERGY USES AND DEMANDS

Energy is used to operate equipment and machines to extend human capabilities to perform tasks that cannot be performed by humans alone, e.g., lifting and transporting materials, transforming and joining materials, converting energy from one source to other sources, creating comfortable living environments, communications and data processing, and so forth. Energy is used by different types of customers located at different geographic locations to perform different activities over different time periods of the day and year in different amounts.

CUSTOMERS OF ENERGY SYSTEMS

The energy is needed and consumed in homes, business, commercial, institutional, and public locations for many different purposes. The energy consumers expect the energy providers to provide the required amount of energy at any given location and to respond to changes in energy demands in terms of quantity (e.g., rate of power supply [kWh] and quality – i.e., quickly with minimum response time, without unexpected interruptions or slowdown [brown-outs]) and at affordable costs.

    The consumers of the energy can be classified as follows:

Residential Customers

Residential customers primarily use electric power and natural gas for purposes such as lighting, cooking, running appliances, pumps, machines and computers, heating or cooling water and air, operating ventilating and air-conditioning equipment, and so forth. Heating oil is also used for residential heating purposes. According to EIA (2009), the average annual U.S. household energy use was about 88 MBtu, which included annual electricity consumption of about 11,000 kWh. As electric vehicles gain more acceptance, the electric power demands for charging their batteries from home will also increase.

Commercial Customers

Commercial customers primarily use electric power and natural gas for heating, air-conditioning, airflow (fans and blowers), lighting, running specialized equipment, e.g., pumps, computers, office equipment, and transportation are the primary energy using devices of commercial customers.

Industrial Customers

Industrial customers use energy in companies, factories, and workstations with heating, air-conditioning, airflow (fans and blowers), lighting, running specialized machines and equipment (operating on electricity, natural gas, petroleum, pressurized air, water, or other fluids).

One major industrial use of energy application involves raising the temperature of components in the manufacturing process, which is called process heating. Refining crude oil, where heat is used to separate various distillates, is an example of this. Another common use of energy in industry is to heat a boiler that generates steam or hot water.

Industry and manufacturing rely heavily on natural gas (30% of all energy consumed by the industrial sector in 2015), petroleum and other liquids (26%), and electricity (10%), with coal, renewables, and biofuels making up the rest.

Institutional Customers

Institutions such as schools and universities, government facilities, hospitals, and other nonprofit organization also purchase energy to meet their functions (e.g., lighting, heating and air-conditioning, running computer and other equipment).

Transportation Customers

The United States uses 28% of its total energy each year to move people and goods from one place to another. The transportation sector includes many modes, from personal vehicles and large trucks to public transportation (buses, trains) to airplanes, freight trains, ships and barges, and pipelines. By far the largest share is consumed by cars, light trucks and motorcycles – about 58% in 2013, followed by other trucks (23%), aircraft (8%), boats and ships (4%), and trains and buses (3%). Pipelines account for 4%. The sources of energy used for transportation include fuels such as gasoline, diesel, kerosene, jet fuel, and electricity.

EIA (2020a) websites provide detailed information on the fact sheets that cover many areas of interest such as:

a. Overall energy use, electricity use, and expenditures
b. Residential consumption by end use (air-conditioning, heating, appliances)
c. Main heating fuel
d. Use of cooling equipment
e. Housing types and year of construction
f. Numbers of TVs and refrigerators
g. *Lighting energy needs and LEDs*: Lighting accounts for one-eighth of total U.S. electrical consumption and 15% of the electricity consumed by the residential and commercial sectors of the economy. There are enormous opportunities for efficiency gains in those sectors by widespread adoption of the technology most Americans first encountered as the glowing lights on appliances: the light-emitting diode (LED). LEDs are the most efficient technology for conventional forms of lighting, such as home and office illumination, and for street lighting. The LED light sources now produce about 5–8 times light flux per Watt as compared with the outgoing filament bulbs. LEDs also have an extremely long life span relative to every other lighting technology (including low-pressure sodium and fluorescent lights but especially compared with incandescent lights). New LEDs can

last 50,000–100,000 hours or more. The typical life span for an incandescent bulb, by comparison, is 1%–5% as long at best (roughly 1,200 hours).

RELATIONSHIP OF CUSTOMER SATISFACTION, ENERGY AVAILABILITY, AND SAFETY

Most energy users expect and assume that the energy will be available anytime of the day or night. The availability of energy is thus what is called an "unspoken customer wants," i.e., the customer does not have to say that they need energy to satisfy their many different needs. According to the Kano model of quality, meeting the unspoken needs of the customers is very important as if the unspoken want is not met, then the customers are very dissatisfied (Yang and El-Haik. 2003; Bhise, 2012).

The customers of energy systems are dissatisfied when the energy companies fail to deliver the required amount of energy due to causes such as adverse weather and resulting electric energy outages or brownouts. Accidents and unsafe situations caused by energy providers also become a customer concern when unexpected energy failures occur. Hospitals, emergency responders, and security providers losing power is major concern, and thus, backup power generators are used by a number of facilities that cannot tolerate interruptions in electric power. Environmental pollution and its effect on health of the customers are also a major concern. Other concerns are from the viewpoint of protection of the environment and wildlife that can be affected by the energy generation and distribution systems.

The major goal of energy providers must be (a) to meet the energy demands of its customers, (b) to deliver the energy without causing accidents and unsafe situations, (c) to reduce pollution created during energy generation and distribution, and (d) to avoid changes in environment that can disrupt other human and natural systems such as wildlife habitats. (Note: Chapter 5 provides information on safety aspects of the energy systems.)

COMMUNICATION BETWEEN CUSTOMERS AND ENERGY PROVIDERS

The customers communicate with the energy providers using telephones, internet, and mail to establish services, pay bills, and obtain service during outages and repairs. With smart phones and internet availability, the providers also keep customers informed about changes in services, rates, planned interruptions, and so forth. With future implementation of smart grid features, the customers will have more information and choices available in consuming energy in terms of time, quantity, and prices.

POWER SECTOR ISSUES

RELIABILITY OF POWER SECTOR

According to DOE (2016) report, while there are numerous standards and regulations that govern reliability of the power sector, they can be consolidated into the following four "rules":

1. Power generation and transmission capacity must be sufficient to meet peak demand for electricity.
2. Power systems must have adequate flexibility to address variability and uncertainty in demand (load) and generation resources.
3. Power systems must be able to maintain steady frequency.
4. Power systems must be able to maintain voltage within an acceptable range.

According to the DOE (2016) report, the following changes from traditional practices need to be considered to meet the above four rules:

The power grid must have sufficient capacity available to meet the demand for electricity. Because there are uncertainties in forecasting demand and the potential for generation and transmission outages, the total amount of capacity is required to exceed the expected level of demand by a given fraction, termed the reserve margin, often about 15%. Traditionally, the need for flexible generation has been met with natural gas generators, which are capable of ramping their output up and down rapidly. Demand response has also played a growing role. Analyses conducted by DOE (2016) indicated that the current level of flexibility on the grid can accommodate variable generation levels of up to 35% of all generation. Many grid operators are planning to or are already implementing policies to increase the flexibility of their systems. New, modern gas generators have been designed to provide very fast ramp rates. Expanded use of demand response also provides more flexibility. Lastly, it is possible to add technology to allow variable resources to decrease generation and, potentially, to increase it if they are not using all available power. This ability to dispatch variable generation is already being used to provide flexibility across the country.

Large spinning generators are traditionally used to arrest any change in frequency because it takes time for them to change their rate of rotation. Generators can have governors that detect any change in their rate of rotation and increase or decrease power to compensate. On longer timescales, generation that can rapidly respond is kept in reserve to match supply and demand.

Studies have shown that increased levels of variable generation on the grid increase reserve requirements necessary to maintain a steady frequency, but these increases are quite modest. Transmission can be used to average out some of the variability and reduce the need for additional reserves. Even as they retire, large spinning generators can be used as "synchronous condensers" that spin synchronously with the grid, not consuming fuel, but serving to arrest changes in frequency. In addition, it is possible to make a variable resource act like a large spinning generator through the use of advanced power electronics. Demand response and storage to balance supply and demand also will likely play a growing role in maintaining a steady frequency.

Traditionally, large spinning generators that are synchronized with the grid can control voltage levels and reactive power by adjusting their output. Various electrical devices such as shunt capacitors are used to control reactive power throughout the transmission and distribution networks. As with frequency control, advanced power electronics can give variable generation resources such as wind and solar the ability to control reactive power and voltage. Many types of storage can also use this sort of power electronics. In addition, synchronous condensers can be used to provide reactive power. Lastly, there is a class of relatively inexpensive electronic devices called Flexible AC Transmission Systems (FACTS) that have existed for a while but are becoming less expensive and more widely deployed and can solve many voltage control problems that historically would have required larger and more costly generators, transmission lines, or electromechanical devices.

ELECTRICITY ENERGY SOURCES

The United States uses many different energy sources and technologies to generate electricity. The sources and technologies have changed over time, and some are used more than others. Three major categories of energy for electricity generation are fossil fuels (coal, natural gas, and petroleum), nuclear energy, and renewable energy sources. Most electricity is generated with steam turbines using fossil fuels, nuclear, biomass, geothermal, and solar thermal energy. Other major electricity generation technologies include gas turbines, hydro turbines, wind turbines, and solar photovoltaics.

Fossil fuels are the largest sources of energy for electricity generation.

a. Natural gas was the largest source – about 38% – of U.S. electricity generation in 2019. Natural gas is used in steam turbines and gas turbines to generate electricity.
b. Coal was the second-largest energy source for U.S. electricity generation in 2019 – about 23%. Nearly all coal-fired power plants use steam turbines. A few coal-fired power plants convert coal to a gas for use in a gas turbine to generate electricity.
c. Petroleum was the source of less than 1% of U.S. electricity generation in 2019. Residual fuel oil and petroleum coke are used in steam turbines. Distillate or diesel fuel oil is used in diesel-engine generators. Residual fuel oil and distillates can also be burned in gas turbines.

Nuclear energy provides one-fifth of U.S. electricity. Nuclear energy was the source of about 20% of U.S. electricity generation in 2019. Nuclear power plants use steam turbines to produce electricity from nuclear fission.

Renewable energy sources provide an increasing share of U.S. electricity. Many renewable energy sources are used to generate electricity and were the source of about 17% of total U.S. electricity generation in 2019.

ENERGY PROVIDERS

The immediate providers of energy to the above types of customers are the utility and fuel supply companies that deliver the electricity through cables, fluids, and gases (e.g., oil, natural gas, steam, hot water) through pipelines. These energy providers either have their own facilities such as plants to generate energy (or convert energy), distribution networks (power lines and pipelines), and storage facilities (tanks and reservoirs), or they have abilities to purchase the energy from other larger energy providers. The locations of the energy conversion plants are generally selected closer to the population centers (where the customers are located to minimize transportation and transmission costs and losses) or where the energy sources are naturally available (e.g., hot water sources for geothermal plants, rivers for dams and hydroelectric projects, nuclear power plants closer to large amount of cooling water source, and so forth).

The locations of fossil fuel mining facilities (crude oil, coal, and natural gas) are dependent upon exploration capabilities and historic data from past geological surveys. For example, the U.S. EIA website (EIA, 2020b) provides the following facts and data for each coal-producing region as follows:

Appalachian Coal Region
a. The Appalachian coal region includes Alabama, Eastern Kentucky, Maryland, Ohio, Pennsylvania, Tennessee, Virginia, and West Virginia.
b. About 26% of the coal produced in the United States came from the Appalachian coal region.
c. West Virginia is the largest coal-producing state in the region and the second-largest coal-producing state in the United States.
d. Underground mines supplied 77% of the coal produced in the Appalachian region.
e. Underground mines in the Appalachian region produced 56% of U.S. total underground coal mine production.

Interior Coal Region
a. The Interior coal region includes Arkansas, Illinois, Indiana, Kansas, Louisiana, Mississippi, Missouri, Oklahoma, Texas, and Western Kentucky.
b. About 18% of total U.S. coal was mined in the Interior coal region.
c. Illinois was the largest coal producer in the Interior coal region, accounting for 36% of the region's coal production and 7% of total U.S. coal production.
d. Underground mines supplied 63% of the region's coal production, and surface mines supplied 37%.

Western Coal Region
a. The Western coal region includes Alaska, Arizona, Colorado, Montana, New Mexico, North Dakota, Utah, Washington, and Wyoming.
b. About 55% of total U.S. coal production was mined in the Western coal region.
c. Wyoming, the largest coal-producing state in the United States, produced 40% of total U.S. coal production and 73% of the coal mined in the Western coal region.
d. Six of the top 10 U.S. coal-producing mines were in Wyoming, and all those mines are surface mines.
e. Surface mines produced 92% of the coal in the Western coal region.

LOCATIONS OF ENERGY SYSTEM FACILITIES

The U.S. mapping system provide locations of individual facilities such as mines, power plants, oil mines and refineries, gas and oil well platforms, pipelines, electric transmission lines, and so forth (EIA, 2020c).

ENERGY PRICES

EIA publications (EIA, 2020a, e) that provide historical, recent, and projected retail gasoline prices (including federal, state, and local taxes) are as follows:

a. *Gasoline and diesel fuel update* – most recent available and historical U.S. weekly (and average monthly and annual) nominal prices.
b. *Short-term energy outlook* – monthly forecasts for average monthly and annual U.S. retail regular grade gasoline prices for the current year and the next year. The Real Prices Viewer shows historical and forecasted nominal and real prices.
c. *Annual energy outlook* – projections for average annual U.S. retail gasoline prices out to the year 2050 are available in a tabular format. The projected prices are the average for all gasoline grades.

In 2019, the U.S. annual average retail price of electricity was about 10.60¢ per kWh. The annual average prices by major types of utility customers (EIA, 2020d, e) in 2019 were:

a. Residential: 13.04¢per kWh.
b. Commercial: 10.66¢per kWh.
c. Industrial: 6.83¢per kWh.
d. Transportation: 9.73¢per kWh.

LIABILITY OF ENERGY PROVIDERS

Just like the product manufacturers who face the risk of being sued for injuries caused by their products, the energy companies spend considerable resources in defending legal cases. The energy companies need to defend against liabilities from accidents and injuries caused by malfunction of their equipment during construction and operations. The principles of negligence, strict liability, and breach of warranty that are applicable to the product manufacturers (Hammer, 1980, 1989; Goetsch, 2007; Bhise 2014) would also apply to injuries caused during provision of energy services. The energy providers also need to provide warnings for unsafe and hazardous situations that may arise from dangerous situations such as down power lines, gas line ruptures, accidents during construction of power plants, and so forth. In addition to the injuries caused during their energy supply, the energy providers are also responsible for supplying clean energy as mandated by state and federal regulatory agencies (see Chapter 11).

Examples of Lawsuits

Typically, a landowner is required to enter into written contractual agreements before a wind turbine (aerogenerator) is constructed on the land. It is important to keep in mind that tort liability may be assessed in cases where harm results as a result of a party's negligence with respect to the construction or maintenance of wind aerogenerators. The fossil fuel companies are already grappling with the risks posed by climate change, from the physical threats of extreme weather to the challenge of switching to cleaner energy. Over the past few years, a growing number of legal cases in the United States – brought by cities, counties, and states, are seeking damages from energy companies for climate-related problems. For example, Baltimore wants compensation for the cost of retrofitting storm drains to prepare for worsening storms. In San Francisco, the city says it will cost $5 billion to upgrade the city's sea wall to prepare for higher sea levels. Meanwhile, Rhode Island expects coastal properties worth $3.6 billion to be under threat by the end of the century (Financial Times, 2020).

Hydraulic fracturing is a process used to extract natural gas from shale formations. This process has been used commercially since the 1940s, but in past decade it has become prevalent as more shale formations have been discovered, specifically the Marcellus Shale formation in Pennsylvania. Although natural gas is a relatively clean source of domestic energy, there have been numerous allegations of water contamination caused by hydraulic fracturing, and several lawsuits have been filed as a result (Coman, 2012).

SAFETY, COMFORT, AND CONVENIENCE-RELATED SPECIAL NEEDS

Comfort, convenience, and mobility needs, especially of mature/older and persons with disabilities, need special considerations during design and implementation of new technology devices and equipment, especially in implementing smart grid and electric vehicles with autonomous features. Special attention should be placed in meeting customer needs in residential, office, and public places related to safety, comfort, and convenience in areas such as interior environments (e.g., comfortable levels of temperature and humidity, reduced exposures of toxic and cancer-causing pollutants) and safety (e.g., incorporating accident prevention countermeasures in using energy consuming devices).

The energy companies also need to work closely with their customers in commercial, industrial, and institutional customers to achieve common goals of the customers and energy companies and support social needs along with the government and energy regulating agencies to create safe, healthful, convenient, and affordable energy systems and operating processes.

CONCLUDING REMARKS

The access to plentiful supply of reliable energy at affordable prices and operating in safe and healthful operating conditions are expected and unspoken wants of all customers of energy systems. Meeting the customer expectations and needs and social responsibilities are especially important for energy systems designers because the well-being, safety, and health of the users depend upon how the energy systems operate. Thus, the designers of energy systems should carefully consider the customer needs and balance many conflicting challenges in their decision-making by meeting their social, regulatory, and legal responsibilities.

REFERENCES

Bhise, V.D. 2012. *Ergonomics in the Automotive Design Process*. Boca Raton, FL: CRC Press. ISBN 978-1-4398-4210-2.

Bhise, V.D. 2014. *Designing Complex Products with Systems Engineering Processes and Techniques*. Boca Raton, FL: CRC Press, Taylor and Francis Group. ISBN: 978-1-4665-0703-6.

Coman, H. 2012. Balancing the Need for Energy and Clean Water: The Case for Applying Strict Liability in Hydraulic Fracturing Suits. *Boston College Environmental Affairs Law Review*, 39(1). Article 5. Website: https://lawdigitalcommons.bc.edu/cgi/viewcontent.cgi?article=2074&context=ealr (Accessed: December 29, 2020).

DOE. 2016. *Maintaining Reliability of Modern Power System*. Department of Energy Office of Energy Policy and Systems Analysis, Prepared under the direction of Aaron Bergman with substantial input from Paul Denholm and Daniel C. Steinberg of the National Renewable Energy Laboratory and David Rosner of the Department of Energy. December 2016.

EIA. 2009. State Fact Sheets on Household Energy Use. Website: https://www.eia.gov/consumption/residential/reports/2009/state_briefs/ (Accessed: October 10, 2020).

EIA. 2020a. EIA Websites. Website: https://www.eia.gov/outlooks/aeo/data/browser/ (Accessed: August 8, 2020).

EIA. 2020b. Where the United States Gets Its Coal. Website: https://www.eia.gov/energyexplained/coal/where-our-coal-comes-from.php (Accessed: August 8, 2020).

EIA. 2020c. U.S. Energy Mapping System. Website: https://www.eia.gov/state/maps.php (Accessed: August 8, 2020).

EIA. 2020d. Factors Affecting Electricity Prices. Website: https://www.eia.gov/energyexplained/electricity/prices-and-factors-affecting-prices.php (Accessed August 8, 2020).

EIA. 2020e. Electric Sales, Revenue, and Average Price. Website: https://www.eia.gov/electricity/sales_revenue_price/ (Accessed: August 8, 2020).

Financial Times. 2020. Oil Majors Gear Up for Wave of Climate Change Liability Lawsuits. Website: https://www.ft.com/content/d5fbeae4-869c-11e9-97ea-05ac2431f453 (Accessed: December 29, 2020).

Goetsch, D.L. 2007. *Occupational Safety and Health for Technologist, Engineers, and Managers*, 6th Edition. ISBN: 0132397609. Englewood Cliffs, NJ: Prentice Hall.

Hammer, W. 1980. *Product Safety Management and Engineering*. Englewood Cliffs, NJ: Prentice-Hall

Hammer, W. 1989. *Occupational Safety Management and Engineering*, 4th edition. Englewood Cliffs, NJ: Prentice Hall.

Yang, K. and B. El-Haik. 2003. *Design for Six Sigma*. ISBN 0071412085. New York: McGraw-Hill.

11 Government Regulations

INTRODUCTION

Whenever a new project such as building a new power plant or developing a new product or a system begins, it is very important to ensure that the system and its outputs will meet all applicable government requirements. Otherwise, the government agency will not allow operation of the system within its enforcement region. In the United States, the Environmental Protection Agency (EPA) is responsible for ensuring that the environment, primarily air and water, is safe, i.e., it does not contain toxic or harmful substances above certain specified concentration levels for human activities. Similarly, the National Highway Traffic Safety Administration (NHTSA) and EPA have jointly created requirements on minimum allowable levels of corporate average fuel economy (CAFE) [in miles per gallon of fuel consumed (mpg)] and maximum allowable levels of emissions requirements [in equivalent grams of CO_2 per mile] on light automotive products (passenger cars and light trucks). The objective of this chapter is to review these important government agencies, their initiatives, and key requirements.

U.S. GOVERNMENT AGENCIES RELATED TO ENERGY

US DEPARTMENT OF ENERGY

The Department of Energy Organization Act of 1977 created the Department of Energy and activated it on October 1, 1977, as the 12th cabinet-level department brought together for the first time within one agency with two programmatic traditions that had long coexisted within the federal establishment:

a. defense responsibilities that included the design, construction, and testing of nuclear weapons dating from the Manhattan Project effort to build the atomic bomb
b. a loosely knit amalgamation of energy-related programs scattered throughout the federal government.

The DOE currently has the following offices:

1. Office of Technology Transitions
2. Office of Energy Efficiency and Renewable Energy
3. Office of Electricity
4. Office of Fossil Energy
5. Office of Nuclear Energy
6. Loan Programs Office
7. Office of Science

DOI: 10.1201/9781003107514-14

8. Artificial Intelligence & Technology Office
9. Office of Cybersecurity, Energy Security, and Emergency Response
10. Advanced Research Projects Agency – Energy
11. Office of Environmental Management
12. Office of Indian Energy Policy and Programs
13. Office of Legacy Management

ENVIRONMENTAL PROTECTION AGENCY (EPA)

In early 1970, because of heightened public concerns about deteriorating city air, natural areas littered with debris, and urban water supplies contaminated with dangerous impurities, President Richard Nixon presented the House and Senate a groundbreaking 37-point message on the environment. The highlights of the message included the following:

a. requesting 4 billion dollars for the improvement of water treatment facilities
b. asking for national air quality standards and stringent guidelines to lower motor vehicle emissions
c. launching federally funded research to reduce automobile pollution
d. ordering a cleanup of federal facilities that had fouled air and water
e. seeking legislation to end the dumping of wastes into the Great Lakes
f. proposing a tax on lead additives in gasoline
g. forwarding to Congress a plan to tighten safeguards on the seaborne transportation of oil
h. approving a National Contingency Plan for the treatment of oil spills

Around the same time, President Nixon also created a council in part to consider how to organize federal government programs designed to reduce pollution, so that those programs could efficiently address the goals laid out in his message on the environment. Following the council's recommendations, the president sent to Congress a plan to consolidate many environmental responsibilities of the federal government under one agency, a new EPA.

This reorganization now permits the EPA to undertake the following (EPA, 2020a):

1. Conduct research on important pollutants irrespective of the media in which they appear and on the impact of these pollutants on the total environment
2. Both by itself and together with other government agencies, the EPA would monitor the condition of the environment – biological as well as physical.
3. With these data, the EPA would be able to establish quantitative "environmental baselines" – critical for efforts to measure adequately the success or failure of pollution abatement efforts.
4. The EPA would be able, in concert with the states, to set and enforce standards for air and water quality and for individual pollutants.
5. Industries seeking to minimize the adverse impact of their activities on the environment would be assured of consistent standards covering the full range of their waste disposal problems.

6. As states developed and expanded their own pollution control programs, they would be able to look to one agency to support their efforts with financial and technical assistance and training.

NATIONAL HIGHWAY TRAFFIC SAFETY ADMINISTRATION (NHTSA)

The NHTSA is an agency of the U.S. federal government, part of the Department of Transportation. It describes its mission as "Save lives, prevent injuries, reduce vehicle-related crashes." The National Traffic and Motor Vehicle Safety Act was enacted in the United States in 1966 to empower the federal government to set and administer new safety standards for motor vehicles and road traffic safety. In 1966, Congress passed legislation to make installation of seat belts mandatory and created the U.S. Department of Transportation. This legislation created several predecessor agencies, which became NHTSA, including the National Traffic Safety Agency, the National Highway Safety Agency, and the National Highway Safety Bureau. Congress established NHTSA with the enacting of the Highway Safety Act of 1970.

NHTSA issues Federal Motor Vehicle Safety Standards (FMVSS) to implement laws created by the congress. The vehicles not certified by the maker or importer as compliant with the FMVSS are not legal to use (or import) into the United States. In 1972, the Motor Vehicle Information and Cost Savings Act (enacted October 20, 1972) expanded NHTSA's scope to include consumer information programs. Since then, automobiles have become far better at avoiding accidents and protecting their occupants in vehicle impacts. These regulations allow NHTSA to fulfill its mission to prevent and reduce vehicle crashes and resulting injuries and fatalities.

NHTSA also administers the Corporate Average Fuel Economy (CAFE), which is intended to incentivize the production of fuel-efficient vehicles.

ENERGY SAVING INITIATIVES

DOE INITIATIVES

More than 40% of the total energy consumed in the United States is used for operating buildings, and most of that energy goes toward appliances and building-related equipment. In accordance with the Energy Policy and Conservation Act of 1975 (EPCA), as amended, the U.S. Department of Energy (DOE) implements minimum efficiency standards for a wide range of appliances and equipment used in residential and commercial buildings. Within the parameters of technical feasibility and cost-effectiveness, federal efficiency standards compel product designers and manufacturers to reduce the amount of energy and water necessary for the proper operation of appliances and other building equipment. Operational efficiency means less waste of natural and financial resources. The legal limit on energy/water consumption for designated products – applied equally to all manufacturers of those products – makes energy efficiency a priority instead of an afterthought or a competitive disadvantage. Regular updates of the standards ensure continuous improvement.

Other important strategies and policies that make buildings and building products more energy efficient include building codes; tax credits; utility rebates; industry product promotions; and award or certification programs. ENERGY STAR, for example, is a voluntary program administered jointly by the EPA and DOE to promote products and buildings that are even more energy-efficient than those that meet the minimum federal standards. As another example, DOE's partnerships with lighting manufacturers to research, develop, and test solid-state lighting technologies have resulted in market-ready LED (light-emitting diodes) light bulbs that are vastly more energy-efficient than old incandescent technology. An LED bulb can last up to 25 times longer than an ordinary incandescent bulb, uses 75% less energy to produce the same amount of light flux, and saves operational costs for each bulb for consumers over the course of its lifetime.

NHTSA and EPA Initiatives

CAFE requirements created jointly by NHTSA and EPA: On October 15, 2012, the EPA announced the final greenhouse gas (GHG) emissions standards for model years 2017–2025. And the NHTSA has announced the final Corporate Average Fuel Economy (CAFE) for MYs 2017–2021 and augural standards for MYs 2022–2025 (EPA and NHTSA, 2012). These standards apply to passenger cars, light-duty trucks, and medium-duty passenger vehicles (i.e., sport utility vehicles, crossover utility vehicles, and light trucks). These standards apply to each manufacturer's fleet (total numbers of vehicles produced by different size and type) and not on an individual vehicle.

The National Program is estimated to save approximately 4 billion barrels of oil and to reduce GHG emissions by the equivalent of approximately 2 billion metric tons over the lifetimes of those light-duty vehicles produced in MYs 2017–2025. Although the agencies estimate that technologies used to meet the standards will add, on average, about $1,800 to the cost of a new light-duty vehicle in MY 2025, consumers who drive their MY2025 vehicles for its entire vehicle lifetime will save, on average, $5,700–$7,400 (based on 7% and 3% discount rates, respectively) in fuel, for a net lifetime savings of $3,400–$5,000.

CLEAN AIR ACT

The Clean Air Act (CAA) is the law that defines EPA's responsibilities for protecting and improving the nation's air quality and the stratospheric ozone layer (DOE, 2020). The last major change in the law enacted by Congress was the Clean Air Act Amendments of 1990. Legislation passed since then has made several minor changes.

The CAA includes the following titles and parts:

    Title I – Air Pollution Prevention and Control
        Part A – Air Quality and Emission Limitations
        Part B – Ozone Protection
        Part C – Prevention of Significant Deterioration of Air Quality
        Part D – Plan Requirements for Nonattainment Areas
    Title II – Emission Standards for Moving Sources
        Part A – Motor Vehicle Emission and Fuel Standards

Part B – Aircraft Emission Standards
Part C – Clean Fuel Vehicles
Title III – General (CAA § 301–328; USC § 7601–7627)
Title IV – Noise Pollution
Title IV-A – Acid Deposition Control
Title V – Permits
Title VI – Stratospheric Ozone Protection

CONTROL OF COMMON POLLUTANTS

To protect public health and welfare nationwide, the CAA requires EPA to establish national ambient air quality standards for certain common and widespread pollutants based on the latest science. EPA has set air quality standards for six common "criteria pollutants": particulate matter (PM) (also known as particle pollution), ozone, sulfur dioxide, nitrogen dioxide, carbon monoxide, and lead.

States are required to adopt enforceable plans to achieve and maintain air quality meeting the air quality standards. State plans also must control emissions that drift across state lines and harm air quality in downwind states.

Other key provisions are designed to minimize pollution increases from growing numbers of motor vehicles and from new or expanded industrial plants. The law calls for new stationary sources (e.g., power plants and factories) to use the best available technology and allows less stringent standards for existing sources.

The CAA also contains specific provisions to address:

a. Hazardous or toxic air pollutants that pose health risks such as cancer or environmental threats such as bioaccumulation of heavy metals
b. Acid rain that damages aquatic life, forests, and property
c. Chemical emissions that deplete the stratospheric ozone layer, which protects us from skin cancer and eye damage
d. Regional haze that impairs visibility in national parks and other recreational areas

NATIONAL AIR TOXICS ASSESSMENT (NATA)

The National Air Toxics Assessment (NATA) is EPA's ongoing review of air toxics, i.e., pollutants known to cause or suspected of causing cancer or other serious health effects (also known as toxic air pollutants or hazardous air pollutants) in the United States. EPA developed NATA as a screening tool for state, local, and tribal air agencies (EPA, 2020c). NATA's results help further to better understand any possible risks to public health from air toxics.

NATA gives a snapshot of outdoor air quality with respect to emissions, i.e., pollutants released into the air or air toxics. It suggests the long-term risks to human health if air toxics emissions are steady over time. NATA estimates the cancer risks by providing the probabilities that adverse health effects will occur from exposure to a hazard from breathing air toxics over many years at different locations in the country. It also estimates noncancer health effects for some pollutants, including diesel PM. NATA calculates these air toxics concentrations and risks at the census tract, i.e., land area defined by the U.S. Census Bureau. A tract usually contains from 1,200 to 8,000 people, with most having close to 4,000 people. Census tracts are usually smaller than 2 square miles in cities but are much larger in rural areas. It only includes outdoor sources of pollutants.

Air quality specialists use NATA results to learn which air toxics and emission source types may raise health risks in certain places. They can then study these places in more detail, focusing where the risks to people may be highest.

In the 2014 NATA results, the results for stationary sources are broken into two groups: "point" and "nonpoint" sources. This reflects the way each source was modeled. Each point source's exact latitude and longitude coordinate is in the NATA source inventory. Large industrial complexes often have many individual point sources. Other smaller sources may not have an exact location in the NATA inventory. They were modeled as nonpoint sources. Emissions from homes, such as wood-burning stoves and fireplaces or solvent emissions, are examples of nonpoint sources.

Table 11.1 provides an example of cancer risk (per million population) data provided by EPA region and county. Risk level of "N" in a million implies a likelihood that up to "N" people, out of 1 million equally exposed people would contract cancer if exposed continuously (24 hours/day) to the specific concentration over 70 years (an assumed lifetime). This risk would be an excess cancer risk that is in addition to any cancer risk borne by a person not exposed to these air toxics. The probability of contracting cancer over the course of a lifetime (assumed to be 70 years for the purposes of NATA risk characterization).

GOVERNMENT REQUIREMENTS IN SAFETY, EMISSIONS, AND FUEL ECONOMY

GOVERNMENT SAFETY REQUIREMENTS

The NHTSA has a legislative mandate under Title 49 of the United States Code, Chapter 301, Motor Vehicle Safety, to issue Federal Motor Vehicle Safety Standards (FMVSS) and Regulations to which manufacturers of motor vehicle and equipment items must conform and certify compliance (NHTSA, 2015). These Federal safety standards are regulations written in terms of minimum safety performance requirements for motor vehicles or items of motor vehicle equipment. These requirements are specified in such a manner "that the public is protected against unreasonable risk of crashes occurring as a result of the design, construction, or performance of motor vehicles and is also protected against unreasonable risk of death or injury in the event crashes do occur."

The Federal Motor Vehicle Safety Standards (FMVSS) can be accessed through the NHTSA website (NHTSA, 2015). The FMVSS are numbered by following categories in Title 49 Code of Federal Regulations (CFR): (a) crash avoidance standards (FMVSS 101–133), (b) crashworthiness standards (FMVSS 201–224), (c) post-crash standards (FMVSS 301–500), and (d) other regulations are included in parts 531–591.

TABLE 11.1

Partial Listing of 2014 NATA Cancer Risk by Source Group Excel File

State	EPA Region	Country	FIPS	Tract	Population	Total Cancer Risk (per Million)	PT-Stationary Point Cancer Risk (per Million)	OR-Light Duty-off Network-Gas Cancer Risk (per Million)	OR-Light Duty-off Network-Diesel Cancer Risk (per Million)	OR-Heavy Duty-off Network-Gas Cancer Risk (per Million)	OR-Heavy Duty-off Network-Diesel Cancer Risk (per Million)
US	Entire US	Entire US	00000	00000000000	312,572,412	31.6890	1.6837	2.0366	0.0487	0.0198	0.0163
AL	EPA Region 4	Entire State	01000	0100000000	4,779,690	43.3141	1.1179	1.5912	0.0325	0.0139	0.0083
AL	EPA Region 4	Autauga	01001	0101000000	54,571	49.5114	0.5092	1.6119	0.0445	0.0183	0.0121
AL	EPA Region 4	Autauga	01001	01001020100	1,912	49.3770	0.4001	1.7251	0.0513	0.0209	0.0150
AL	EPA Region 4	Autauga	01001	01001020200	2,170	50.3207	0.5641	1.9060	0.0587	0.0235	0.0171
AL	EPA Region 4	Autauga	01001	01001020300	3,373	50.7721	0.4650	1.9811	0.0610	0.0239	0.0173
AL	EPA Region 4	Autauga	01001	01001020400	4,386	51.5959	0.4784	1.9789	0.0614	0.0242	0.0173
AL	EPA Region 4	Autauga	01001	01001020500	10,766	51.7300	0.5065	1.9501	0.0591	0.0233	0.0164
AL	EPA Region 4	Autauga	01001	01001020600	3,668	50.4676	1.5207	1.6314	0.0441	0.0205	0.0139
AL	EPA Region 4	Autauga	01001	01001020700	2,891	51.0658	1.1144	1.6726	0.0475	0.0206	0.0143
AL	EPA Region 4	Autauga	01001	01001020801	3,081	48.6344	0.8959	1.2994	0.0252	0.0162	0.0084
AL	EPA Region 4	Autauga	01001	01001020802	10,435	48.8757	0.2316	1.6750	0.0461	0.0176	0.0114
AL	EPA Region 4	Autauga	01001	01001020900	5,675	46.4156	0.1852	1.0658	0.0236	0.0082	0.0040
AL	EPA Region 4	Autauga	01001	01001021000	2,894	45.9419	0.2487	0.9515	0.0172	0.0080	0.0033
AL	EPA Region 4	Autauga	01001	01001021100	3,320	46.6365	0.2807	0.9236	0.0155	0.0105	0.0046
AL	EPA Region 4	Baldwin	01003	01003000000	182,265	35.6106	0.7775	0.8523	0.0235	0.0102	0.0043
AL	EPA Region 4	Baldwin	01003	01003010100	3,804	41.8121	0.5412	0.5704	0.0120	0.0038	0.0013
AL	EPA Region 4	Baldwin	01003	01003010200	2,902	40.8647	0.3765	0.7028	0.0182	0.0066	0.0026
AL	EPA Region 4	Baldwin	01003	01003010300	7,826	40.5079	0.4950	0.8549	0.0224	0.0083	0.0034
AL	EPA Region 4	Baldwin	01003	01003010400	4,736	39.7490	0.3781	0.7597	0.0190	0.0071	0.0025
AL	EPA Region 4	Baldwin	01003	01003010500	7,815	42.3954	0.4215	1.0693	0.0337	0.0133	0.2264
AL	EPA Region 4	Baldwin	01003	01003010600	3,325	41.8856	0.4193	1.0263	0.0318	0.0124	0.0058
AL	EPA Region 4	Baldwin	01003	01003010701	7,882	38.0075	0.4200	1.0199	0.0281	0.0097	0.0041
AL	EPA Region 4	Baldwin	01003	01003010703	13,166	37.0575	0.3862	1.0037	0.0284	0.0112	0.0047

Source: https://www.epa.gov/national-air-toxics-assessment/2014-nata-assessment-results.

EPA's GHG Emissions and NHTSA's CAFE Standards

On October 15, 2012, the U.S. EPA announced the final greenhouse gas (GHG) emissions standards for vehicles with model years 2017–2025. And the NHTSA has announced the final Corporate Average Fuel Economy (CAFE) for MYs 2017–2021 and augural standards for MYs 2022–2025 (EPA and NHTSA, 2012). These standards apply to passenger cars, light-duty trucks, and medium-duty passenger vehicles (i.e., sport utility vehicles, crossover utility vehicles and light trucks – GVWR less than 8,500 lbs.). These standards apply to each manufacturer's fleet (total numbers of vehicles produced by different size and type) and not on an individual vehicle.

Rationale Behind Footprint-based Standard

The requirements in these standards are illustrated in Figures 11.1–11.4. The requirements are based on the footprint of the vehicle. The footprint is the area covered under the four tire touch points on the ground, and it is defined as the product of the wheelbase and the tread width. With this footprint-based standard approach, EPA and NHTSA continue to believe that the rules will not create significant incentives to produce vehicles of any particular size or type, and thus there should be no significant effect on the relative availability of different vehicle sizes in the fleet. These standards will also help to maintain consumer choice during the MY 2017–MY 2025 rulemaking timeframe.

Figures 11.1 and 11.2 present fuel economy requirements (in miles per gallon) for passenger cars and light trucks, respectively. These figures show that a smaller footprint vehicle will need to have higher fuel economy relative to a larger footprint

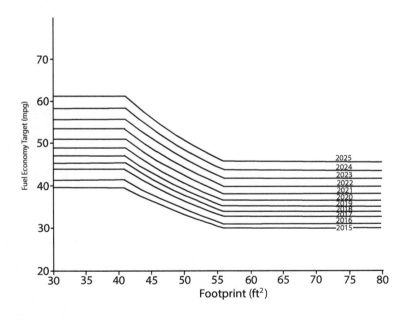

FIGURE 11.1 Passenger car fuel economy requirements. (Redrawn from: EPA and NHTSA, 2012.)

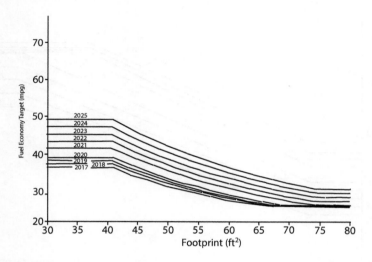

FIGURE 11.2 Light truck fuel economy requirements. (Redrawn from: EPA and NHTSA, 2012.)

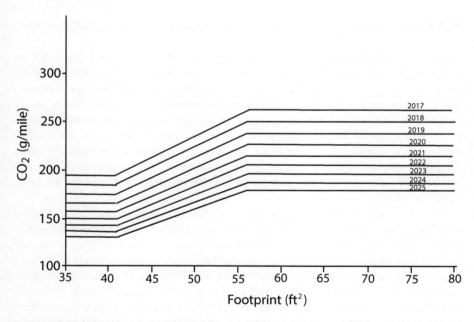

FIGURE 11.3 Passenger car emission requirements. (Redrawn from: EPA and NHTSA, 2012.)

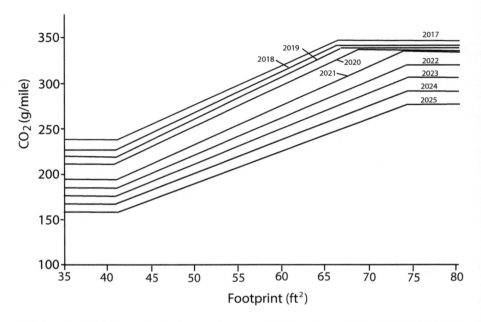

FIGURE 11.4   Light truck emission requirements. (Redrawn from: EPA and NHTSA, 2012.)

vehicle (when both vehicles have a comparable level of fuel efficiency improvement technology). Conversely, Figures 11.3 and 11.4 (for passenger car and light truck emission requirements) show that a smaller footprint vehicle will have lower equivalent CO_2 emissions relative to a larger footprint vehicle (when both vehicles have a comparable level of fuel efficiency improvement technology). The standards apply to a manufacturer's overall passenger car fleet and overall light truck fleet, not to an individual vehicle. Thus, if one of a manufacturer's fleets is dominated by small footprint vehicles, then that fleet will have a higher fuel economy requirement and a lower CO_2 requirement than a manufacturer whose fleet is dominated by large footprint vehicles.

A wide range of technologies are available for automakers to meet these standards. The technologies include advanced gasoline engines and transmissions, vehicle mass reduction, improved aerodynamics, lower rolling resistance tires, diesel engines, more efficient accessories, improvements in air conditioning systems, and so forth. The automakers will increase electric technologies, such as stop–start systems, mild and strong hybrids, plug-in hybrids (PHEVs), and all electric vehicles (EVs). However, NHTSA also projected that automakers would meet the standards largely through advancements in internal combustion engines. The rulemaking analysis showed that automakers would only need to produce about 1%–3% of the 2025 new vehicle fleet as EVs/PHEVs to meet the 2025 standards.

The National Program is estimated to save approximately 4 billion barrels of oil and to reduce GHG emissions by the equivalent of approximately 2 billion metric tons over the lifetimes of those light-duty vehicles produced in MYs 2017–2025.

Although the agencies estimate that technologies used to meet the standards will add, on average, about $1,800 to the cost of a new light-duty vehicle in MY 2025, consumers who drive their MY2025 vehicles for its entire vehicle lifetime will save, on average, $5,700–$7,400 (based on 7% and 3% discount rates, respectively) in fuel, for a net lifetime savings of $3,400–$5,000.

On August 1, 2018, EPA and NHTSA signed (published) a notice of proposed rulemaking entitled, "The Safe and Fuel-Efficient (SAFE) Vehicles Rules for Model Years 2021–2026 Passenger Cars and Light Trucks" to amend existing CAFE Rule. The proposed SAFE vehicle rule was estimated to save $500 billion in social costs and 12,700 lives over the lifetime of the vehicles through MY2029. On June 28, 2020, EPA and NHTSA issued a Final Rule on SAFE. The final rule increases stringency of both fuel economy and emissions by 1.5% per year from MY 2020 levels over MYs 2021–2026 (EPA and NHTSA, 2020).

Greenhouse Gas Equivalencies Calculator

Since the exhaust gases from automotive internal combustion engines contain a mixture of gases, the EPA created a GHG equivalency calculator to convert emission potential of different gases and gas mixtures into equivalent CO_2 emitted. This calculator may be useful in understanding toxicity effects and in communicating GHG reduction tasks, reduction targets, or other initiatives aimed at reducing GHG emissions.

The GHG equivalencies calculator can help in translating abstract measurements into equivalent terms, such as the annual emissions from cars, households, or power plants (EPA, 2020b). Some equivalency values obtained from the calculator are provided below:

1 lb. of Carbon or carbon equivalent $= 3.7$ lbs. of CO_2 equivalent
1 lb. of Methane $(CH_4) = 25$ lbs. of CO_2 equivalent
1 lb. of Nitrous Oxide $(NO_2) = 298$ lbs. of CO_2 equivalent
1 lb. of Hydrofluorocarbon gases $(HCFC-22) = 14,800$ lbs. of CO_2 equivalent
1 lb. of Perfluorocarbon gases $(CF4) = 7,390$ lbs. of CO_2 equivalent
1 lb. of Sulfur Hexafluoride $(SF_6) = 22,800$ lbs. of CO_2 equivalent

CONCLUDING REMARKS

U.S. government departments and its agencies have undertaken many initiatives and tasks to reduce motor vehicle accidents, pollutants in air, water, and the environment and to reduce energy consumption. They have also created a number of databases, conducted and sponsored research projects, generated valuable information, and created requirements on a variety of systems, equipment, and vehicles that have been used by researchers, engineers, designers, and administrators of various organizations to reduce accidents, pollutants, and/or energy consumption of equipment used by variety of customers (e.g., residential, commercial, industrial, institutional). Knowledge of this work and meeting applicable requirements is essential in conducting analyses and decision-making related to energy systems problems.

REFERENCES

DOE. 2020. Clean Air Act Overview. Website: https://www.epa.gov/clean-air-act-overview/-clean-air-act-text (Accessed: November 2, 2020).

EPA. 2020a. The Origins of EPA. Website: https://www.epa.gov/history/origins-epa (Accessed: August 9, 2020).

EPA. 2020b. Greenhouse Gas Equivalencies Calculator. Website: http://www.epa.gov/cleanenergy/energy-resources/calculator.html#results (Accessed: August 9, 2020).

EPA. 2020c. National Air Toxics Assessment (NATA). Website: https://www.epa.gov/-national-air-toxics-assessment (Accessed: October 11, 2020).

EPA and NHTSA. 2012. 2017 and Later Model Year Light-Duty Vehicle Greenhouse Gas Emissions and Corporate Average Fuel Economy Standards. *Federal Register*, 77(199): 62623–63200, October 15. Environmental Protection Agency, 40 CFR Parts 85, 86, and 600. Department of Transportation National Highway Traffic Safety Administration, 49 CFR Parts 523, 531, 533, 536, and 537. [EPA-HQ-OAR-2010-0799; FRL-9706-5; NHTSA-2010-0131]. RIN 2060-AQ54; RIN 2127-AK79.

EPA and NHTSA. 2020, April 30. The Safer Affordable Fuel-Efficient (SAFE) Vehicles Rule for Model Years 2021–2026 Passenger Cars and Light Trucks. Final Rule on SAFE. *Federal Register*, 85(84).

12 Meeting Future Automotive Fuel Economy and Emissions Requirements

INTRODUCTION

Implementing new technologies always creates a lot more work. The teams involved in the applications of new technologies to improve performance or capabilities of one or more vehicle systems have to first clearly understand a number of issues related to functions of different systems, how they affect different vehicle attributes and other interfacing vehicle systems. The important issues that need be considered here are as follows:

1. Level of improvement in performance or capability of the vehicle that could be achieved (Note: Proving that an improvement is achievable and assessing the level of improvement generally requires considerable amount of research.)
2. Ability and willingness of the customers to adapt to the new changes, accept the changes, and maintain (i.e., service and upgrade) the new features as needed
3. Effect of the changes on other vehicle systems and the resulting trade-offs between various affected vehicle attributes
4. Technical resources (e.g., availability of specialists, analysis techniques, and test equipment) readily available to analyze effects of changes in the vehicle design
5. Packaging space available in the vehicle to incorporate the hardware needed to incorporate the changes
6. Effect of the changes on the overall cost of the vehicle
7. Effect of the changes on the curb weight of the vehicle
8. Effect of the changes on the fuel consumption characteristics of the vehicle
9. Time and costs associated with implementing each new technology (i.e., making sure that each new technology is feasible, effective, and is ready for implementation)
10. Effect of the changes on vehicle quality in short and long terms (i.e., to ensure that the new technologies do not introduce defects in the vehicle)

DOI: 10.1201/9781003107514-15

11. Effect of the changes on "make vs. buy" decisions related to new entities on the company's production resources versus the capabilities of potential suppliers to produce the new or affected vehicle systems

Major Reasons for Changes Affecting Future Vehicle Designs

1. Meeting government requirements (e.g., NHTSA's CAFE and EPA's GHE requirements; see Chapter 11 for more details)
2. Advances in technologies related to vehicle attributes and features that can provide many advantages such as improved functionality, efficiency, safety, comfort, convenience, packaging space, weight reduction, emission reduction, and cost reduction. Some examples of implementation areas are as follows:
   a. Incorporation of advanced driver aids and safety technologies
   b. Improving driver comfort and convenience during ride and operation of the vehicle
   c. Incorporation of advanced driver information and communication systems
   d. Incorporation of new more fuel-efficient and less polluting powertrains
   e. Incorporating new lightweight materials in vehicle body structure and redesigning components with lightweight materials
   f. Improvements in manufacturing and assembly methods and equipment (e.g., material joining techniques, robotics systems, material handling systems)
   g. Advances in global communication, sourcing, and project management methods
   h. Improving vehicle reliability, durability, and quality

Another major consideration in implementing the new technologies is to ensure that the new vehicle will be perceived by its customers to be "improved" (or advanced) as compared with the outgoing older vehicle. This is especially important if the major competitors of the vehicle have already adopted or are in a process of adopting many of the new technologies.

CREATING A TECHNOLOGY PLAN

Development of a new vehicle begins with determining its specifications and creating a list of improvements (e.g., features and technologies) to be incorporated. A technology plan is generally initiated in parallel with the development of specification and vehicle attribute requirements. The technology plan should consider inclusion of every major vehicle system, and it should include descriptions of changes and technologies, risks in incorporating the changes, and major open issues that need to be resolved.

Table 12.1 presents an example of an early technology plan illustrating how changes in various vehicle systems and subsystems can be planned to meet weight reduction and fuel economy objectives of a future electric vehicle. The last two

TABLE 12.1
Illustration of a Technology Plan for an Automotive Product

Vehicle System	Major Changes Planned	Brief Description of Major Technological Challenges	Comments
Body System	Use lightweight and recyclable materials for the exterior body and for vehicle interior	Lightweight materials are expensive and their properties such as strength and stiffness need improvements. Lighter weight vehicle can be less safe. Reduce aerodynamic drag.	Using lightweight materials like aluminum, magnesium or glass fiber-reinforced polymer composites will increase cost of the vehicle. Recyclable materials are environmental friendly. Most parts of the body are cheaper to make internally than buy from outside suppliers. Reduce aerodynamic drag by improved exterior shape, reduce vehicle height and incorporating underbody panels.
Electric Powertrain System	160kW permanent magnet electric traction motor 60 kWh lithium-ion battery or more advance replaceable battery	Increasing motor power would increase its power consumption, and also increase cost and weight of the vehicle.	Some body modifications will be required to accommodate the new motor and the replaceable battery. It is recommended to buy the motor and the battery from suppliers. If a major breakthrough occurred in battery technology in the future, the electric powertrain would be more appealing to the customers.
Chassis System	Low friction bearings Low rolling resistance tires Quick change battery exchange (release and installation)	Safety must be considered because low rolling resistance tires may increase braking distance and time response for braking. Also adding a new battery releasing and installing system needs special consideration about how the new system will interface with all other vehicle systems.	Low rolling resistance tire and low friction bearings reduce the energy use to roll tires, thus increase the range. Some modifications are required to accommodate the battery releasing and installing system.
Climate Control System	Automatic climate control system	More sensors and wiring needed to measure sun load at different body locations.	This climate control system will adjust the interior temperature according to the sun illumination and sun angle. The automatic climate control system will help to reduce the power consumption by adjusting the internal environment according to the outside environment.

(Continued)

TABLE 12.1 (*Continued*)
Illustration of a Technology Plan for an Automotive Product

Vehicle System	Major Changes Planned	Brief Description of Major Technological Challenges	Comments
Infotainment System	Voice recognition, voice controls and voice displays Reconfigurable displays and multi-function controls Navigation systems with real time traffic alerts Text-to-speech conversion systems Projected displays	Since there are many voice activated systems, highly sophisticated software and hardware (e.g., voice filters) are needed to differentiate human voice vs. road noise. Incorporation of many advanced systems in vehicle will increase costs.	Some modifications are required to accommodate all these systems. Buying these systems from a supplier is recommended.
Steering System	Tilt/telescoping steering column	May increase the overall price of the vehicle.	The cross-car beam around the steering column need to be reconfigured.
Braking System	Electronic braking system (EBS)	All weather reliability need to be evaluated.	This braking system activates all components electronically and brake force distribution adapts to load distribution. It is recommended to get this system from a supplier.
Lighting System	Smart lighting and visibility systems LED lamps, fiber optics, smart headlamps Night vision system	Adding new systems to the vehicle will increase costs.	Smart lighting can help to reduce power used for lighting by turning on and off the light sources according to the environment around the vehicle.
Solar Power System	Solar panels or more advance technology for reducing battery load	Solar panels and additional electronics will add more weight and cost to the vehicle.	Modifications needed on the roof to accommodate the solar panels. Solar panels must be curved to be compatible with the body style of the vehicle. Explore breakthrough in solar panel construction.

(*Continued*)

TABLE 12.1 (Continued)
Illustration of a Technology Plan for an Automotive Product

Vehicle System	Major Changes Planned	Brief Description of Major Technological Challenges	Comments
Safety System	Smart airbags and belts	Additional sensors may increase costs. Need more extensive verification tests with large range of dummies to ensure protection to increased percentage of population.	Extensive verification testing required.
	Drowsy driver or alertness monitoring and lane departure warning systems	Adding all these safety system will increase the overall price of the vehicle. Accuracy and reliability need additional verification tests.	
	Collision avoidance systems	Higher costs, false alarms and reliability concerns.	
	Driver assistance systems	Higher costs, false alarms and reliability concerns.	
	Blind spot warning	Higher cost to incorporate sensors and warning displays.	

columns of the table provide brief descriptions of the technological challenges in designing the major changes in various vehicle systems and comments describing details that need to be considered in incorporating the changes.

RISKS IN TECHNOLOGY IMPLEMENTATION

Implementing any new technology involves risks. It is important for the vehicle development team members to consider all types of risks and make sure that the higher management of the company understands the risks, consequences, and challenges facing the company. The risks can be classified in the following three categories:

1. *Technical risks*: Technical risks result due to inability of the auto company in developing the technology or perfecting it to create the required functionality within the available time or resources. The technology readiness may be overestimated during the development of the business and technology plans. The new features may not be debugged thoroughly to remove all possible errors or defects, which may force costly product recalls to fix the problems and/or defending potential liability situations.
2. *Schedule risk*: The vehicle program may be delayed due to additional technology development needed for its implementation.
3. *Cost overruns*: The costs to develop the required technology and its implementation may increase well above the budgeted amounts.

Cost sharing with other vehicle models or with other vehicle manufacturers in joint development projects (e.g., Ford and GM jointly developed new nine and ten-speed transmissions (Wernle and Colias, 2013)) is a possible approach considered in undertaking new technology development projects. A number of auto companies have been developing electric vehicles (e.g., Ford with Volkswagen, Honda with GM) and vehicle batteries (e.g., BYD with Toyota, GM with LG Chem) jointly to share and reduce development and production costs.

NEW TECHNOLOGIES

This section provides descriptions of leading design trends and technologies considered during the early phases of the vehicle development process.

DESIGN TRENDS IN POWERTRAIN DEVELOPMENT

Smaller, Lighter, and More Fuel-Efficient Gasoline Engines

With the increasingly stringent fuel economy and emissions requirements (see Chapter 11, Figures 11.1–11.4), the average vehicle engine weights are decreasing and new engine technologies are being introduced at a greater pace. The current improvements in automotive engines involve the following:

1. *Forced induction/ turbo charging/ turbo-boost*: Forced induction is the process of delivering compressed air to the intake port of an internal combustion engine (ICE). A forced induction engine uses a gas compressor (e.g.,

a turbo charger, which is an exhaust powered or an electric motor-driven turbine) to increase the pressure, temperature, and density of the air. An engine without forced induction is considered a naturally aspirated engine. Turbo charging has helped in downsizing engines and maintaining or even increasing its output. For example, many of the currently available turbo-boost gasoline engines are providing about 120 hp/L output as compared with about 80–100 hp/L outputs provided by the naturally aspirated gasoline engines. Turbo chargers also help recycle exhaust energy and reduce the energy loss when hot exhaust gases are released in the atmosphere. The energy loss is typically about 25%–30% of the energy in the fuel consumed.

It should be noted that a supercharger is another method to obtain higher output from and ICE. The supercharger does not work off the exhaust gas as it is attached to and powered by the engine. When the crankshaft spins the supercharger, it forces air into the engine. The turbo is more efficient as it does not require engine power to spin it, so it makes more power per boost. A supercharger also does not create full boost until the redline (near the top end of engine speed), which is when the engine is spinning the supercharger as fast as possible.

Thus, with the implementation of forced induction techniques, the number of cylinders used in automotive engines has been decreasing, which has resulted in an increase in percentage use of four cylinder engines with turbo-boost; and the eight cylinder engines are being replaced with six cylinder engines with turbo-boost. Electric assist from motor generators attached to the turbine shafts can further assist in recovering the electrical energy.

2. *Direct fuel injection vs. carburetor-based engines*: Fuel injection engines are more efficient and reduce emissions as compared with engines with carburetors. The carburetor contains jets that inject the fuel (e.g., gasoline) into the combustion chambers. The amount of fuel that can flow through these jets depends completely on the amount of air that can be pulled into the carburetor intake. The main disadvantage with obtaining the best performance using a carburetor is that it cannot adjust the air-to-fuel ratio for each individual cylinder. Fuel injection systems, which can inject precise amount of fuel into the engine, are now more popular in obtaining the best performance from the engines.

There are two different versions of fuel injection, namely port fuel injection and direct injection. Port fuel injection is the most commonly used, and direct fuel injection is the latest developed fuel injection system. Both systems use computer-controlled electric injectors to spray fuel into the engine, but the difference is where they spray the fuel. Port injection sprays the fuel into the intake ports where it mixes with the incoming air. The injectors are often mounted in the intake manifold runners, and the fuel sits in the runners till the intake valve opens and the mixture is pulled into the engine cylinder. The port injection systems are much cheaper to manufacture than injectors mounted in the cylinders. The port injectors are not exposed to the high heat and pressure of the combustion chamber, and they do not have to handle the high fuel pressures. Port injection systems typically operate in the 30–60 PSI range, which is dramatically lower than direct injection systems. Support systems such as fuel pumps are also cheaper because fuel pressures are lower.

In the direct injection, the injectors are mounted in the cylinder head and the injectors spray fuel directly into the engine cylinders, where it then mixes with the air. Only air passes through the intake manifold runners and past the intake valves with direct injection. Direct injection can meter the amount of fuel exactly into each cylinder for optimum performance, and it is sprayed in under very high pressure – up to about 15,000 PSI in some vehicles, so the fuel atomizes well and ignites almost instantly. With current computer controls, the injectors can be pulsed several times for each combustion stroke so the fuel can be injected over a longer time frame to maximize the power output of the cylinder.

Thus, the main advantage of using direct injection is that the amount of fuel and air can be precisely injected into the cylinder according to the engine load conditions. The electronics used in the system will calculate the fuel needs and constantly adjust timings of the fuel injection. The controlled fuel injection results in a higher power output, greater fuel efficiency, and much lower emissions. Improvements of about 15% are not uncommon just by changing from port to direct injection.

Disadvantages of direct injection are its cost and complexity. Because the injector tips are mounted right into the combustion chamber, the materials in the injector have to withstand both high temperatures and high pressures, and thus, they are more expensive. Also, the high pressure needed to inject fuel directly into the cylinders means that more expensive high-pressure fuel pumps are required. These are typically mechanically driven from the engine, and thus, they increase the engine complexity.

3. *Cylinder deactivation*: This method involves deactivation of some cylinders (typically 2–4 cylinders in 6–8 cylinder vehicles) when the vehicle is cruising at constant speed and the demand on power is lower as compared with when the vehicle is accelerating. Under light driving load conditions, the cylinder deactivation will reduce pumping losses from deactivated cylinders and thus, improve fuel economy.

4. *Stop/start*: Stop/start method involves stopping the internal combustion engine when the vehicle comes to a full stop and restarting it immediately when the driver presses the gas pedal to accelerate the vehicle. The system requires a larger starter motor and battery capacity to handle frequent stop/start cycles. The stop/start method can reduce energy consumption in city traffic conditions where the vehicle makes frequent stops in traffic and at intersections.

5. *Alternate fuel sources*: To conserve demand on gasoline, engines using a number of alternative fuel sources have been developed. These include (a) natural gas (compressed natural gas [CNG], (b) liquefied natural gas [LNG]), (c) diesel (e.g., turbo-diesel), (d) biomass fuels, and (e) hydrogen (i.e., hydrogen-powered fuel cell vehicles).

Each of the alternate sources has some disadvantages and advantages over the gasoline-powered internal combustion engines. For example, since the energy density of CNG is much lower than gasoline, large on-board CNG tanks are required. To carry LNG, refrigeration unit is needed to store the fuel at low temperature in its liquefied state before its use. Diesel engines are more expensive than gasoline engines. The biomass fuels (developed from organic materials, e.g., lumber, crops, manure) are not very common and not standardized. Hydrogen-powered vehicle would need large hydrogen tank or need to carry a hydrogen fuel cell to generate hydrogen.

6. *Hybrid powertrains*: The hybrid powertrains involve an ICE along with one or more electric motors. In series configuration, the drive wheels are powered by an electric motor, and the internal combustion engine drives an alternator, which charges the battery. The electric motor is driven by the battery through an electronic module. In parallel powertrain configuration, both the ICE and the electric motor provide power to the drive wheels. Some hybrid powertrains have two or more electric motors (e.g., each wheel motor directly drives a wheel). The hybrid powertrain consumes less fuel because energy is supplied by the electric motor that is more efficient than the ICE engine. Further, during vehicle deceleration, the electric motor acts like a generator and recovers dynamic energy of the vehicle and uses it to recharge the battery.

7. *Electric vehicles*: Vehicles driven purely on electric power through energy stored in batteries or electric power generated by on-board sources (e.g., hydrogen fuel cells or lithium-ion batteries) are available in steadily increasing numbers. Future advances in abilities to increase energy storage capabilities and reduction in battery weight and volume will increase driving distance range and thus, accelerate their market share.

Higher Efficiency Transmissions

Share of 8–10-speed transmissions, continuous variable transmissions (CVT), and dual clutch transmissions is slowly increasing. These transmissions can improve fuel efficiency by about 2%–10% over the 5- or 6-speed transmissions. The added weight and complexity of these newer higher-speed transmissions however can increase the cost and may not provide substantial improvements in fuel economy. However, several manufacturers have produced vehicles with such complexity and claimed improvements in fuel consumption.

DRIVER AIDS AND SAFETY TECHNOLOGIES

These features are incorporated in the vehicles to perform certain functions to aid the drivers in performing the driving tasks safely. Such features typically include sensors that monitor the vehicle motion and other variables related to road, traffic, and weather and warn the driver and activate vehicle controls (e.g., braking or steering the vehicle) to avoid the drivers getting into unsafe situations. Some important features are described below.

1. *Lane departure warning systems*: A lane departure warning system provides a warning to a driver when his vehicle begins to move out of its lane (unless a turn signal is activated in that direction of the lane deviation) on

freeways and arterial roads (typically while driving over about 40 mph). These systems are designed to minimize run-off-the-road accidents by addressing the main causes of collisions such as driver error, distractions, and drowsiness. There are two main types of systems: (a) systems that warn the driver (i.e., provides lane departure warning signal) if the vehicle is leaving its lane by providing visual, audible, and/or vibratory warning (e.g., vibrating the steering wheel), and (b) systems that warn the driver and, if no action is taken by the driver, it automatically takes steps to ensure that the vehicle stays in its lane.

2. *Driver monitoring or alertness warning systems*: If the driver is not paying attention to the road ahead and a dangerous situation is detected, the system will warn the driver by flashing lights, warning sounds, and/or vibratory warning. If no action is taken by the driver, the vehicle will apply the brakes (e.g., a warning alarm will sound followed by a brief automatic application of the braking system).

3. *Adaptive cruise control system*: Adaptive cruise control (also called autonomous or radar cruise control) is an optional cruise control system that automatically adjusts the vehicle speed to maintain a safe distance from the vehicle ahead. The control is based on sensor information from on-board sensors (radar or laser-based). Most systems provide steering wheel mounted controls for setting maximum cruising speed and safe headway distance from the leading vehicle.

4. *Automated braking system*: This system applies vehicle brakes when the sensor and processor in the vehicle determine if the vehicle is headed on a collision course with a stationary or a moving object. The unit applies brakes automatically if the vehicle is on a collision course and the driver has not executed a collision avoidance maneuver.

5. *Backup camera system*: A backup camera is a special type of video camera that is produced specifically for attaching to the rear of a vehicle to aid in backing up and to alleviate the rear blind area. The backup camera is alternatively known as the "reversing camera" or "rear view camera." It is specifically designed to avoid a backup collision by providing the driver a view of the projected path of the vehicle with color-coded distance markers in the rear camera view. The rear facing video camera is typically mounted at the vehicle centerline, and above the rear license plate, or near the top or bottom edge of the backlite (rear window). During backing maneuvers (as soon as the gear shifter is placed in the reverse gear), the camera output along with the projected color-coded markings is displayed in a screen located in the center stack. The red, yellow, and green color-coded zones, respectively, indicate that an object to the rear of the vehicle is very close, somewhat close, or far from the vehicle. In some vehicles, the rear camera display with the color-coded zones is integrated within the inside rearview mirror.

6. *Blind spot monitoring system*: A blind spot monitor is a vehicle-based sensor device that detects other vehicles located to the driver's both sides (i.e., adjacent lanes) and the rear. Warnings can be displayed via visual, audible, vibrating, or tactile signals. While driving in forward direction, the most

common warning signal is by activating an amber-colored LED warning lamp mounted near the outboard edge of each outside rearview mirror. In backup sensing mode, the system provides auditory warning beeps when the detected targets are approaching close to the collision zone. The system can be also integrated with the backup camera system.

7. *Night vision systems*: This system allows the driver to see further than what the driver could see with the vehicle headlamp system during night driving. The night vision system typically uses an infrared camera that can detect objects on the roadway far beyond what a driver can see with the low beam of the vehicle headlamps. The output of the infrared camera is provided to the driver through a separate display in the front of the driver or in an augmented screen of a head-up display. The detected objects are typically shown as augmented superimposed images on the view of the forward road scene captured by the camera.

8. *Adaptive forward lighting systems*: The adaptive forward lighting systems offer the most potential for improving night driving safety performance. The system monitors the forward road scene for oncoming drivers and road features such as curves, grades, and intersections (e.g., through integrated GPS and map database system) and alters the beam pattern to provide more illumination in target areas and reducing glare illumination into the oncoming driver's eyes. Some of these functions are already allowed under FMVSS 108 by allowing for a portion of the emitted light to move within a compliant headlamp beam and/or through an automatic re-aim of a headlamp beam pattern.

9. *Active rollover protection/stability system*: Active rollover protection (ARP) systems involve sensors and microprocessors to recognize impending rollover and selectively apply brakes to resist the rollover. ARP builds on an electronic stability control and its three chassis control systems, namely the vehicle's anti-lock braking system, traction control, and yaw control. ARP adds another function, i.e., detection of an impending rollover. Excessive lateral force, generated by excessive speed in a turn, may result in a rollover. ARP automatically responds whenever it detects a potential rollover. ARP rapidly applies the brakes with a high burst of pressure to the appropriate wheels and in some situations decreases the engine torque to interrupt the rollover before it occurs.

10. *Advanced automatic collision notification system*: The system is also known as advanced automatic crash notification and is the successor to automatic collision notification (ACN) system. It alerts emergency medical responders and provides data to more quickly determine if a vehicle occupant needs care at a trauma center after a vehicle crash. The real-time crash data from the advanced automatic crash notification (AACN) vehicle telematics system and similar systems can be used to determine whether injured patients need care at a trauma center. By using a collection of sensors, the vehicle telemetry systems such as AACN send crash data to an advisor if a vehicle is involved in a moderate or severe front, rear, or side-impact crash. Depending on the type of system, the data include information about crash

severity, the direction of impact, air bag deployment, multiple impacts, and rollovers (if equipped with appropriate sensors). Advisors can relay this information to emergency dispatchers, helping them to quickly determine the appropriate response involving combination of emergency personnel, equipment, and medical facilities.

CONNECTED VEHICLES OR V2X TECHNOLOGIES

These wireless technologies (called "vehicle-to-X" as "V2X") allow two-way communication between the vehicle and other entities (X) outside the vehicle, such as:

1. *V2V = Vehicle-to-vehicle*: The subject's vehicle communicates its position, motion, and state of control activations (e.g., turning, accelerating, decelerating) to other vehicles.
2. *V2H = Vehicle-to-home*: The subject's vehicle communicates information with his/her home-related programming or controlling of functions related to the vehicle (e.g., charging of an electric vehicle) or home systems (e.g., security system, appliances).
3. *V2I = Vehicle-to-infrastructure*: The subject's vehicle can communicate with roadside infrastructure such as traffic signals at intersections, state of road, traffic conditions, and so forth.
4. *V2P = Vehicle-to-person or pedestrian communication*: The subject's vehicle can communicate with nearby persons or pedestrians (e.g., by sending them warning messages through wireless devices about the vehicle location, direction of approach, or arrival time).
5. *V2C = Vehicle-to-cloud-based data sources*: The driver can access information from other cloud-based databases for personnel needs (e.g., looking for the nearest bank, gas station, electric vehicle charging station, restaurant, and so forth).

Thus, the V2X technologies allow connected vehicles to wirelessly communicate with each other and other locations. The communicated information can be used to assist the driver to provide warning messages related to different unsafe situations or even initiate certain maneuvers to avoid accidents. The V2X technologies have potential to reduce fuel consumption and emissions by providing the drivers information traffic congestions, rerouting to avoid the congestions, and providing locations and status of electric vehicle charging stations.

SELF-DRIVING VEHICLES

A number of vehicle manufacturers have demonstrated vehicles that have capabilities to drive without any inputs or interventions from the drivers. These vehicles have sensing capabilities to continuously monitor the roadway and traffic situations and make necessary lateral (steering) and longitudinal (accelerator and brake pedal actions) control actions. With integrated GPS support, the vehicles can also select routes and reach preprogrammed destinations. With the implementation of such technologies, the vehicle becomes "autonomous" (i.e., acting separately from other things or people; having the power or right to govern itself).

Many of the currently available driver assistance systems such as automatic braking, adaptive cruise control, lane keeping systems will be integrated over time to create the self-driving cars.

The future of such technologies is currently debated because the drivers may not be ready to trust such systems. Further, the problem of hacking into such cars needs to be solved to the highest degree of confidence because if the hackers can get into the electronic systems of such vehicles, they can alter output actions of the vehicle. It is expected that in the near future, the automakers will integrate many of the driver assistance capabilities and offer vehicles with limited capabilities (semiautomated and not fully automated self-driving vehicles) such as (a) adaptive cruise control with lane changing capabilities, (b) self-parking vehicles, (c) auto-pilot features that allow drivers to take their hands off the wheel under certain pre-approved conditions (Naughton, 2015).

Self-driving trucks are another important application area for this technology. Many commercial trucks including those of the army can use self-driving trucks. Sedgwick (2016) describes how the army can benefit from having a convoy of self-driving trucks that can follow a lead truck with a human driver. The potential for reducing driver workload and number of human drivers is also very appealing for many commercial delivery applications. The self-driving trucks can operate over long distances with much less breaks (no coffee breaks and only stop to refuel) and thus can transport cargo in shorter delivery times.

Use of autonomous features also has a potential to reduce fuel and emissions by reducing speed variations, reducing traffic density, following most economic route, and so forth (NHTSA, 2013).

LIGHT-WEIGHTING TECHNOLOGIES

The lightweight materials and new structural optimization technologies are used to reduce the weight of the vehicle. Reducing vehicle weight requires less power to accelerate and maintain a given speed of the vehicle; and thus, weight reduction reduces the fuel consumption of the vehicle. During early stages of product development, all vehicle systems are studied to evaluate weight reduction possibilities. Automakers have been experimenting for decades with light-weighting technologies, but the effort has gained urgency with the adoption of tougher fuel economy standards (EPA and NHTSA, 2012). To meet the government's goal of nearly doubling average fuel economy to 45 mpg by 2025, the light vehicles (less than 8,500 lbs. GVWR) need to reduce some weight.

The weight reduction possibilities generally involve combinations of the following approaches: (a) use of different lightweight materials (e.g., high-strength steels, aluminum, magnesium, composites/plastics/carbon fiber, etc.), (b) new structural designs and mechanisms (e.g., space-frame designs with composite body panels and hollow coil springs), (c) different production techniques (e.g., hydro-formed body and chassis components, titanium suspension links, spray-painted metal circuits), (d) joining methods (e.g., riveting of steel and aluminum body parts, adhesives, laser welding of dissimilar materials), and (e) smaller lower weight and more efficient fuel-saving powertrains. Many technological advances such as turbo-boost engines, eight-speed transmissions, stop–start, and cylinder deactivation have been attempted to improve fuel saving capabilities of powertrains. Hybrid and electric power plants provide improved fuel economy; however, the need to carry heavy batteries generally increases the weight of the vehicle. All the above approaches generally increase costs and development time and add challenges in maintaining high levels of reliability and durability in achieving desired performance levels.

Following is a short summary of various materials currently used in the auto industry (Helms, 2014).

1. *High-strength steel (HSS)*: The HSSs are lighter and stronger steels, and they are mixtures with other elements such as nickel and titanium. Currently, the HSS makes up at least 15% of the car's weight. Some vehicles (e.g., 2014 Cadillac ATS) are using nearly 40% HHSs. The HSS costs about 15% more than regular steel, but less than aluminum. HSS weighs more than aluminum. However, with continuing advances in structural designs, the vehicle weight can be further reduced with HSS.
2. *Aluminum*: The typical vehicle already contains around 340 lbs. of aluminum, or about 10% of the weight of a midsize car. The 2013 Range Rover dropped around 700 lbs. with its all-aluminum body, while the 2014 Acura MDX shed 275 lbs. with increased use of HSS, aluminum, and magnesium. The 2015 F-150 pickup reduced up to 700 lbs. as compared with its earlier version. Aluminum is most commonly used in engines, wheels, hoods, and trunk lids. Aluminum is lighter than steel and easy to form into a variety of parts. It is also more corrosion-resistant than steel. The supply of steel is many times greater than that of aluminum. Aluminum costs about 30% more than conventional steel, and a rapid increase in demand could make aluminum prices volatile. Some projections have estimated that the aluminum's use in the auto industry will triple by 2030.
3. *Carbon fiber*: It is a high-strength material made from woven fibers. The specific weight of carbon fiber is about half as that of steel. The carbon fiber is resistant to dents and corrosion, and it offers high design flexibility, as it can be shaped in ways as compared with the stamped steel. However, the high cost of carbon fiber and the longer part forming (manufacturing) times are substantial drawbacks for the auto industry. Carbon fiber parts are made from petroleum-based strands, which must go through several stages before they are woven into carbon fiber. After that, it takes about 5 minutes to form the material into a part, compared with about 1 minute for steel or aluminum. The carbon fiber is about 5–6 times more expensive than steel.

   Many organizations are experimenting with cheaper materials for the fibers and faster-curing resins that could shorten the time and costs to form parts. The carbon fiber is expected to be used in limited amounts on low-volume or luxury cars until major advances occur.

AERODYNAMIC DRAG REDUCTION

Many improvements in aerodynamics are constantly developed and incorporated in new vehicle designs. The improvements range from coming up with basic more aerodynamic vehicle shape to introducing active aerodynamic elements to alter air flow around the vehicle (Gehm, 2015). Some recently introduced aerodynamic improvements include:

1. Lowering vehicle height at higher driving speeds by use of adjustable suspensions
2. Lightweight underbody panels to reduce underbody turbulence
3. Active shutters in grills and front bumpers to reduce and deflect air into the engine compartment
4. Active deflector elements that move outward and rearward to reduce drag around wheels
5. Flush (non-cupped) wheels or active wheel rims (e.g., Intelligent Aerodynamic Automobile (IAA) concept introduced by Mercedes-Benz, which changes their cupping from 50 mm to 0 – from five-spoke to flat-disc wheels)
6. Extendable rear ends and spoilers
7. Low profile or flush-mounted door handles
8. Smaller rear view mirrors or replacing outside mirrors with rear facing video cameras displaying their view on a screen mounted in the instrument panel. Note: FMVSS 111 requires flat (unit magnification) inside and left outside mirrors (NHTSA, 2016).

CONCLUDING REMARKS

Reducing fuel consumption and emissions is a major goal facing the automobile industry. Reducing vehicle weights, incorporating new powertrains, and reducing aerodynamic drag are some of the solutions currently considered to accomplish these goals. Advances in new technologies are also allowing auto manufacturers to incorporate new safety technologies. In addition, advances in autonomous vehicles will allow further reductions in fuel economy and pollution potential. Increase in vehicle costs as these technologies are incorporated in future vehicles is one major concern facing the automobile manufacturers.

REFERENCES

Environmental Protection Agency and National Highway Traffic Safety Administration 2012, October 15. 2017 and Later Model Year Light-Duty Vehicle Greenhouse Gas Emissions and Corporate Average Fuel Economy Standards. *Federal Register*, 77(199): 62623–63200. Environmental Protection Agency, 40 CFR Parts 85, 86, and 600. Department of Transportation National Highway Traffic Safety Administration, 49 CFR Parts 523, 531, 533, 536, and 537. [EPA-HQ-OAR-2010-0799; FRL-9706-5; NHTSA-2010-0131]. RIN 2060-AQ54; RIN 2127-AK79.
Gehm, R. 2015, November. *Active in Aero*. Automotive Engineering. Warrendale, PA: SAE International.

Helms, J.H. 2014. Advanced Engineered Material Technologies for a Challenging Environment. *SAE Off-Highway Engineering.* Website: http://articles.sae.org/13054/ (Accessed June 16, 2015).

National Highway Traffic Safety Administration. 2013, May. NHTSA's Preliminary Statement of Policy on Vehicle Automation. Website: www.nhtsa.gov/staticfiles/rulemaking/pdf/ Automated_Vehicles_Policy.pdf (Accessed: January 22, 2014).

Naughton, K. 2015. Self-driving Cars Are a Lot Closer Than You Think. *Automotive News.* http://www.autonews.com/article/20150507/OEM06/150509895/self-driving-cars-are-a-lot-closer-than-you-think (Accessed: May 7, 2015).

Sedgwick, D. 2016, February 6. Army Marches with Self-Driving Trucks. *Automotive News.* Website: http://www.autonews.com/article/20160206/OEM06/302089997/army-marches-forward-with-self-driving-trucks (Accessed: February 6, 2016).

Wernle, B. and M. Colias. 2013, April 14. Ford, GM Work Together on New Nine-, 10-Speed Transmissions. *Autoweek.* Website: http://autoweek.com/article/car-news/ford-gm-work-together-new-nine-10-speed-transmissions (Accessed: October 23, 2014).

Section IV

Current Issues Facing the Energy Industries

13 Smart Grid

INTRODUCTION

The objectives of this chapter are to provide background information on the power grid and the smart grid and discuss issues related to the operation of the smart grid and its advantages and disadvantages. The chapter also discusses issues related to the future of the smart grid.

WHAT IS AN ELECTRIC GRID?

An electrical grid (or power grid) is an interconnected network for delivering electricity from producers to consumers. It consists of:

1. Generating sources that produce electric power (e.g., coal-fired power plants, natural gas power plants, wind turbines, solar power plants)
2. Electrical substations for stepping electrical voltage up for transmission and stepping down the voltage for distribution
3. High-voltage transmission lines to carry power from distant power sources to demand centers
4. Controllers to ensure that power is routed from sources to demand centers efficiently with minimum power disruptions
5. Distribution lines to connect individual customers to enable use of electrical devices (e.g., lighting systems, climate control systems, appliances, and so forth).

An electrical grid can vary in size from covering a single building to a national grid (which covers a whole country) to transnational grids (which can cross continents). A large power grid thus is a system of synchronized power providers and consumers connected by transmission and distribution lines and operated by one or more control centers. In the continental United States, the electric power grid consists of three systems (a) the Eastern Interconnect, (b) the Western Interconnect, and (c) the Texas Interconnect. In Alaska and Hawaii, several systems encompass areas smaller than the state (e.g., the interconnect serving Anchorage, Fairbanks, and the Kenai Peninsula; individual islands). According to the U.S. Energy Information Administration (EIA), the U.S. power grid is made up of over 7,300 power plants, nearly 160,000 miles of high-voltage power lines, and millions of miles of low-voltage power lines and distribution transformers, connecting 145 million customers throughout the country (EIA, 2016).

Electricity in the United States is generated using a variety of resources and technologies. The majority of electricity consumed in the United States is produced using conventional sources, such as natural gas, oil, coal, and nuclear. EIA provides an

DOI: 10.1201/9781003107514-17

hourly electric grid monitor that provides online hourly demand by regions. It also provides data on the power supplied by different types of power sources (generation mix) and interchanges between United States and neighboring countries – Canada and Mexico (EIA, 2020a).

CHALLENGES FACING THE POWER GRID

Construction of electricity infrastructure in the United States began in the early 1900s, and investment was driven by new transmission technologies, central station generating plants, and growing electricity demand, especially after World War II. Now, some of the older, existing transmission and distribution lines have reached the end of their useful lives and must be replaced or upgraded. New power lines are also needed to maintain the electrical system's overall reliability and to provide links to new renewable energy generation resources, such as wind and solar power, which are often located far from where electricity demand is concentrated.

Several challenges exist for improving the infrastructure of the grid (EIA, 2020b):

a. Planning new transmission lines (getting approval of new routes and obtaining rights to the necessary land)
b. Determining an equitable approach for recovering the construction costs of a new transmission line built in one state when the line provides benefits to consumers in other states
c. Addressing the uncertainty in federal regulations regarding who is responsible for paying for new transmission lines, which affects the private sector's ability to raise money to build transmission lines
d. Expanding the network of long-distance transmission lines to renewable energy generation sites where high-quality wind and solar resources are located, which are often far from where electricity demand is concentrated
e. Protecting the grid from physical and cyber attacks

WHAT IS A SMART GRID?

A smart grid is an electricity supply network that uses digital communications technology to detect and react to local changes in power demand. It monitors grid loads, anticipates changes in loads, and automatically switches/diverts powers to match demands (DOE, 2020; Wikipedia, 2020). And thus, it reduces risks due to power failures. Thus, a smart grid is the implementation of (a) digital technology to the power grid to allow for two-way communication between the utility and its customers, (b) sensing electricity parameters along the transmission lines, and (c) controlling the electricity flows in the transmission and distribution lines. Like the Internet, the smart grid will consist of application of new technologies using computer-operated communications between the electricity providers and its customers, controls (or actuators) to direct electricity flows, and sensors to monitor electricity flows at many points along the transmission and distribution lines to meet the continually changing electric demands within the power grid – automatically and quickly.

The smart grid represents an unprecedented opportunity to move the energy industry into a new era of increased reliability, availability, and efficiency that will contribute to achieving higher levels of economic benefits and environmental health. During the transition period, it will be critical to carry out testing, technology improvements, consumer education, development of standards and regulations, and information sharing between projects to ensure that the envisioned benefits from the smart grid become a reality.

The benefits associated with the smart grid include:

a. More efficient transmission of electricity (i.e., provide needed electric power at lower cost). It also reduces operations and management costs for utilities and ultimately lowers power costs for consumers.
b. Reduces peak power demand, which will also help lower electricity rates
c. Quicker restoration of electricity after power disturbances (e.g., blackouts and brownouts). Thus, provides more reliable electric power and improved security.
d. Increased integration of large-scale conventional and renewable energy generation systems and customer–owner power generating systems (e.g., wind turbines in farmland) in providing reliable power to customers.

MINIMIZE ELECTRICITY DISRUPTION

An electricity disruption such as a blackout can have a domino effect – a series of failures that can affect banking, communications, traffic, and security. This is a particular threat in the winter when homeowners can be left without heat. A smarter grid will add resiliency to the electric power system and make it better prepared to address emergencies due to events such as severe storms, earthquakes, large solar flares, and terrorist attacks.

Because of its two-way interactive capacity, the smart grid will allow for automatic rerouting when equipment fails, or outages occur. This will minimize outages and minimize the effects when they do happen. When a power outage occurs, smart grid technologies will detect and isolate the outages, containing them before they become large-scale blackouts.

The new technologies will also help ensure that electricity recovery resumes quickly and strategically after an emergency – routing electricity to emergency services first, for example. In addition, the smart grid will take greater advantage of customer-owned power generators to produce power when it is not available from utilities. By combining these "distributed generation" resources, a community could keep its health center, police department, traffic lights, phone system, and grocery store operating during emergencies.

In addition, the smart grid is a way to address an aging energy infrastructure that needs to be upgraded or replaced.

GIVING CONSUMERS CONTROL

The smart grid is not just about incorporating technologies to improve responding capabilities and efficiencies of the utility companies. It will provide the information and tools to the customers to make choices about their energy use. The customers can manage their electric needs and transactions similar to how they manage their banking activities from their home computers. A smarter grid will enable an unprecedented level of consumer participation.

For example, the customer will no longer have to wait for his monthly statement to know how much electricity is used. "Smart meters," and other mechanisms will allow the customers to see how much electricity they use, when they use it, and its cost. Combined with real-time pricing, this will allow the customers to save money by using less power when electricity rate is most expensive. While the potential benefits of the smart grid are usually discussed in terms of economics, national security, and renewable energy goals, the smart grid has the potential to help the customers to save money by helping them to manage their electricity use and choose the best times to purchase electricity. And they can save even more by generating their own power through their generators such as roof-mounted solar panels or wind turbines in farmlands.

BUILDING AND TESTING THE SMART GRID

The smart grid will include a number of devices such as controls, computers, power lines, and new technologies and equipment. It will take some time for all the technologies to be perfected, equipment installed, and systems tested before it comes fully online. And it will not happen all at once – the smart grid is evolving over the next decade or so. Once mature, the smart grid will likely bring the same kind of transformation that the Internet has already brought to the way we live, work, play, and learn.

GRID OPERATION CENTERS

Today's electrical transmission system operates much like a system of interconnected streams. Power flows through the transmission system along the path of least resistance, finding multiple paths between the power plants and the cities that are demanding the power. Grid operators actually have very little control over today's system. Their primary task is to make sure that as much power is being generated as is being used. Otherwise, the grid's voltage could drop, causing the grid to become unstable. Operators generally know which lines are in service and when relays have opened to protect lines against faults, but they have limited control capabilities. Unfortunately, like water in a bathtub, power can "slosh around" within the grid, developing oscillations that, under the worst of conditions, could lead to widespread blackouts. To compound the problem, grid operators also have limited information about how the power is flowing through the grid. The smart grid will help solve this problem by adding new capabilities for measurement and control of the transmission system. These technologies will make the grid much more reliable and will minimize the possibility of widespread blackouts.

SMART METERS

Incorporation of smart meters in homes and businesses is another key element to the smart grid system. Replacing the traditional analog meters, these new digital devices are capable of two-way communication – relaying information about both supply and demand between producers and consumers.

The data collected via smart meters, too, is essential to the function of the smart grid. By analyzing these data, power generation plants can better predict and respond to periods of peak demand. This allows them to reduce production when less power is needed and quickly ramp up generation when peak periods approach.

By harnessing the power of computers, communications, and data analysis technologies, the smart grid can improve the flexibility and efficiency of the traditional grid and can open up new opportunities for more intermittent generation methods, e.g., wind and solar, and new stresses to the network, such as charging electric cars and trucks.

SMART GRID OPERATION AND BENEFITS

The power demand sensors will continuously sample and monitor state of grid power flow to control and stabilize the transmission of electricity. Thus, the benefits of the smart grid will be as follows:

1. Quicker restoration of electricity after power disturbances
2. Reduced operations and management costs and ultimately lower power costs for consumers
3. Reduced peak demand, which will also help lower electricity rates
4. Increased integration of large-scale renewable energy systems
5. Better integration of customer–owner power generation systems, including renewable systems
6. Reduce cascading effects of blackouts (e.g., isolating outages, automatic rerouting, and prioritization of power restoration)
7. Customers can manage their energy needs like an online bank account (e.g., charge cars during off-peak times)
8. Improved security (e.g., avoided blackouts)

SAMPLING FOR MONITORING AND CONTROLLING THE POWER GRID

Smart grid technologies offer a new solution to the problem of monitoring and controlling the grid's transmission System. New technologies called Phasor Measurement Units (PMU) sample voltage and current many times per second at a given location, providing a snapshot of the power system at work. PMUs provide a new monitoring tool for the smart grid. In our current electric grid, measurements are taken once every 2 or 4 seconds, offering a steady-state view into the power system behavior. Equipped with smart grid communications technologies, measurements can be taken many times a second, offering dynamic visibility into the power system. This makes it easier to detect the types of oscillations that led to the 2003 blackout.

DATA SHARING WITH SMART GRID

Another contributor to the 2003 blackout was the limited situational awareness of the various grid operators involved. At the time, there was limited data sharing and transparency among the grid operators in different regions of North America, making it hard for the individual grid operators to see the big picture. By including new standards that make it easier for grid systems to interact with one another, the smart grid will make data sharing among regional grid operators easier to accomplish. Potentially, grid operators will be able to explore the state of the grid at the national level and switch within seconds to explore specific details at the local level. These technologies will provide rapid information about blackouts and power quality as well as insights into system operations for utilities.

THE "SELF-HEALING" GRID

Smart grid technologies also offer new means of controlling the transmission system. New high-power electronics function essentially as large-scale versions of transistors, adding a new level of control to the transmission system. New technologies could also help dampen unwanted power oscillations and avoid unproductive flows of current through the grid that only serves to waste energy. The combination of new measurement and control technologies also enables a new automated approach to controlling the grid. Software could potentially monitor the grid in real time for potential disturbances that could lead to blackouts, and it could take actions to check the disturbances. Such monitoring software could act to dampen out oscillations in the power grid, or it could even reroute power through the grid to avoid overloading a transmission line. In the event that a power line needs to be removed from service, control software could reroute the power in a way that causes minimal disruptions to the grid. This approach is often referred to as the "self-healing" grid. The ideal self-healing grid will involve a combination of transmission system monitoring and control software and comparable measures for the local distribution systems that deliver the power to individual homes and businesses. These distribution system measures are sometimes referred to as distribution intelligence.

SMART GRID COSTS

The additional costs to implement features of the smart grid into existing power grid will include (a) updating equipment with sensors, computers, communication (wireless) and switching equipment (e.g., smart meters, interfacing power loads, two-way communications between customers and utility companies), (b) integration of renewable energy systems, (c) maintenance of equipment and software, (d) training of personnel to operate new features, and (e) additional security considerations to reduce hacking and protection against malicious intrusion, vandalism, and attacks (Horowitz et al., 2018).

The question of who should pay for the added costs and who should benefit from the lowered costs due to increased efficiencies is also a major topic for discussions. However, the customers of the electric power and providers of the power are both expected to benefit in the long term.

Currently, many of the additional costs are charged back by the utility companies to the customers. Therefore, the future electric bills from a typical utility company will most likely include the following costs:

1. Customer service charges to recover cost of monthly billing, meter and equipment operation, and meter reading
2. Generation and transmission charges
3. Distribution charges (e.g., cost of power transmission and distribution lines, voltage step-up and step-down substations)
4. System improvement cost
5. Electricity usage charge (charge for actual electric energy (kWh) consumed)
6. Capacity charge (capacity charges are based on the highest amount of energy you are estimated to use or consume during a month (or year in some locations). Essentially, you pay a fee to ensure that the electricity you might use is there for you when you need to use it, whenever you need to use it.
7. Surcharges for costs incurred due to energy optimization programs, renewable energy [cost of the company's efforts to harness solar and wind power], nuclear [for increased security at nuclear-generating facilities], and state-mandated low-income energy assistance fund (LIEAF)
8. Residential state sales tax

CONCLUDING REMARKS

As many features of the smart grid are implemented, the customers will benefit from the grid's increased reliability and flexibility to obtain electricity in a more informed way. The customers can also plan their electricity use (especially of heavy electricity equipment such as air-conditioners and electric vehicles) and reduce their electricity bill. The utility companies can also benefit from reducing variability in distribution systems, reducing peak demands and downtime.

REFERENCES

DOE. 2020. SmartGrid.gov. Website: https://www.smartgrid.gov/the_smart_grid/smart_grid.html (Accessed: October 12, 2020).EIA. 2016. Today in Energy – U.S. Energy Information Administration (EIA) (Website: www.eia.gov/todayinenergy/detail.php?id=27152 (Accessed: September 17, 2021).

EIA. 2020b. Electricity Explained: How Electricity is Delivered to Consumers. Website: https://www.eia.gov/energyexplained/electricity/delivery-to-consumers.php (Accessed: August 11, 2020).

EIA. 2020a. Hourly Electric Grid Monitor. Website: https://www.eia.gov/beta/electricity/-gridmonitor/dashboard/electric_overview/US48/US48 (Accessed: August 11, 2020).

Horowitz, K.A. Fei Ding, W., Mather, B. and B. Palmintier. 2018. *The Cost of Distribution System Upgrades to Accommodate Increasing Penetrations of Distributed Photovoltaic Systems on Real Feeders in the United States.* Golden, CO: National Renewable Energy Laboratory. NREL/TP-6A20-70710. Website: https://www.nrel.gov/docs/fy18osti/70710.pdf (Accessed: November 1, 2020).

Wikipedia. 2020. Electrical Grid. Website: https://en.wikipedia.org/wiki/Electrical_grid (Accessed: March 30, 2020).

14 Electricity Storage Technologies

INTRODUCTION

Solar plants and wind turbines operate intermittently as they depend on the presence of sun light and wind velocity, respectively. Lack of electric energy from these intermittent sources needs to be compensated by supplying electric power generated by other standby power sources that can run when the solar and wind-powered turbines cannot produce sufficient energy. The standby power sources can be nuclear, natural gas, and even coal-fired plants. Other possibility is that if excess electric energy generated is stored in devices such as batteries, the stored energy can be retrieved and used when sufficient electric power cannot be generated by the intermittent sources. The objective of this chapter is to present relevant energy storage technologies and other related issues, problems, and information available in this area.

For example, other issues related to electricity storage are as follows:

a. Increasing usage of portable electronic devices such as smart phones, laptop computers, and other equipment operated by rechargeable batteries that produce discontinuous energy flows. These discontinuous energy demands need to be smoothened by use of energy from electricity storages rather than changing the output levels of other power plants, which may not be able to respond quickly to changes in demands.
b. Low-cost batteries are needed as lithium batteries are not yet cheap.
c. Other storage systems such as flywheels, hydro and chemical storage systems have limitations.

ENERGY STORAGE

The electric power grid operates on a delicate balance between supply (generation) and demand (consumer use). One way to help balance fluctuations in electricity supply and demand is to store electricity during periods of relatively high power generation and low demand, then release it back to the electric power grid during periods of lower generation and higher demand. When the generated energy is greater than the electricity demanded, the excess generated energy can be stored in devices such as batteries; then it could be used when other power sources are not available. Energy storage is the capture of energy produced for use at a later time. A device that stores energy is generally called an accumulator or battery. Energy comes in multiple forms including radiation, chemical, gravitational potential, electrical potential, electricity, elevated temperature, latent heat, and kinetic. Energy storage involves converting energy from forms that are difficult to store to more conveniently or economically storable forms. Depending on the extent to

DOI: 10.1201/9781003107514-18

which it is deployed, electricity storage could help the utility grid operate more efficiently, reduce the likelihood of brownouts during peak demand, and allow for more renewable resources to be built and used.

Some technologies provide short-term energy storage, while others can endure for much longer. Bulk energy storage is currently dominated by hydroelectric dams, both conventional at higher elevation and systems that use pumps. Common examples of energy storage are (a) the rechargeable battery, which stores chemical energy readily convertible to electricity to operate an electric device (e.g., a mobile phone), (b) the hydroelectric dam, which stores energy in a reservoir as gravitational potential energy, and (c) ice storage tanks, which store ice frozen by cheaper energy at night to meet peak daytime demand for cooling. Fossil fuels such as coal and gasoline store ancient energy derived from sunlight by organisms that were buried and were converted over time into these fuels. Food (which is made by the same process as fossil fuels) is a form of energy stored in chemical form.

According to the U.S. Department of Energy, the United States had more than 25 GW of electrical energy storage capacity as of March 2018. Of that total, 94% was in the form of pumped hydroelectric storage, and most of that pumped hydroelectric capacity was installed in the 1970s. The 6% of other storage capacity is in the form of battery, thermal storage, compressed air, and flywheel (EIA, 2020).

ATTRIBUTES OF AN ENERGY STORAGE SYSTEM

Attributes are characteristics that an energy storage system must have to be acceptable to its customers.

1. Storage capacity or rated power output [kW]
2. Capital cost [$/kW]
3. Operating and maintenance costs (fixed costs [$/kW] and variable [$/kWh])
4. Efficiency (output energy [kW]/ Input energy [kW])
5. Accessibility (ease, response time, distance to nearest power hookup)
6. Life (years, number of charging cycles)
7. Environmental impact (pollution potential, wildlife danger)
8. Safety (accident rate, risk priority number)
9. Security (likelihood of hackers' attack, loss due to security breach [$])
10. Storage time [h]

Different customers have different needs. Large consumers would most likely have large power demands as compared with small residential customers. Selecting a technology to build a new energy storage system (from other available energy storage systems) will depend upon the results of analyses such as a cost–benefit analysis expressed in measures of net present value of benefits minus costs and/or benefit-to-cost ratio over the life cycle of the storage system.

TRADE-OFFS BETWEEN ATTRIBUTES RELATED TO ENERGY STORAGE SYSTEMS

Many trade-offs between important attributes of a storage system must be considered during initial decision to develop a storage system. Some examples of the trade-offs are as follows:

1. Capital costs vs. environmental impact (e.g., reduction in environmental impact will require higher capital costs to capture pollutants)
2. Operating and maintenance costs vs. safety (e.g., higher operating and maintenance costs can reduce accidents and hence less accident-related costs)
3. Storage capacity vs. accessibility (e.g., large-capacity storage systems may be in remote areas that are not easily accessible)
4. Capital costs vs. safety (e.g., higher design and safety protection devices costs can lead to lower accident costs)
5. Storage time vs. efficiency (e.g., higher storage time will reduce efficiency due to higher losses during storage)

CURRENTLY AVAILABLE ENERGY STORAGE TECHNOLOGIES

Many technologies are available to provide different types of energy storage systems (EIA, 2020; DOE, 2020; Wikipedia, 2020). The technologies currently being deployed can be divided into the following seven main categories:

1. *Chemical* – a range of electrochemical storage solutions, including advanced chemistry batteries and capacitors
2. *Electrical and magnetic* – capacitor, supercapacitor, and superconducting magnetic energy storage
3. *Thermal* – capturing materials at high temperature (or low temperature) to create energy on demand by heat transfer (e.g., molten salt storage, hot water from thermal solar plant)
4. *Mechanical storage* – using technologies that harness kinetic energy (e.g., using flywheels), gravitational energy (e.g., mass storage at higher elevations), and energy stored in compressed materials (e.g., compressed air or compressed springs)
5. *Hydrogen* – excess electricity generation can be converted into hydrogen via electrolysis and stored in cylinders or tanks
6. *Pumped hydropower* – creating large-scale reservoirs of energy by pumping water into the reservoirs located at higher elevation
7. *Biological* – converting biomass into materials such as biofuels, glycogen, and starch

Brief descriptions of the above technologies are provided below.

MECHANICAL

Energy is used to raise a mass through a height. Thus energy can be stored as gravitational potential energy. The amount of gravitational energy stored in a mass (m) raised at height (h) is equal to "mgh," where "g" is the gravitational acceleration. Energy can be stored in water pumped to a higher elevation using pumped storage methods or by moving solid matter to higher locations (called gravity batteries). Other commercial mechanical methods include compressing air and storing it in tanks, using flywheels rotating at high speeds (revolutions per minute), or deflecting or compressing springs. The electrical energy can be used to compress air (or accelerate a flywheel at higher speed) and then converted back again into electric energy when electrical demand peaks.

Pumped Hydro

The most common large-scale use of gravity energy storage in current use is pumped hydro storage. Electricity powers a pump that raises water from a low elevation reservoir to a high elevation reservoir, thus storing energy as gravitational potential energy. Subsequently energy is recovered as water flows down and drives a water turbine connected to a generator producing electricity.

At times of low electrical demand, excess generation capacity is used to pump water from a lower elevation into a reservoir located at higher elevation. When demand grows, water is released back into a lower reservoir (or waterway or body of water) through a turbine, generating electricity. Reversible turbine-generator assemblies act as a motor–pump combination.

Compressed Air

Compressed-air energy storage (CAES) uses surplus energy (electrical or mechanical) to compress air for subsequent electricity generation. Small-scale systems have long been used in such applications as propulsion of mine locomotives. Here the compressed air typically drives a piston engine like an internal combustion engine (but without any combustion because there is no fuel to burn, only compresses air). The compressed air can be stored in an on-board cylinder (tank) or an underground reservoir, such as a salt dome.

CAES plants can bridge the gap between production volatility and load. CAES storage addresses the energy needs of consumers by effectively providing readily available energy to meet demand. CAES plants can take in the surplus energy output of renewable energy sources during times of energy overproduction. This stored energy can be used later when demand for electricity increases or energy resource availability decreases.

Compression of air creates heat. The air is warmer after compression. Expansion of air reduces the air temperature. If no extra heat is added, the air will be much colder after expansion. If the heat generated during compression can be stored and used during expansion, efficiency improves considerably. A CAES system can deal with the heat in three ways. Air storage can be adiabatic (without the transfer of heat), diabatic (involving transfer of heat), or isothermal (temperature remains constant).

Flywheel

Flywheel energy storage (FES) works by accelerating a flywheel (mounted on a shaft or rotor) to a very high speed, holding energy as rotational energy. A *flywheel* is a mechanical device specifically designed to efficiently store rotational energy (kinetic energy), which is proportional to the square of its rotational speed and its mass. When energy is added, the rotational speed of the flywheel increases and when energy is extracted its speed declines, due to conservation of energy. Most FES systems use electricity to accelerate and decelerate the flywheel, but mechanical devices can be coupled to the flywheel for increasing output or to store excess energy.

High-efficiency FES systems have rotors made of high-strength carbon-fiber composites, suspended by magnetic bearings and spinning at speeds from 20,000 to over 50,000 revolutions per minute (rpm) in a vacuum enclosure. Such flywheels can reach maximum speed in a matter of minutes. The flywheel system is connected to a combination electric motor/generator system.

SOLID MASS GRAVITATIONAL

Changing the altitude of solid masses can store or release energy via a raising and elevating system (e.g., using an electric motor/generator). Potential energy storage or gravity energy storage systems have used various methods such as (a) the movement of earth-filled hopper rail cars driven by electric locomotives from lower to higher elevations and then recouping the energy as the rail cars are rolled back to lower elevation, (b) cranes moving concrete weights up and down using solar-powered winches, (c) raising and lowering concrete in mine shafts of recently closed mines, (d) using winches mounted on an ocean barge for taking advantage of a 4 km (13,000 ft) elevation difference between the surface and the seabed. Efficiencies of the mass gravitational systems can be as high as 85% recovery of stored energy.

HYDROELECTRICITY

Hydroelectric dams with reservoirs can be operated to provide electricity at times of peak demand. Water is stored in the reservoir during periods of low demand and released when demand is high. The net effect is like the pumped storage, but without the pumping loss.

While a hydroelectric dam does not directly store energy from other generating units, it behaves equivalently by lowering output in periods of excess electricity from other sources. In this mode, dams are one of the most efficient forms of energy storage, because only the timing of its generation changes. Hydroelectric turbines have a start-up time on the order of a few minutes.

THERMAL

Seasonal thermal energy storage (STES) allows heat or cold to be used months after it was collected from waste energy or natural sources. The material can be stored in contained aquifers, clusters of boreholes in geological substrates such as sand or crystalline bedrock, in lined pits filled with gravel and water, or water-filled mines. STES projects often have paybacks in 4–6 years. For example, a heat pump can be run when there is surplus wind power available. It is used to raise the temperature to 80°C (176°F) for distribution. When surplus wind generated electricity is not available, a gas-fired boiler can be used.

LATENT HEAT THERMAL

Latent heat thermal energy storage systems (LHTES) work by transferring heat to or from a material to change its phase. A phase change is the melting, solidifying, vaporizing, or liquifying. Such a material is called a phase change material (PCM). Materials used in LHTESs often have a high latent heat so that at their specific temperature, the phase change absorbs a large amount of energy, much more than sensible heat. A steam accumulator is a type of LHTES where the phase change is between liquid and gas and uses the latent heat of vaporization of water.

ELECTROCHEMICAL

Rechargeable battery: A rechargeable battery comprises one or more electrochemical cells. It is known as a "secondary cell" because its electrochemical reactions are electrically reversible. Rechargeable batteries come in many shapes and sizes, ranging from button cells to megawatt grid systems.

Rechargeable batteries have lower total cost of use and environmental impact than non-rechargeable (disposable) batteries. Some rechargeable battery types are available in the same form factors as disposables. Rechargeable batteries have higher initial cost but can be recharged very cheaply and used many times.

Common rechargeable battery chemistries include:

1. *Lead–acid battery*: Lead acid batteries hold the largest market share of electric storage products. Lead–acid battery technology has been developed and used extensively. Upkeep requires minimal labor, and its cost is low. The battery's available energy capacity is subject to a quick discharge resulting in a low life span and low energy density. A single cell produces about 2V when charged. In the charged state, the metallic lead negative electrode and the lead sulfate positive electrode are immersed in a dilute sulfuric acid (H_2SO_4) electrolyte. In the discharge process, electrons are pushed out of the cell as lead sulfate is formed at the negative electrode while the electrolyte is reduced to water.
2. *Nickel–cadmium battery (NiCd)*: Uses nickel oxide hydroxide and metallic cadmium as electrodes. Cadmium is a toxic element and was banned for most uses by the European Union in 2004. Nickel–cadmium batteries have been almost completely replaced by nickel–metal hydride (NiMH) batteries.
3. *Nickel–metal hydride battery (NiMH)*: First commercial types were available in 1989. These are now a common consumer and industrial type. The battery has a hydrogen-absorbing alloy for the negative electrode instead of cadmium.
4. *Lithium-ion battery*: The choice in many consumer electronics and has one of the best energy-to-mass ratios and a very slow self-discharge when not in use.
5. *Lithium-ion polymer battery*: These batteries are light in weight and can be made in any shape desired.

Energy density is a measure of how much energy a battery can hold. The battery energy density is the proportion of energy that can be included in a unit of mass or volume. The energy density of a battery is generally expressed in two ways: (a) The gravimetric energy density of a battery is a measure of how much energy a battery contains in comparison to its mass, and it is typically expressed in Watt-hours/kilogram (Wh/kg), and (b) The volumetric energy density of a battery is a measure of how much energy a battery contains in comparison to its volume and is typically expressed in Watt-hours/liter (Wh/L).

The energy densities of the battery technologies have been improving steadily over the past 30 years. The lithium-ion batteries have high energy densities (about 100–265 Wh/kg or 250–670 Wh/L). In addition, Li-ion battery cells can deliver up to 3.6 V, three times higher than technologies such as Ni-Cd or Ni-MH. The higher the energy density, the longer will be its runtime. Lithium ion with cobalt cathodes offers the highest energy densities. Typical applications are cell phones, laptops, and digital cameras.

The electric vehicles currently use lithium-ion batteries, which have capacities in the range of 17 kWh (in Smart EQ) to 100 kWh (in Tesla Models S and X). Tesla's 100 kWh battery weighs about 625 kg. Tesla also sells a Powerwall system involving a 14 kWh lithium-ion battery home storage system. The battery can be charged by roof-mounted solar panels when the electricity demand during the daytime is low. The power from the battery can be used to provide electricity to the home when the solar panels do not produce sufficient power. The homeowners can also sell excess stored power to the utility companies.

Capacitor: A capacitor (originally known as a "condenser") is a passive two-terminal electrical component used to store energy electrostatically. Practical capacitors vary widely, but all contain at least two electrical conductors (plates) separated by a dielectric (i.e., insulator). A capacitor can store electric energy when disconnected from its charging circuit, so it can be used like a temporary battery, or like other types of rechargeable energy storage system. Capacitors are commonly used in electronic devices to maintain power supply while batteries are disconnected for change. (This prevents loss of information in volatile memory.) Conventional capacitors provide less than 360 joules per kilogram (J/kg), while a conventional alkaline battery has a density of 590 kJ/kg.

Capacitors store energy in an electrostatic field between their plates. Given a potential difference across the conductors (e.g., when a capacitor is attached across a battery), an electric field develops across the dielectric, causing positive charge (+Q) to collect on one plate and negative charge (−Q) to collect on the other plate. If a battery is attached to a capacitor for enough time, no current can flow through the capacitor. However, if an accelerating or alternating voltage is applied across the leads of the capacitor, a displacement current can flow. Besides capacitor plates, charge can also be stored in a dielectric layer.

Capacitance is greater given a narrower separation between conductors and when the conductors have a larger surface area. In practice, the dielectric between the plates emits a small amount of leakage current and has an electric field strength limit, known as the breakdown voltage. However, the effect of recovery of a dielectric after a high-voltage breakdown holds promise for a new generation of self-healing capacitors. The conductors and leads introduce undesired inductance and resistance.

Supercapacitor: They bridge the gap between conventional capacitors and rechargeable batteries. They store the most energy per unit volume or mass (energy density) among capacitors. They support up to 10,000 farads/1.2 Volt – up to 10,000 times that of electrolytic capacitors, but deliver or accept less than half as much power per unit time (power density). Supercapacitors have a charging time from 1 to 10 seconds, compared with 10–60 minutes to reach a full charge on a battery. While supercapacitors have specific energy and energy densities that are approximately 10% of batteries, their power density is generally 10–100 times greater. This results in much shorter charge/discharge cycles. The life of a supercapacitor can reach up to a 1 million cycles, whereas typical batteries have 500–1,000 charge–discharge cycles.

Supercapacitors have many applications, including: (a) low supply current for memory backup in static random-access memory (SRAM), and (b) power for cars, buses, trains, cranes, and elevators, including energy recovery from braking, short-term energy storage, and burst-mode power delivery.

OTHER CHEMICAL

Conversion of electricity to gaseous fuel: This technology involves converting electricity to a gaseous fuel, such as hydrogen or methane. There are three commercial methods that use electricity to reduce water into hydrogen and oxygen by means of electrolysis.

In the first method, hydrogen is injected into the natural gas grid or is used for transportation. The second method is to combine the hydrogen with carbon dioxide to produce methane using a methanation reaction such as the Sabatier reaction, or biological methanation, resulting in an extra energy conversion loss of 8%. The methane may then be fed into the natural gas grid. The third method uses the output gas of a wood gas generator or a biogas plant, after the biogas upgrader is mixed with the hydrogen from the electrolyzer, to upgrade the quality of the biogas.

Hydrogen: The element hydrogen can be a form of stored energy. Hydrogen can produce electricity via a hydrogen fuel cell. A *fuel cell* is a device that converts chemical potential energy (energy stored in molecular bonds) into electrical energy. A PEM (Proton Exchange Membrane) *cell* uses *hydrogen* gas (H_2) and oxygen gas (O_2) as *fuel*. The products of the reaction in the *cell* are water, electricity, and heat.

Hydrogen is an alternative fuel that has very high energy content by weight. It is locked up in enormous quantities in water, hydrocarbons, and other organic matter. Hydrogen can be produced from diverse, domestic resources including fossil fuels, biomass, and water electrolysis with wind, solar, or grid electricity. The environmental impact and energy efficiency of hydrogen depend on how it is produced. Energy losses involved in the hydrogen storage cycle come from the electrolysis of water, liquification or compression of the hydrogen, storage and transportation and conversion to electricity.

Methane: Methane is the simplest hydrocarbon with the molecular formula CH_4. Methane is more easily stored and transported than hydrogen. Storage and combustion infrastructure (pipelines, gasometers, power plants) are mature.

Synthetic natural gas (syngas or SNG) can be created in a multistep process, starting with hydrogen and oxygen. Hydrogen is then reacted with carbon dioxide in a Sabatier process, producing methane and water. Methane can be stored and later used to produce electricity. The resulting water is recycled, reducing the need for water. In the electrolysis stage, oxygen is stored for methane combustion in a pure oxygen environment at an adjacent power plant, eliminating nitrogen oxides.

Conversion of electric power to liquids: Conversion of electric power to liquid is similar to conversation of electric power to gas except that the hydrogen is converted into liquids such as methanol or ammonia. These are easier to handle than gases and require fewer safety precautions than hydrogen. They can be used for transportation, including aircraft, but also for industrial purposes or in the power sector.

Biofuels: Various biofuels such as biodiesel, vegetable oil, alcohol fuels, or biomass can replace fossil fuels. Various chemical processes can convert the carbon and hydrogen in coal, natural gas, plant and animal biomass, and organic wastes into short hydrocarbons suitable as replacements for existing hydrocarbon fuels. Examples are Fischer–Tropsch diesel, methanol, dimethyl ether, and syngas. This diesel source was used extensively in World War II in Germany, which faced limited access to crude oil supplies. South Africa produces most of the country's diesel from coal for similar reasons. A long-term oil price above US$35/bbl may make such large-scale synthetic liquid fuels economical.

CONCLUDING REMARKS

While a number of different possible energy storage technologies presented in the literature, the practical implementation of many of the technologies has not progressed due a number of reasons such as inefficiencies in energy conversion processes, low cost-effectiveness (i.e., lower benefit-to-cost ratios), safety concerns in their operations, high capital, and operating costs. Hydroelectric power plants, electrochemical batteries (rechargeable and non-rechargeable), flywheels, and capacitors have probably the highest number of applications.

REFERENCES

DOE. 2020. U.S. Department of Energy Global Energy Storage Database (Accessed January 1, 2021).
EIA. 2020. Electricity Storage. Website: https://www.epa.gov/energy/electricity-storage (Accessed: December 31, 2020).
Wikipedia. 2020. Energy Storage. Website: https://en.wikipedia.org/wiki/Energy_storage (Accessed: October 12, 2020).

15 Infrastructure Standardization for Electric Vehicles

INTRODUCTION

To accommodate the electric vehicles (EVs) within the nation's existing highway transportation and economic system, the infrastructure must be expanded to serve the needs of the growing EV users. The needs of the customers are to allow them to (a) recharge their EVs quickly, economically, and safely, (b) provide additional convenience and comfort features and facilities offered by existing gas stations and highway plazas such as restrooms, food shops, restaurants, and ATM machines. Or the charging stations should be integrated with the existing facilities that provide fuel filling and other convenience features.

WHAT IS INFRASTRUCTURE FOR THE ELECTRIC VEHICLES?

The term infrastructure normally includes all the systems and structures including access roads and utilities required to provide the required service (with all necessary functions) to the area it supports. For example, the electrical grid across a city, state, or country is the infrastructure based on the power generation, transmission, and distribution equipment and supporting facilities to meet electricity demand to the areas it supports. Similarly, the physical cabling, wi-fi communications and components making up the data network operating within a specific location are also the infrastructure for a business. Thus, the infrastructure can be classified into the following the three categories:

1. *Hard infrastructure*: These make up the physical systems that make it necessary to run a modern, industrialized nation. Examples of physical systems for EV-based transportation here will include vehicle charging equipment, electric power transmission and generating capabilities, data communication and processing equipment, associated buildings, roads, bridges, as well as the capital/assets needed to make them operational and user-friendly.
2. *Soft infrastructure*: These types of infrastructure make up institutions that help maintain the economy. These usually require jobs and people to provide and maintain the required services.
3. *Critical infrastructure*: These are assets defined by a government as being essential to the functioning of a society and economy, such as safe transportation and facilities to support-related functions such as surveillance and policing, shelter and heating, telecommunication, and public health.

DOI: 10.1201/9781003107514-19

From the viewpoint of convenience of users of the EVs, the customers should have similar, or even more advanced, facilities and services available when they use their EVs as they use their conventional internal combustion engine (ICE)-powered vehicles. The charging ports for the EV will be different from the fuel filling ports. Standardizing charging interfaces between EVs and charging stations is one of the topics covered in the infrastructure standardization. The equipment should be easy to use, i.e., not require many different types of interfaces, adapters, and other equipment/systems. The equipment and processes involved in charging and billing should be easy to learn and use. On the overall, the whole processes of charging should thus reduce user difficulties and frustrations and keep the charging time and costs as low as possible. Other issues-related standardization are transaction processing, e.g., billing and payments for services.

As the EV industry matures, interoperability (i.e., open communication and exchange of data; see next section) will continue to remain important to the development of vehicle hook-up and communications with the chargers, the grid, and the power company. We can expect to see further advancements in the communication systems between EVs and grid connected assets. Intelligent power supply is an emerging technical and commercial opportunity carrying many benefits across the vehicle electrification landscape and will undoubtedly demand open and harmonized communication standards (Bablo, 2016, DOE, 2020a).

While EV infrastructure is still a relatively new and quickly evolving space, regardless of which vehicles and charging innovations will capture and drive the market, open standardization will always be the optimal approach for rolling out the most future-proof and reliable charging infrastructure. The interoperability strategy can deliver the most returns for those who will fund, deploy, operate, and use these critical assets in the years to come – for the most convenient, reliable, and clean transportation.

As more EVs replace the ICE-powered vehicles, the demand for electricity will also increase. Additional electricity generating capacity along with rules and incentives to redistribute the available electric supply (e.g., by charging EV's at off-peak and nighttime hours at lower rates when electricity demand from other equipment is less) are possible solutions always under consideration (Gold, 2020). When too many drivers want to fast-charge their EVs, the load on the grid can increase substantially. And the power grid may need upgrading with additional power distribution and power generation capabilities. For example, since majority of people who own EVs usually charge them at homes, it would mean changes in substations and distribution circuits to accommodate multiple homes in a neighborhood drawing power to recharge the EV batteries.

Physical Infrastructure-Related Considerations
a. Number of charging stations
b. Parking spaces for vehicles to be charged including waiting areas for vehicles. Number of parking spaces will depend upon time required for full charge and traffic flow in the road network near the charging station.
c. Electrical grid (with available sources, their capacity, distribution network, and characteristics of power demands) in the charging area and in

the vicinity of the charging area. The capacity of the grid in kWh to provide electric power required to charge all vehicles at the charging stations. Backup electrical sources (e.g., standby generators) and emergency electrical storage systems in case of grid failures.

d. Restaurants and comfort stations with communication capabilities (e.g., telephone, wi-fi routers)
e. Access to tow trucks, service vehicles for disabled vehicles, firefighting equipment, and emergency vehicles

Critical Infrastructure
a. Local management personnel for service, safety, shops, and restaurants personnel
b. Level of automation involved in minimizing charging time and communicating charging progress with the customers

Soft Infrastructure
a. Number of full-time jobs by each needed job classification (e.g., maintenance personnel, electrical and computer, safety, and security personnel) dependent upon the charging facility
b. Incentives from local, state, and federal governments
c. Incorporation of customer convenience and satisfaction features with the whole charging experience (e.g., ability to reserve a charging spot within a specified time interval)

WHAT IS INTEROPERABILITY?

Interoperability, in the most universal terms, is the open communication and exchange of data between and among devices and/or software systems. Interoperability is a key issue for many industries such as software development, home automation, healthcare, telecommunications, and public safety. Interoperability has many benefits such as varied mobile devices to work across different cellular networks in different regions or when our communities rely on police and fire departments to communicate with each other using common platforms during emergencies.

The term is often used to describe multiple aspects of EV charging and can include form factor, communication, and compatible ratings among any of the following entities in a charging system:

a. The vehicle
b. The charging station hardware for conductive as well as the wireless power transfer technologies and battery swapping capabilities
c. The charging station connectivity software
d. The back-office or payment back end
e. The network operator
f. The energy management system
g. The power supply

CHARGING STATIONS AND CHARGERS FOR THE ELECTRIC VEHICLES

CHARGING STATIONS

There are essentially three types of charge stations that are widely in use, but they are designated differently around the globe.

In North America, there is the Portable EV Cord Set, which is a portable device intended to stay with the vehicle and be used with any convenient receptacle. In the IEC document, this was referred to as a Mode 2 cable assembly (IEC, 2017).

Second, in North America, there is a fastened in place charge station, which is a device that can be moved but is not intended to be moved often. It typically is "hung on a hook" in a residential or commercial garage for use with EVs that are parked in the vicinity. In the IEC document, this is a Mode 2 charge station.

The last device in North America is a fixed charge station, which is typically a public access charge station that is permanently fixed in one location and is hardwired. In the IEC document, this is designated as a Mode 3 charge station. The designations are not all that important to the discussion except for the portable EV cord set/Mode 2 cable assembly.

TYPES OF CHARGERS

To get the most out of a plug-in EV, it must be charged on a regular basis. Charging frequently maximizes the range of all-EVs and the electric-only miles of plug-in hybrid EVs. Drivers can charge at home, at work, or in public places. While most drivers do more than 80% of their charging at home and it is often the least expensive option, workplace and public charging can complement.

Conductive Charging

Conductive charging can be accomplished by delivering AC power to a vehicle with an on-board charger or by delivering DC power to a vehicle for directly charging the battery (no on-board charger needed). In some cases, the vehicle may contain both an AC and DC connection, and in such cases, the on-board charger is used, as necessary. AC delivery is done through what will be called a charging station. DC delivery is done through what will be called a quick charger. Charge stations and quick chargers, although different internally and perhaps using different communication protocols, all must be provided with a means to connect conductively to the vehicle. This is done through an output cable and an EV connector. Standards covering these products are also included in the discussion within the SAE and IEC committees. Lastly, internal to the charge station or the quick charger is a system of protection, designated and treated differently by different standards, that is provided to protect the user when recharging the electric vehicle. The standards associated with this protection system are also included in the discussion.

Charging an EV requires plugging into a charger connected to the electric grid, also called electric vehicle supply equipment (EVSE). There are three major categories of chargers, based on the maximum amount of power the charger provides to the battery from the grid (DOE, 2020c). They are SAE level 1, level 2, and level 3 chargers (SAE J1772, SAE, 2020).

a. *Alternate-current (AC) charging* (also known as level 1 or level 2)

In this system, an in-car inverter converts AC to direct current (DC), which then charges the battery at either level 1 (equivalent to a U.S. household outlet) or level 2 (240 volts). It operates at powers up to roughly 20 kW.

Level 1: Provides charging through a 120 V AC plug and does not require installation of additional charging equipment. Can deliver 2–5 miles of range per hour of charging. Most often used in homes, but sometimes used at workplaces.

Level 2: Provides charging through a 240 V (for residential) or 208 V (for commercial) plug and requires installation of additional charging equipment. It can deliver 10–20 miles of range per hour of charging, and it is used in homes, workplaces, and for public charging.

The kilowatt capacity of a charger determines the speed at which the battery receives electricity. AC level 1 and level 2 are most applicable for homes and workplaces because of the long parking (recharging) periods and their lower cost: a simple level 2 for a home can cost as little as $500.

Basic AC level 1 and level 2 power will overwhelmingly remain the dominant charging technology through 2030, providing from 60% to 80% of the energy consumed. Most of this charging will take place at homes, workplaces, and via slow-charge public stations.

b. *DC charging*

This charging system converts the AC from the grid to DC before it enters the car and charges the battery without the need for an inverter. Usually called direct-current fast charging or level 3, it operates at powers from 25 kW to more than 350 kW.

DC fast charge: It provides charging through 480 V AC input and requires highly specialized, high-powered equipment as well as special equipment in the vehicle itself. (Plug-in hybrid EVs typically do not have fast-charging capabilities.) It can deliver 60–80 miles of range in 20 minutes of charging. It is used most often in public charging stations, especially along heavy traffic corridors.

Charging times range from less than 30 minutes to 20 hours or more based on the type of EVSE, as well as the type of battery, how depleted it is, and its capacity. All-EVs typically have more battery capacity than plug-in hybrid EVs, so charging a fully depleted all-EV takes longer. Direct Current Fast Charging (DCFC) chargers are most applicable in situations where time matters, such as on highways and for fast public charging. DCFC will likely play a much larger role in China, which requires more public-charging infrastructure.

c. *Types of plugs*

Most modern chargers and vehicles have a standard connector and receptacle, called the SAE J1772 connector (SAE, 2020). Any vehicle with this plug receptacle can use any Level 1 or Level 2 EVSE. All major vehicle and charging system manufacturers support this standard, so the EV should be compatible with nearly all non-fast charging workplace and public chargers.

Fast charging currently does not have a consistent standard connector. The SAE International, an engineering standards-setting organization, has passed a standard for fast charging that adds high-voltage DC power contact pins to the SAE J1772 connector currently used for Level 1 and Level 2. This connector enables use of the same receptacle for all levels of charging and is available on certain models such as the Chevrolet Spark EV. However, other EVs (the Nissan Leaf and Mitsubishi i-MiEV in particular) use a different type of fast-charge connector called CHAdeMO. Fortunately, an increasing number of fast chargers have outlets for both SAE and CHAdeMO fast charging. Lastly, Tesla's Supercharger system can only be used by Tesla vehicles and is not compatible with vehicles from any other manufacturer. Tesla vehicles can use CHAdeMO connectors through a vehicle adapter.

Wireless Charging

In addition to the three types above, wireless charging uses an electromagnetic field to transfer electricity to an EV without a cord. This system uses electromagnetic waves to charge batteries. There is usually a charging pad connected to a wall socket and a plate attached to the vehicle. Current technologies align with level 2 chargers and can provide power up to 11 kW. The Department of Energy is supporting research to develop and improve wireless charging technology. Wireless chargers are currently available for use with certain vehicle models.

SAE International recently published two new documents, SAE J2954 and SAE J2847/6, which ensure a safe and efficient method for transferring power from charging stations to EVs. SAE J2954 Standard: "Wireless Power Transfer & Alignment for Light Duty Vehicles" establishes the first standard for wireless power transfer (WPT) for both EV and electric vehicle supply equipment (EVSE). This enables light-duty EVs and infrastructure to safely charge up to 11 kW, over an air gap of 10 inches (250 mm), achieving up to 94% efficiency.

The SAE J2954 standard is a gamechanger by giving a "cook-book" specification for developing both the vehicle and charging infrastructure for wireless power transfer, as one system, compatible to 11 kW. The SAE J2954 alignment technology gives additional parking assistance, even allowing for vehicles to park and charge themselves autonomously. The SAE task force coordinated with industry and international standards organizations to ensure global WPT harmonization.

CURRENT STATUS OF CHARGING STATIONS

Currently, the U.S. Department of Energy has a database of charging station locations in the United States (DOE, 2020b). The DOE website claims that about tens of thousands of EV charging stations are available in the United States. When a user inputs "Electric" (as the fuel type of his vehicle) and his location (city, state, and street address), the website provides a map and list of charging stations with available chargers and connectors.

ADVANTAGES AND DISADVANTAGES OF ELECTRIC VEHICLES

This section provides brief descriptions of advantages and disadvantages of EVs to help the reader in understanding issues related to the EVs.

Advantages

1. *Convenience*: The vehicle can be charged at home. Thus, no need to find and go to a gas station for recharging. EVs can be charged using wired plug-in chargers, which provide more flexibility as compared with the gas stations used for refueling ICE-equipped vehicles. They can be charged at home, in charging stations at work, in public, and on highways for long-distance trips. People tend to follow a charging hierarchy that starts at home.

2. *Reduced energy cost*: EVs are cheaper to use as compared with the ICE-equipped vehicles. They cost typically about 2–3 cents/mile as compared with about 10–15 cents/mile for gasoline-powered vehicles. Regenerative braking used in the EVs recovers energy used in braking. Thus, EVs are more energy-efficient. Further, most individual passenger cars remain parked for 8–12 hours at night, and home charging can be easy and often cheaper than charging elsewhere. Also, most charging can happen overnight when off-peak electricity prices are lower.

3. *Cleaner nonpolluting*: EVs do not generate exhaust gases. Thus, they will reduce environmental pollution and resulting health-related (e.g., respiratory and cardiovascular) problems. More renewable electric sources can be added to the grid as the power demand on the grid will increase.

4. *Low sound*: EVs produce much lower sound levels than vehicles with their exhaust systems. EVs do not produce sound when stopped. Thus, quieter under idling situation than ICE-equipped vehicles (except with stop–start capability).

5. *Higher acceleration*: Newer EVs have more powerful electric motors that can provide higher accelerations than most gasoline-powered vehicles. Thus, they are fun to drive.

6. *Package efficient*: EVs do not need transmissions and exhaust systems. The electric motor in an EV is much smaller than a gas engine of similar power, and it can be placed right between the driven wheels (a few companies are even building prototype EVs with motors inside the wheel hubs). An EV needs no transmission as such, so there is no need for the central tunnel that takes up so much space in rear-wheel-drive gas engine vehicles. There is also no need for an exhaust system, thermal shielding, or a catalytic converter. The EV batteries are typically packaged under vehicle floor between the front and rear wheels. This battery location also reduces center of gravity of the vehicle, which also improves vehicle stability and handling.

7. *Cheaper batteries*: As the demand for the EVs will increase, the prices of batteries are also more likely to decrease. The increased availability of cheaper batteries will also increase sales of EVs and the use batteries to store excess energy produced from renewable energy sources.

Disadvantages

1. *Longer recharging time*: The battery recharging time for an EV is longer than refueling time for an ICE-equipped vehicle.
2. *Too few recharging stations*: Recharging stations for EVs are currently not as easily available as the gas stations. This results into "range anxiety" for many drivers.
3. *Need for recharging stations*: Approximately 3%–6% of total miles driven involve long-distance trips that average more than 100 miles. Even with a full charge leaving home, most of today's EVs cannot make that round-trip without recharging. This makes the case for long-distance chargers. Drivers without chargers at home or work must charge in public; drivers who exceed their battery range on a given day may need to visit fast-charge stations; and drivers who forget to charge at home or who do not have home chargers must rely on other options, making the case for public charging.
4. *Battery weight and size*: The battery in EV weighs the most as compared with other systems in the vehicle. The battery volume is also large. Thus, batteries are typically packaged under vehicle floor between the front and rear wheels.
5. *Reduced battery output at low and high temperatures*: The battery output reduces as the ambient (battery operating) temperature decreases. Thus, the batteries provide lower range under low ambient temperatures. Further, greater use of electric heaters and air-circulating fans in colder weather conditions consumes additional electric power. Under higher ambient temperatures (e.g., summertime in southwestern states), air-conditioning systems are used heavily. The constant use of air-conditioning systems will also substantially reduce the range of EVs.

CONCLUDING REMARKS

This chapter covered some details associated with the problems in the standardization of infrastructures associated in implementation of future EVs. With improvements in EV technologies and reduction in anticipated prices of EVs, the demand and uses of EVs are more likely to increase rapidly over the next few decades. The increased energy efficiencies and reduction in pollution potential are also major advantages. Further with the implementation of additional features of smart grid technologies will also provide many benefits over the future years. As these new technologies are developing, coordination and integration between different systems and products used in the EV transportation and its infrastructure are needed for increased convenience. The tasks associated in the coordination and successful implementation of the EVs are huge because they are spread between many industry sectors such as transportation, energy generation, energy distribution, and modernization of the smart grid. Thus, coordinated efforts between the auto industry, energy companies, and government organization are needed.

REFERENCES

Bablo, J. 2016, June 19–22. Electric Vehicle Infrastructure Standardization. *EVS29 Symposium*, Montreal, Canada.

DOE. 2020a. *Vehicle Charging*. Energy Efficiency and Renewable Energy Office, Department of Energy. Website: https://www.energy.gov/eere/electricvehicles/vehicle-charging (Accessed: August 20, 2020).

DOE. 2020b. Electric Vehicle Charging Station Locations. Website: https://afdc.energy.gov/fuels/electricity_locations.html#/find/nearest?fuel=ELEC (Accessed: October 13, 2020).

DOE. 2020c. Developing Infrastructures to Charge Plug-in Electric Vehicles. Website: https://-afdc.energy.gov/fuels/electricity_infrastructure.html (Accessed: October 13, 2020).

Gold, R. 2020, September 26–27. For Electric Cars, California Needs a Bigger Grid. *The Wall Street Journal.*

International Electrotechnical Commission (IEC). 2017. Electric vehicle conductive charging system - Part 1: General requirements. IEC 61851-1:2017. Website: https://webstore.ansi.org/Standards/IEC/IEC61851Ed2017 (Assessed: September 18, 2021).

SAE. 2020. *SAE Standards*. *SAE Handbook*. Warrendale, PA: SAE International (formerly Society of Automotive Engineers, Inc.).

Section V

Applications of Methods: Examples and Illustrations

16 Selection of Power Generation Alternatives

INTRODUCTION

The objective of this chapter is to present a detailed cost–benefit analysis. The students in the author's course on risk analysis in energy system were asked to conduct a homework project in conducting a detailed cost–benefit analysis to evaluate five different alternatives to generate additional electric power for a power company. The analysis conducted for the project by Emmen (2020) was modified and edited here to provide a better understanding of the data used to determine the best alternative among the five given alternatives and four possible outcomes are described below.

THE PROBLEM

A detailed cost–benefit analysis was conducted to evaluate five different alternatives to supply an additional 500 MW of electricity distribution capacity to customers of a power company located in Michigan. It was assumed that the power company had decided to study the following five alternatives:

A1: Build new land-based wind turbines to generate the additional capacity
A2: Build a natural-gas-fueled power plant to generate the additional capacity
A3: Build a geothermal plant to generate the additional capacity
A4: Build a concentrating solar plant to generate the additional capacity
A5: Do not build a new plant and purchase the needed energy from other utility companies

The company had also decided to study the problem by considering the following four outcomes:

O1: The economy will grow at current 1% annual rate of increase in electricity demand.
O2: The economy will accelerate to 2% annual rate of increase in electricity demand.
O3: The economy will be stagnant with no increase in electricity demand.
O4: The economy will get worse and decelerate at 2% annual rate of electricity demand.

The following input conditions were assumed:

1. Probability of occurrences of outcomes O1, O2, O3, and O4 were 0.25, 0.50, 0.20, and 0.05, respectively.

DOI: 10.1201/9781003107514-21

2. The capacity factors of wind turbines (A1), natural-gas-fueled plant (A2), geothermal plant (A3), and concentrating solar plant (A4) were assumed to be 35%, 87%, 92%, and 25%, respectively.
3. The utility company can finance 100% of the project at 3.0% interest over the next 33 years (first 3 years using a construction loan with annual interest only payments, and 30 years of annual payments for principal and interest during the plant operation).
4. In the first year of the plant operation, the company will begin selling 400 MW at $0.13/kWh rate, and the rate will increase at 1% per year after the first year of operation.
5. Additional energy, if needed, can be purchased from other utility companies at a pre-negotiated fixed rate of $0.12/kWh over the 30 years of new plant operation (4th–33rd years after plant construction).
6. The utility company can gain benefit (or rebate from the state government) equal to 5% of economic impact generated during the construction and operation of the plant. The economic benefit will be estimated by using the Jobs and Economic Development Impact (JEDI) models developed by the National Renewable Energy Laboratory (NREL, 2020a).
7. The power plant will operate 24 hours/day and over all 365 days per year.
8. All power plants will be assumed to be built in a midwestern state of the United States.
9. The 3% discount rate will be used for present value calculations.
10. Annual safety costs were assumed to be as follows:
    a. Wind turbines at 2% of its capital
    b. Natural gas plant at 0.5% of its capital cost
    c. Geothermal plant at 1.75% of its capital cost
    d. Solar plant at 0.5% of its capital

METHODOLOGY

A decision matrix with five alternatives and four outcomes was set up to conduct a cost–benefit analysis (see Chapter 6 for the decision matrix). The values of the evaluation measures were the net present values of benefits minus costs and benefit-to-cost ratios computed over the total 33 years (first 3 years of plant construction time and next 30 years of electricity generation) for each combination of alternatives and outcomes.

The steps used in analysis are as follows:

1. *Cost data*: Most cost data for the different power plants were obtained from the available NREL developed JEDI models (NREL, 2020a). Specific inputs for this analysis included construction costs (annual payments of principal and interest), annual operation and maintenance (O&M) costs, property taxes, land lease or purchase costs, and the economic impact during construction as well as the normal operation time.
2. *Nameplate capacity*: The nameplate capacity for each power plant required for the application of JEDI models was calculated by dividing the 500 MW required capacity by the capacity factor (for each plant).

Outputs of JEDI Models

The power plant characteristics, costs data provided in the outputs of the JEDI models and other inputs for alternatives A1–A4 are summarized in Table 16.1.

The data for last alternative, A5, were not included in Table 16.1 because it did not involve a power plant. It involved purchasing the needed additional electric power from other utility companies at the wholesale price of $0.12/kWh and selling the electricity directly to the customers. The benefit will be from the profit generated by reselling the purchased electricity at a higher rate through the power company.

The outputs of the JEDI models for the four plants (wind, natural gas, geothermal, and concentrating solar) are provided in Tables 16.2–16.5. It should be noted that the dollar values of economic impact provided in the lower half of these tables are in millions and in 2009–2010 dollars. Thus, a multiplier of 1.20 was used to adjust the values to 2020 dollars, and these adjusted values are shown in Table 16.1.

COMPUTATIONS OF COSTS AND BENEFITS FOR COMBINATIONS OF ALTERNATIVES AND OUTCOMES

Similar approach for calculating costs and benefits for the five alternatives was used to allow comparisons. However, unique issues related to each plant technology such as safety and accident cost, cost of carbon, and economic impact were considered during cost calculations. The annual insurance cost for the different alternatives was set as a base percentage of the capital costs. Since annual insurance cost was estimated separately, this was assumed to be 3% for wind (A1), natural gas (A2), and geothermal (A3) of their respective construction costs. The annual insurance cost for concentrating solar was assumed to be 0.5% of the capital cost. The annual safety costs for the four plants were assumed to be 2.0%, 0.5%, 1.75%, and 0.5% of the capital costs for the wind, natural gas, geothermal, and solar plants, respectively. One additional parameter that was necessary for use in the concentrated solar alternative was the figure for solar radiation per day. This value was obtained based on the NREL PVWatts calculator using Detroit, MI, as the project location (NREL, 2020b). This low value of 4.6 kWh/m²/day increases the final size of the reflector array needed to generate the nameplate capacity and ends up being a significant disadvantage for this technology.

Five separate spreadsheets were prepared to estimate costs and benefits generated over the 33 years for the five alternatives. Tables 16.6–16.29 present the spreadsheets for the five alternatives. Since each spreadsheet is large, its width is divided into 4–5 subparts. The length of each subpart is 33 years (3 years for the construction of each plant and 30 years for the operation after the plant construction.)

The variables included in the columns of each subpart of each spreadsheet are identified by an alphanumeric code, such as A1B or A3O2C. The alphanumeric code is presented above each column of the tables. The first two alphanumeric positions of the codes indicate the alternative, such as A1 or A3. The last position indicates the sequential position of the variable by a letter (e.g., A, B, C, and so on) in each subpart of the spreadsheets. The codes for the outcomes, O1 through O5, are inserted after the codes for the alternatives for variables that are used to compute their values

TABLE 16.1

Power Plant Characteristics, Data Provided by the JEDI Models and Other Inputs

Variable	Power Plant			
	Wind	Natural Gas	Geothermal	Concentrating Solar
Capacity factor	35%	87%	82%	25%
Nameplate capacity (MW)	1429	575	543	2000
Total construction cost (from JEDI model)	$2,179,065,918.00	$718,750,000.00	$1,646,772,679.20	$11,411,402,657.11
Annual O&M cost (from JEDI model)	$56,505,388	$154,264,073	$488,648,427	$51,323,763
Economic impact during construction (Value added from JEDI model)	$472,262,848	$234,355,665	$337,090,403	$6,095,777,407
Economic impact during operation (Value added from JEDI model)	$72,701,016	$9,588,038	$18,456,904	$72,250,457
$/kWh for the cost to purchase electricity	$0.12	$0.12	$0.12	$0.12
$/kWh for the price of electricity in year 4 (increases at 1% per year)	$0.13	$0.13	$0.13	$0.13
Insurance cost (set at % of capital cost)	3%	3%	3%	0.50%
Assumed safety cost (% of capital cost)	2%	0.50%	1.75%	0.50%
Fuel heat rate: Btu/kWh		7000		
lbs of CO_2 per Billion Btu		117000		
Social cost of carbon: $/metric ton of CO_2		42		
Solar radiation: peak sun hours/day in kWh/m²/day				4.6

TABLE 16.2
JEDI Model Output for Wind Turbines

Wind Farm – Project Data Summary Based on User Modifications to Default Values

Project location	Michigan		
Year of construction	2020		
Total project size - nameplate capacity (MW)	1,429		
Number of projects (included in total)	1		
Turbine size (kW)	2,300		
Number of turbines	622		
Installed project cost ($/kW)	1,524.8887	1,453	without taxes
Annual O&M cost ($/kW)	39.541909	38	without taxes
Money value (Dollar year)	2017		
Total Installed project cost	2.179E+09		
Local spending	473,339,574		
Total annual operational expenses	404,718,676		
Direct operating and maintenance costs	56,505,388		
Local spending	12,446,609		
Other annual costs	348,213,288		
Local spending	33,054,582		
Debt and equity payments	0		
Property taxes	26,559,394		
Land lease	4,291,800		

Local Economic Impacts - Summary Results

	Jobs	Earnings	Output	Value Added
During construction period				
Project development and onsite labor impacts				
Construction and interconnection labor	620.8746642	36.691943		
Construction related services	65.21832265	5.8638505		
Total	686.0929868	42.555793	48.2028	43.62534801
Turbine and supply chain impacts	2,691.792872	146.15186	455.3964	223.7227258
Induced impacts	1,396.20239	71.767425	217.2063	126.2042994
Total impacts	4,774.088248	260.47508	720.8054	393.5523732
During operating years (annual)				
Onsite labor impacts	64.78133333	4.0681711	4.068171	4.068171083
Local revenue and supply chain impacts	132.1858955	7.415959	51.46346	40.4418203
Induced impacts	164.3109744	9.1427625	27.666	16.07418897
Total impacts	361.2782033	20.626893	83.19763	60.58418035

TABLE 16.3
JEDI Model Output for the Natural Gas Power Plant

Natural Gas Plant – Project Data Summary Based on User Modifications to Default Values

Project location	Michigan
Year construction starts	2020
Project size - nameplate capacity (MW)	575
Capacity factor (Percentage)	0.87
Heat rate (Btu per kWh)	7,000
Construction period (Months)	36
Plant construction cost ($/KW)	1,250
Cost of fuel ($/mmbtu)	4.46
Produced locally (Percent)	0
Fixed operations and maintenance cost ($/kW)	8.25
Variable operations and maintenance cost ($/MWh)	2.9
Money value (Dollar year)	2010
Project construction cost	718,750,000
Local spending	200,108,455.9
Total annual operational expenses	235,187,141.4
Direct operating and maintenance costs	154,264,072.8
Local spending	5,722,829.906
Other annual costs	80,923,068.59
Local spending	5,993,967.088
Debt and equity payments	0
Property taxes	0

Local Economic Impacts – Summary results

	Jobs	Earnings	Output	Value Added
During construction period				
Project development and onsite labor impact	686.8482	79.31199939	113.0974265	83.51486283
Construction and interconnection labor	686.8482	79.31199939		
Construction related services	0	0		
Power generation and supply chain impacts	732.6295	33.28221827	118.1967369	62.99761245
Induced impacts	641.868	29.21439371	85.73755893	48.78391227
Total impacts	2061.346	141.8086114	317.0317223	195.2963875
During operating years (annual)				
Onsite labor impacts	29.12262	1.842613007	1.842613007	1.842613007
Local revenue and supply chain impacts	49.87764	2.511620616	8.216450254	4.491400995
Induced impacts	21.89673	0.991677747	2.910500099	1.656017714
Total impacts	100.897	5.34591137	12.96956336	7.990031716

TABLE 16.4
JEDI Model Output for Geothermal Power Plant

Geothermal Plant – Project Data Summary (based on advanced analysis)

Project location	Michigan		
Year of construction	2020	Installed project cost ($/kW)	$3,294
Construction period (months)	36	Annual O&M cost ($/kW)	$60
Nominal plant size (MW net output)	500.0	Money value (Dollar year)	2010
Nameplate capacity (MW)	611.3	Installed project cost	$1,646,772,679
Technology	Hydrothermal	Local spending	$330,380,446
Plant type	Binary	Total annual operational expenses	$488,648,427
Resource temperature (°C)	200	Direct operating and maintenance costs	$54,967,548
Resource depth (m)	2250	Local spending	$10,132,438
Number of exploration wells	2	Other annual costs	$433,680,879
Number of production wells	93	Local spending	$0
Number of injection wells	47	Debt and equity payments	$0
Production well flow rate (kg/s)	80	Property taxes	$0
Well flow (lb/h)	634,931	Land lease	$0
Total flow (lb/h)	59,174,789		

Local Economic Impacts – Summary Results

	Jobs (FTE)	Earnings (Millions of $2010)	Output (Millions of $2010)	Value Added (Millions of $2010)
During construction period				
Project development and onsite labor impacts	2,488	$86.00	$128.99	$86.00
Construction labor	2,488	$86.00		
Construction related services	0	$0.00		
Turbine and supply chain impacts	1,404	$83.14	$337.49	$136.08
Induced impacts	796	$35.23	$103.40	$58.83
Total impacts	4,689	$204.36	$569.87	$280.91
During operating years (annual)				
Onsite labor impacts	67	$10.13	$10.13	$10.13
Local revenue and supply chain impacts	27	$1.45	$14.34	$2.37
Induced impacts	37	$1.72	$5.06	$2.88
Total impacts	130	$13.31	$29.53	$15.38

TABLE 16.5
JEDI Model Output for the Concentrating Solar Power Plant

CSP Trough Plant – Project Data
  Summary

Project location	Michigan
Year of construction	2020
Solar direct normal resource (kWh/m²/day)	4.65
Project size - nameplate capacity (MW)	2,000
Solar field aperture area (square meters)	15,527,272.73
Plant capacity factor	0.25
Construction cost ($/KW)	5,705.701329
Annual direct O&M cost ($/KW)	25.6618817
Money value (Dollar year)	2009
Project construction cost	11,411,402,657
Local spending	4,530,440,839
Local sales tax	477,790,036.2
Total annual operational expenses	1,159,445,573
Direct operating and maintenance costs	51,323,763.41
Local spending	32,009,895.31
Other annual costs	1,108,121,810
Local spending	2,531,345.953
Debt and equity payments	0
Property taxes	0
Sales tax	1,023,259.589
Insurance	0
Land purchase	1,508,086.364
Land lease	0

Local Economic Impacts
  – Summary Results

	Jobs	Earnings	Output	Value Added
During construction period		Million 2009$	Million 2009$	Million 2009$
Project development and onsite labor impacts	18,457.52176	1,957.124637	2,681.143431	2,194.564126
Construction and interconnection labor	12,146.50845	1,691.936729		
Construction related services	6,311.013317	265.1879079		
Equipment and supply chain impacts	15,901.55649	836.8926558	3,580.200318	1,798.78957
Induced impacts	13,585.82869	600.1545465	1,761.335826	1,086.460809
Total impacts	47,944.90694	3,394.171839	8,022.679576	5,079.814505

(Continued)

TABLE 16.5 (Continued)
JEDI Model Output for the Concentrating Solar Power Plant

During operating years (annual)	Annual Jobs	Annual Earnings Million 2009$	Annual Output Million 2009$	Annual Output Million 2009$
Onsite labor impacts	460.748873	24.24189739	24.24189739	24.24189739
Local revenue and supply chain impacts	242.0313258	12.54135713	42.81216883	23.27094401
Induced impacts	150.560803	6.781112547	19.90219406	12.69587276
Total impacts	853.3410017	43.56436708	86.95626028	60.20871416

related to outcomes (e.g., A3O2B indicates a variable for alternative A3, outcome O2, and second variable (B) in the subpart related to A3O2).

ALTERNATIVE 1: BUILDING A 1429 MW WIND TURBINE PLANT

Tables 16.6–16.10 present the spreadsheet for alternative 1 of building a wind turbine plant that can generate 500 MW of power. The computations are made over 33 years (see variable A1A in Table 16.6). During the first three years, the plant is assumed to be under construction and therefore O&M costs (variable A1D) are set to zero. The safety costs (variable A1H) were assumed to be 2% of the capital costs and constant throughout the construction and operation years. The variable, A1B shows the price at which the electricity is sold to the retail customers. During the fourth year (i.e., first year of plant operation), the price is assumed to be $0.13/kwh. After the fourth year, the price is assumed to increase 1% every year.

Determining the Costs

All cost values provided in the tables are shown as negative values. The total capital cost of the plant obtained from the JEDI model was $2,178,065,918. It is assumed that the utility company obtained a construction loan at 3% interest rate over the 3 years (1/3rd the capital cost per year), and only the interest was to be paid over the 3 years. From 4th through 33 years, the annual payment to cover principal and interest during each year was included in the value of variable A1C. The annual land lease cost obtained from the JEDI model was included in the variable A1F. The total cost for each year was shown in column A1I. The present value of total costs for each year was shown in the column A1J. It was calculated by assuming 3% discount rate, and the interest period used is shown in column A1A. The sum of present values (shown in column A1J) over all the 33 years of the costs shown at the bottom of the column is $6,213,899,811.

The variable A1O1A shows the amount of annual kW capacity needed to sell the required amount of energy in years 4–33 (see Table 16.7). When the plant functions at the beginning of fourth year, it is expected to produce 400 MW capacity. To meet

TABLE 16.6
Alternative 1 (Wind Turbines) Costs over 33 Years

A1A	A1B	A1C	A1D	A1E	A1F	A1G	A1H	A1I	A1J
Years	Elect Selling Price ($/kwh)	Capital Cost Related Interest or Principal + Interest	Operating & Maintenance Cost	Property Taxes	Land Lease Cost	Insurance Cost	Safety Cost	Total Cost	Present Value of Total Cost
1		-$21,790,659	$0	-$26,559,394	-$4,291,800	-$65,371,978	-$43,581,318	-$161,595,149	-$161,595,149
2		-$43,581,318	$0	-$26,559,394	-$4,291,800	-$65,371,978	-$43,581,318	-$183,385,808	-$178,044,474
3		-$65,371,978	$0	-$26,559,394	-$4,291,800	-$65,371,978	-$43,581,318	-$205,176,467	-$193,398,499
4	$0.130	-$111,174,329	-$56,505,388	-$26,559,394	-$4,291,800	-$65,371,978	-$43,581,318	-$307,484,207	-$281,391,607
5	$0.131	-$111,174,329	-$56,505,388	-$26,559,394	-$4,291,800	-$65,371,978	-$43,581,318	-$307,484,207	-$273,195,735
6	$0.133	-$111,174,329	-$56,505,388	-$26,559,394	-$4,291,800	-$65,371,978	-$43,581,318	-$307,484,207	-$265,238,578
7	$0.134	-$111,174,329	-$56,505,388	-$26,559,394	-$4,291,800	-$65,371,978	-$43,581,318	-$307,484,207	-$257,513,183
8	$0.135	-$111,174,329	-$56,505,388	-$26,559,394	-$4,291,800	-$65,371,978	-$43,581,318	-$307,484,207	-$250,012,799
9	$0.137	-$111,174,329	-$56,505,388	-$26,559,394	-$4,291,800	-$65,371,978	-$43,581,318	-$307,484,207	-$242,730,872
10	$0.138	-$111,174,329	-$56,505,388	-$26,559,394	-$4,291,800	-$65,371,978	-$43,581,318	-$307,484,207	-$235,661,041
11	$0.139	-$111,174,329	-$56,505,388	-$26,559,394	-$4,291,800	-$65,371,978	-$43,581,318	-$307,484,207	-$228,797,127
12	$0.141	-$111,174,329	-$56,505,388	-$26,559,394	-$4,291,800	-$65,371,978	-$43,581,318	-$307,484,207	-$222,133,133
13	$0.142	-$111,174,329	-$56,505,388	-$26,559,394	-$4,291,800	-$65,371,978	-$43,581,318	-$307,484,207	-$215,663,236
14	$0.144	-$111,174,329	-$56,505,388	-$26,559,394	-$4,291,800	-$65,371,978	-$43,581,318	-$307,484,207	-$209,381,783
15	$0.145	-$111,174,329	-$56,505,388	-$26,559,394	-$4,291,800	-$65,371,978	-$43,581,318	-$307,484,207	-$203,283,284
16	$0.146	-$111,174,329	-$56,505,388	-$26,559,394	-$4,291,800	-$65,371,978	-$43,581,318	-$307,484,207	-$197,362,412
17	$0.148	-$111,174,329	-$56,505,388	-$26,559,394	-$4,291,800	-$65,371,978	-$43,581,318	-$307,484,207	-$191,613,992
18	$0.149	-$111,174,329	-$56,505,388	-$26,559,394	-$4,291,800	-$65,371,978	-$43,581,318	-$307,484,207	-$186,033,002
19	$0.151	-$111,174,329	-$56,505,388	-$26,559,394	-$4,291,800	-$65,371,978	-$43,581,318	-$307,484,207	-$180,614,565

(Continued)

TABLE 16.6 (Continued)
Alternative 1 (Wind Turbines) Costs over 33 Years

A1A	A1B	A1C	A1D	A1E	A1F	A1G	A1H	A1I	A1J
Years	Elect Selling Price ($/kwh)	Capital Cost Related Interest or Principal + Interest	Operating & Maintenance Cost	Property Taxes	Land Lease Cost	Insurance Cost	Safety Cost	Total Cost	Present Value of Total Cost
20	$0.152	−$111,174,329	−$56,505,388	−$26,559,394	−$4,291,800	−$65,371,978	−$43,581,318	−$307,484,207	−$175,353,947
21	$0.154	−$111,174,329	−$56,505,388	−$26,559,394	−$4,291,800	−$65,371,978	−$43,581,318	−$307,484,207	−$170,246,550
22	$0.155	−$111,174,329	−$56,505,388	−$26,559,394	−$4,291,800	−$65,371,978	−$43,581,318	−$307,484,207	−$165,287,913
23	$0.157	−$111,174,329	−$56,505,388	−$26,559,394	−$4,291,800	−$65,371,978	−$43,581,318	−$307,484,207	−$160,473,702
24	$0.159	−$111,174,329	−$56,505,388	−$26,559,394	−$4,291,800	−$65,371,978	−$43,581,318	−$307,484,207	−$155,799,710
25	$0.160	−$111,174,329	−$56,505,388	−$26,559,394	−$4,291,800	−$65,371,978	−$43,581,318	−$307,484,207	−$151,261,855
26	$0.162	−$111,174,329	−$56,505,388	−$26,559,394	−$4,291,800	−$65,371,978	−$43,581,318	−$307,484,207	−$146,856,170
27	$0.163	−$111,174,329	−$56,505,388	−$26,559,394	−$4,291,800	−$65,371,978	−$43,581,318	−$307,484,207	−$142,578,806
28	$0.165	−$111,174,329	−$56,505,388	−$26,559,394	−$4,291,800	−$65,371,978	−$43,581,318	−$307,484,207	−$138,426,025
29	$0.167	−$111,174,329	−$56,505,388	−$26,559,394	−$4,291,800	−$65,371,978	−$43,581,318	−$307,484,207	−$134,394,199
30	$0.168	−$111,174,329	−$56,505,388	−$26,559,394	−$4,291,800	−$65,371,978	−$43,581,318	−$307,484,207	−$130,479,805
31	$0.170	−$111,174,329	−$56,505,388	−$26,559,394	−$4,291,800	−$65,371,978	−$43,581,318	−$307,484,207	−$126,679,422
32	$0.172	−$111,174,329	−$56,505,388	−$26,559,394	−$4,291,800	−$65,371,978	−$43,581,318	−$307,484,207	−$122,989,730
33	$0.173	−$111,174,329	−$56,505,388	−$26,559,394	−$4,291,800	−$65,371,978	−$43,581,318	−$307,484,207	−$119,407,505
									−$6,213,899,811

TABLE 16.7
Alternative 1 under Outcome 1 Revenue Calculations

A1O1A	A1O1B	A1O1C	A1O1D	A1O1E	A1O1F	A1O1G	A1O1H	A1O1I
			Outcome 1 : 1% Increase in Demand/Year					
kW Sold	kWh Sold (24/7 Operation)	kW Produced	kWh Produced (24/7 Operation)	kWh Bought from Other Utility cos.	Income from Sale of Electricity	Economic Impact Benefit	Total Revenue	Present Value of Total Revenue
0	0	0	0	0	$0	$23,613,142	$23,613,142	$23,613,142
0	0	0	0	0	$0	$23,613,142	$23,613,142	$22,925,381
0	0	0	0	0	$0	$23,613,142	$23,613,142	$22,257,651
400,000	3,504,000,000	400,000	3,504,000,000	0	$455,520,000	$3,635,051	$459,155,051	$420,191,915
404,000	3,539,040,000	404,000	3,539,040,000	0	$464,675,952	$3,635,051	$468,311,003	$416,088,260
408,040	3,574,430,400	408,040	3,574,430,400	0	$474,015,939	$3,635,051	$477,650,989	$412,025,939
412,120	3,610,174,704	412,120	3,610,174,704	0	$483,543,659	$3,635,051	$487,178,710	$408,004,500
416,242	3,646,276,451	416,242	3,646,276,451	0	$493,262,887	$3,635,051	$496,897,937	$404,023,495
420,404	3,682,739,216	420,404	3,682,739,216	0	$503,177,471	$3,635,051	$506,812,521	$400,082,484
424,608	3,719,566,608	424,608	3,719,566,608	0	$513,291,338	$3,635,051	$516,926,389	$396,181,034
428,854	3,756,762,274	428,854	3,756,762,274	0	$523,608,494	$3,635,051	$527,243,544	$392,318,713
433,143	3,794,329,897	433,143	3,794,329,897	0	$534,133,024	$3,635,051	$537,768,075	$388,495,099
437,474	3,832,273,195	437,474	3,832,273,195	0	$544,869,098	$3,635,051	$548,504,149	$384,709,774
441,849	3,870,595,927	441,849	3,870,595,927	0	$555,820,967	$3,635,051	$559,456,018	$380,962,325
446,267	3,909,301,887	446,267	3,909,301,887	0	$566,992,968	$3,635,051	$570,628,019	$377,252,344
450,730	3,948,394,906	450,730	3,948,394,906	0	$578,389,527	$3,635,051	$582,024,578	$373,579,429
455,237	3,987,878,855	455,237	3,987,878,855	0	$590,015,157	$3,635,051	$593,650,207	$369,943,183
459,790	4,027,757,643	459,790	4,027,757,643	0	$601,874,461	$3,635,051	$605,509,512	$366,343,213
464,388	4,068,035,220	464,388	4,068,035,220	0	$613,972,138	$3,635,051	$617,607,189	$362,779,132
469,031	4,108,715,572	469,031	4,108,715,572	0	$626,312,978	$3,635,051	$629,948,029	$359,250,558

(Continued)

TABLE 16.7 (Continued)
Alternative 1 under Outcome 1 Revenue Calculations

A1O1A	A1O1B	A1O1C	A1O1D	A1O1E	A1O1F	A1O1G	A1O1H	A1O1I
			Outcome 1 : 1% Increase in Demand/Year					
kW Sold	kWh Sold (24/7 Operation)	kW Produced	kWh Produced (24/7 Operation)	kWh Bought from Other Utility cos.	Income from Sale of Electricity	Economic Impact Benefit	Total Revenue	Present Value of Total Revenue
473,722	4,149,802,728	473,722	4,149,802,728	0	$638,901,869	$3,635,051	$642,536,920	$355,757,114
478,459	4,191,300,755	478,459	4,191,300,755	0	$651,743,796	$3,635,051	$655,378,847	$352,298,425
483,244	4,233,213,762	483,244	4,233,213,762	0	$664,843,847	$3,635,051	$668,478,897	$348,874,124
488,076	4,275,545,900	488,076	4,275,545,900	0	$678,207,208	$3,635,051	$681,842,259	$345,483,846
492,957	4,318,301,359	492,957	4,318,301,359	0	$691,839,173	$3,635,051	$695,474,224	$342,127,233
497,886	4,361,484,373	497,886	4,361,484,373	0	$705,745,140	$3,635,051	$709,380,191	$338,803,930
502,865	4,405,099,216	500,000	4,380,000,000	25,099,216	$716,918,712	$3,635,051	$720,553,762	$334,116,980
507,894	4,449,150,208	500,000	4,380,000,000	69,150,208	$726,103,198	$3,635,051	$729,738,249	$328,520,173
512,973	4,493,641,711	500,000	4,380,000,000	113,641,711	$735,525,682	$3,635,051	$739,160,733	$323,069,973
518,103	4,538,578,128	500,000	4,380,000,000	158,578,128	$745,191,482	$3,635,051	$748,826,533	$317,761,815
523,284	4,583,963,909	500,000	4,380,000,000	203,963,909	$755,106,028	$3,635,051	$758,741,078	$312,591,278
528,516	4,629,803,548	500,000	4,380,000,000	249,803,548	$765,274,863	$3,635,051	$768,909,914	$307,554,081
533,802	4,676,101,583	500,000	4,380,000,000	296,101,583	$775,703,650	$3,635,051	$779,338,701	$302,646,080
								$10,990,632,626
							Benefits − Costs	$4,776,732,815
							Benefits/Costs	1.77

the needs under Outcome 1 after the first year of operation, the plant should increase it capacity by 1% per year. This 1% increase is shown in variable A1O1A from 5th through 33rd year. The next variable, A1O1B presents total kWh of energy generated by the plant by assuming that it produces constant level of capacity shown by variable A1O1A over 24 hours/day and 365 days/year. Since the plant is designed to produce maximum capacity of 500 MW, additional electric energy needed will be purchased by the company when the demand exceeds 500 MW. The additional capacity need is shown in variable A1O1E. The revenue generated by selling the energy (A1O1B at rate shown in A1B) minus additional energy purchased (shown in A1O1E) at $0.12/kwh from other utility companies is shown in variable A1O1F.

The economic impact estimates of the new plant during plant construction and during its operation are provided by the JEDI model. It was assumed that the utility company can gain benefit (or rebate from the state government) equal to 5% of economic impact generated during the construction and operation of the plant. This benefit is shown in column A1O1G.

Column A1O1H presents total revenue, which is the sum of values in columns A1O1F and A1O1G.The column A1O1I presents the present value of A1O1H at 3% discount rate. The present values for all years (1–33) are added, and the sum is shown at the bottom of the column A1O1I as $ $10,990,632,626. Thus, the present value of the benefits minus cost shown at the bottom of column A1O1I is $4,776,732,815 and the benefit-to-cost ratio is 1.77.

Table 16.8 similarly presents data for Alternative 1 under Outcome 2. Outcome 2 represents 2% increase in electricity demand. The column A1O2A shows that during the first year of operation, the power generating capacity is at 400 MW. Under outcome 2, the capacity will increase by 2% per year. The increased capacity is shown in column A1O2A for years 5–33.

Table 16.9 and 16.10 similarly present data for Alternative 1 under Outcomes 3 and 4, respectively.

Determining the Potential Benefit for Each Outcome

Benefits for the different power plants include the income (revenue) from sale of electricity and the 5% kickback from economic input during construction and operation. Each outcome has a change in energy demand over the operational life. As this changes, the amount of energy sold to the customer will also change. When the demand is increasing, it is possible to reach a point where the plant cannot generate enough energy to meet the demand – at which point the remaining energy that needs to be sold to the consumer is purchased and then resold for minimal profit. For the income due to sale of electricity, this will be common to all the alternatives. Thus, under each alternative the same amount of electricity for a given outcome. The 5% kickback from economic input is a result of jobs created to construct and operate each power plant. The values for economic impact during construction or operation are outputs from the JEDI model.

Determining Totals and Present Value of Costs and Benefits

Total values for costs and benefits (revenues) were calculated for each year. These yearly totals were then used to calculate present values for costs and revenues.

TABLE 16.8

Alternative 1 under Outcome 2 Revenue Calculations

Outcome 2 : 2% Increase in Demand/Year

A1O2A	A1O2B	A1O2C	A1O2D	A1O2E	A1O2F	A1O2G	A1O2H	A1O2I
kW Sold	kWh Sold (24/7 Operation)	kW Produced	kWh Produced (24/7 Operation)	kWh Bought	Income from Sale of Electricity	Economic Impact	Total Revenue	Present Value of Total Revenue
0	0	0	0	0	0	$23,613,142	$23,613,142	$23,613,142
0	0	0	0	0	0	$23,613,142	$23,613,142	$22,925,381
0	0	0	0	0	0	$23,613,142	$23,613,142	$22,257,651
400,000	3,504,000,000	400,000	3,504,000,000	0	455,520,000	$3,635,051	$459,155,051	$420,191,915
408,000	3,574,080,000	408,000	3,574,080,000	0	469,276,704	$3,635,051	$472,911,755	$420,175,969
416,160	3,645,561,600	416,160	3,645,561,600	0	483,448,860	$3,635,051	$487,083,911	$420,162,861
424,483	3,718,472,832	424,483	3,718,472,832	0	498,049,016	$3,635,051	$501,684,067	$420,152,508
432,973	3,792,842,289	432,973	3,792,842,289	0	513,090,096	$3,635,051	$516,725,147	$420,144,831
441,632	3,868,699,134	441,632	3,868,699,134	0	528,585,417	$3,635,051	$532,220,468	$420,139,752
450,465	3,946,073,117	450,465	3,946,073,117	0	544,548,697	$3,635,051	$548,183,748	$420,137,197
459,474	4,024,994,579	459,474	4,024,994,579	0	560,994,067	$3,635,051	$564,629,118	$420,137,091
468,664	4,105,494,471	468,664	4,105,494,471	0	577,936,088	$3,635,051	$581,571,139	$420,139,365
478,037	4,187,604,360	478,037	4,187,604,360	0	595,389,758	$3,635,051	$599,024,809	$420,143,949
487,598	4,271,356,448	487,598	4,271,356,448	0	613,370,529	$3,635,051	$617,005,580	$420,150,776
497,350	4,356,783,577	497,350	4,356,783,577	0	631,894,319	$3,635,051	$635,529,370	$420,159,782
507,297	4,443,919,248	507,297	4,443,919,248	0	650,977,527	$3,635,051	$654,612,578	$420,170,904
517,443	4,532,797,633	517,443	4,532,797,633	0	670,637,049	$3,635,051	$674,272,099	$420,184,080
527,792	4,623,453,586	527,792	4,623,453,586	0	690,890,287	$3,635,051	$694,525,338	$420,199,252
538,347	4,715,922,657	538,347	4,715,922,657	0	711,755,174	$3,635,051	$715,390,225	$420,216,360
549,114	4,810,241,111	549,114	4,810,241,111	0	733,250,180	$3,635,051	$736,885,231	$420,235,351

(*Continued*)

TABLE 16.8 (Continued)
Alternative 1 under Outcome 2 Revenue Calculations

Outcome 2 : 2% Increase in Demand/Year

A1O2A	A1O2B	A1O2C	A1O2D	A1O2E	A1O2F	A1O2G	A1O2H	A1O2I
kW Sold	kWh Sold (24/7 Operation)	kW Produced	kWh Produced (24/7 Operation)	kWh Bought	Income from Sale of Electricity	Economic Impact	Total Revenue	Present Value of Total Revenue
560,097	4,906,445,933	560,097	4,906,445,933	0	755,394,336	$3,635,051	$759,029,387	$420,256,168
571,298	5,004,574,852	500,000	4,380,000,000	624,574,852	703,258,263	$3,635,051	$706,893,313	$379,989,989
582,724	5,104,666,349	500,000	4,380,000,000	724,666,349	714,749,142	$3,635,051	$718,384,193	$374,919,323
594,379	5,206,759,676	500,000	4,380,000,000	826,759,676	726,709,557	$3,635,051	$730,344,608	$370,059,587
606,267	5,310,894,869	500,000	4,380,000,000	930,894,869	739,156,140	$3,635,051	$742,791,191	$365,404,046
618,392	5,417,112,766	500,000	4,380,000,000	1,037,112,766	752,106,071	$3,635,051	$755,741,122	$360,946,169
630,760	5,525,455,022	500,000	4,380,000,000	1,145,455,022	765,577,100	$3,635,051	$769,212,151	$356,679,619
643,375	5,635,964,122	500,000	4,380,000,000	1,255,964,122	779,587,565	$3,635,051	$783,222,616	$352,598,250
656,242	5,748,683,405	500,000	4,380,000,000	1,368,683,405	794,156,410	$3,635,051	$797,791,461	$348,696,101
669,367	5,863,657,073	500,000	4,380,000,000	1,483,657,073	809,303,202	$3,635,051	$812,938,253	$344,967,390
682,755	5,980,930,214	500,000	4,380,000,000	1,600,930,214	825,048,155	$3,635,051	$828,683,206	$341,406,509
696,410	6,100,548,818	500,000	4,380,000,000	1,720,548,818	841,412,148	$3,635,051	$845,047,199	$338,008,017
710,338	6,222,559,795	500,000	4,380,000,000	1,842,559,795	858,416,747	$3,635,051	$862,051,797	$334,766,638
								$11,900,335,922
							Benefits – Costs	$5,686,436,111
							Benefits/Costs	1.92

TABLE 16.9
Alternative 1 under Outcome 3 Revenue Calculations

A1O3A	A1O3B	A1O3C	A1O3D	A1O3E	A1O3F	A1O3G	A1O3H	A1O3I
					Outcome 3: No Change in Demand/Year			
kW Sold	kWh Sold (24/7 Operation)	kW Produced	kWh Produced (24/7 Operation)	kWh Bought	Income from Sale of Electricity	Economic Impact	Total Revenue	Present Value of Total Revenue
0	0	0	0	0	$0	$23,613,142	$23,613,142	$23,613,142
0	0	0	0	0	$0	$23,613,142	$23,613,142	$22,925,381
0	0	0	0	0	$0	$23,613,142	$23,613,142	$22,257,651
400,000	3,504,000,000	400,000	3,504,000,000	0	$455,520,000	$3,635,051	$459,155,051	$420,191,915
400,000	3,504,000,000	400,000	3,504,000,000	0	$460,075,200	$3,635,051	$463,710,251	$412,000,552
400,000	3,504,000,000	400,000	3,504,000,000	0	$464,675,952	$3,635,051	$468,311,003	$403,969,185
400,000	3,504,000,000	400,000	3,504,000,000	0	$469,322,712	$3,635,051	$472,957,762	$396,094,680
400,000	3,504,000,000	400,000	3,504,000,000	0	$474,015,939	$3,635,051	$477,650,989	$388,373,965
400,000	3,504,000,000	400,000	3,504,000,000	0	$478,756,098	$3,635,051	$482,391,149	$380,804,027
400,000	3,504,000,000	400,000	3,504,000,000	0	$483,543,659	$3,635,051	$487,178,710	$373,381,915
400,000	3,504,000,000	400,000	3,504,000,000	0	$488,379,096	$3,635,051	$492,014,146	$366,104,732
400,000	3,504,000,000	400,000	3,504,000,000	0	$493,262,887	$3,635,051	$496,897,937	$358,969,642
400,000	3,504,000,000	400,000	3,504,000,000	0	$498,195,515	$3,635,051	$501,830,566	$351,973,862
400,000	3,504,000,000	400,000	3,504,000,000	0	$503,177,471	$3,635,051	$506,812,521	$345,114,666
400,000	3,504,000,000	400,000	3,504,000,000	0	$508,209,245	$3,635,051	$511,844,296	$338,389,378
400,000	3,504,000,000	400,000	3,504,000,000	0	$513,291,338	$3,635,051	$516,926,389	$331,795,378
400,000	3,504,000,000	400,000	3,504,000,000	0	$518,424,251	$3,635,051	$522,059,302	$325,330,097
400,000	3,504,000,000	400,000	3,504,000,000	0	$523,608,494	$3,635,051	$527,243,544	$318,991,015
400,000	3,504,000,000	400,000	3,504,000,000	0	$528,844,579	$3,635,051	$532,479,629	$312,775,663
400,000	3,504,000,000	400,000	3,504,000,000	0	$534,133,024	$3,635,051	$537,768,075	$306,681,619
400,000	3,504,000,000	400,000	3,504,000,000	0	$539,474,355	$3,635,051	$543,109,405	$300,706,510

(Continued)

TABLE 16.9 (Continued)
Alternative 1 under Outcome 3 Revenue Calculations

A1O3A	A1O3B	A1O3C	A1O3D	A1O3E	A1O3F	A1O3G	A1O3H	A1O3I
			Outcome 3: No Change in Demand/Year					
kW Sold	kWh Sold (24/7 Operation)	kW Produced	kWh Produced (24/7 Operation)	kWh Bought	Income from Sale of Electricity	Economic Impact	Total Revenue	Present Value of Total Revenue
400,000	3,504,000,000	400,000	3,504,000,000	0	$544,869,098	$3,635,051	$548,504,149	$294,848,008
400,000	3,504,000,000	400,000	3,504,000,000	0	$550,317,789	$3,635,051	$553,952,840	$289,103,833
400,000	3,504,000,000	400,000	3,504,000,000	0	$555,820,967	$3,635,051	$559,456,018	$283,471,748
400,000	3,504,000,000	400,000	3,504,000,000	0	$561,379,177	$3,635,051	$565,014,227	$277,949,560
400,000	3,504,000,000	400,000	3,504,000,000	0	$566,992,968	$3,635,051	$570,628,019	$272,535,120
400,000	3,504,000,000	400,000	3,504,000,000	0	$572,662,898	$3,635,051	$576,297,949	$267,226,320
400,000	3,504,000,000	400,000	3,504,000,000	0	$578,389,527	$3,635,051	$582,024,578	$262,021,095
400,000	3,504,000,000	400,000	3,504,000,000	0	$584,173,422	$3,635,051	$587,808,473	$256,917,419
400,000	3,504,000,000	400,000	3,504,000,000	0	$590,015,157	$3,635,051	$593,650,207	$251,913,306
400,000	3,504,000,000	400,000	3,504,000,000	0	$595,915,308	$3,635,051	$599,550,359	$247,006,810
400,000	3,504,000,000	400,000	3,504,000,000	0	$601,874,461	$3,635,051	$605,509,512	$242,196,021
400,000	3,504,000,000	400,000	3,504,000,000	0	$607,893,206	$3,635,051	$611,528,257	$237,479,069
								$9,683,113,287
							Benefits - Costs	$3,469,213,475
							Benefits/Costs	1.56

TABLE 16.10
Alternative1 under Outcome 4 Revenue Calculations

Outcome 4 : 2% Decrease in Demand/Year

A1O4A kW Sold	A1O4B kWh Sold (24/7 Operation)	A1O4C kW Produced	A1O4D kWh Produced (24/7 Operation)	A1O4E kWh Bought	A1O4E Income from Sale of Electricity	A1O4F Economic Impact	A1O4G Total Revenue	A1O4H Present Value of Total Revenue
0	0	0	0	0	$0	$23,613,142	$23,613,142	$23,613,142
0	0	0	0	0	$0	$23,613,142	$23,613,142	$22,925,381
0	0	0	0	0	$0	$23,613,142	$23,613,142	$22,257,651
400,000	3,504,000,000	400,000	3,504,000,000	0	$455,520,000	$3,635,051	$459,155,051	$420,191,915
392,000	3,433,920,000	392,000	3,433,920,000	0	$450,873,696	$3,635,051	$454,508,747	$403,825,135
384,160	3,365,241,600	384,160	3,365,241,600	0	$446,274,784	$3,635,051	$449,909,835	$388,096,176
376,477	3,297,936,768	376,477	3,297,936,768	0	$441,722,782	$3,635,051	$445,357,832	$372,980,173
368,947	3,231,978,033	368,947	3,231,978,033	0	$437,217,209	$3,635,051	$440,852,260	$358,453,230
361,568	3,167,338,472	361,568	3,167,338,472	0	$432,757,594	$3,635,051	$436,392,644	$344,492,383
354,337	3,103,991,703	354,337	3,103,991,703	0	$428,343,466	$3,635,051	$431,978,517	$331,075,563
347,250	3,041,911,868	347,250	3,041,911,868	0	$423,974,363	$3,635,051	$427,609,414	$318,181,563
340,305	2,981,073,631	340,305	2,981,073,631	0	$419,649,824	$3,635,051	$423,284,875	$305,790,000
333,499	2,921,452,159	333,499	2,921,452,159	0	$415,369,396	$3,635,051	$419,004,447	$293,881,289
326,829	2,863,023,115	326,829	2,863,023,115	0	$411,132,628	$3,635,051	$414,767,679	$282,436,607
320,293	2,805,762,653	320,293	2,805,762,653	0	$406,939,075	$3,635,051	$410,574,126	$271,437,865
313,887	2,749,647,400	313,887	2,749,647,400	0	$402,788,297	$3,635,051	$406,423,348	$260,867,681
307,609	2,694,654,452	307,609	2,694,654,452	0	$398,679,856	$3,635,051	$402,314,907	$250,709,349
301,457	2,640,761,363	301,457	2,640,761,363	0	$394,613,322	$3,635,051	$398,248,373	$240,946,815
295,428	2,587,946,136	295,428	2,587,946,136	0	$390,588,266	$3,635,051	$394,223,317	$231,564,650
289,519	2,536,187,213	289,519	2,536,187,213	0	$386,604,266	$3,635,051	$390,239,316	$222,548,029

(Continued)

TABLE 16.10 (Continued)
Alternative1 under Outcome 4 Revenue Calculations

Outcome 4 : 2% Decrease in Demand/Year

A1O4A	A1O4B	A1O4C	A1O4D	A1O4E	A1O4E	A1O4F	A1O4G	A1O4H
kW Sold	kWh Sold (24/7 Operation)	kW Produced	kWh Produced (24/7 Operation)	kWh Bought	Income from Sale of Electricity	Economic Impact	Total Revenue	Present Value of Total Revenue
283,729	2,485,463,469	283,729	2,485,463,469	0	$382,660,902	$3,635,051	$386,295,953	$213,882,703
278,054	2,435,754,199	278,054	2,435,754,199	0	$378,757,761	$3,635,051	$382,392,812	$205,554,979
272,493	2,387,039,115	272,493	2,387,039,115	0	$374,894,432	$3,635,051	$378,529,482	$197,551,698
267,043	2,339,298,333	267,043	2,339,298,333	0	$371,070,508	$3,635,051	$374,705,559	$189,860,215
261,702	2,292,512,366	261,702	2,292,512,366	0	$367,285,589	$3,635,051	$370,920,640	$182,468,376
256,468	2,246,662,119	256,468	2,246,662,119	0	$363,539,276	$3,635,051	$367,174,327	$175,364,503
251,339	2,201,728,877	251,339	2,201,728,877	0	$359,831,176	$3,635,051	$363,466,226	$168,537,373
246,312	2,157,694,299	246,312	2,157,694,299	0	$356,160,898	$3,635,051	$359,795,948	$161,976,198
241,386	2,114,540,413	241,386	2,114,540,413	0	$352,528,056	$3,635,051	$356,163,107	$155,670,615
236,558	2,072,249,605	236,558	2,072,249,605	0	$348,932,270	$3,635,051	$352,567,321	$149,610,660
231,827	2,030,804,613	231,827	2,030,804,613	0	$345,373,161	$3,635,051	$349,008,212	$143,786,762
227,190	1,990,188,521	227,190	1,990,188,521	0	$341,850,355	$3,635,051	$345,485,406	$138,189,721
222,647	1,950,384,750	222,647	1,950,384,750	0	$338,363,481	$3,635,051	$341,998,532	$132,810,696
								$7,581,539,099
							Benefits − Costs	$1,367,639,288
							Benefits/Costs	1.22

The present values for each of the costs and revenue totals were calculated per year using the following equation: $P = F[1/(1+i)^n]$; where F is the future value, i is the annual interest rate, and n is the year.

All the present values of cost and benefit over the life of the project were summed separately to give a cumulative costs and cumulative revenue numbers for comparison of each combination of alternatives and outcomes.

Calculating the Profit and Benefit-to-Cost Ratio for Each Alternative Across all Outcomes

The cumulative present values of costs and benefits were used to compute profit for each alternative under each outcome.

Profit was calculated for each outcome where, $Profit = [Total\ Benefit - Total\ Cost]$.

Additionally, the cost-to-benefit ratio can be calculated, where $ratio = [Total\ Benefit\ /\ Total\ Cost]$. Both variables are useful for comparing each alternative to one another for a single outcome. In general, a positive value for profit indicates that the benefits will outweigh the costs. Similarly, a benefit-to-cost ratio value greater than 1.0 indicates that the benefits will outweigh the cost over the life of the project.

Compiling the Results into a Decision Matrix and Compare Alternatives Based on Their Expected Values of Profits and Benefit-to-Cost Ratios

Comparison of each alternative and outcome was done with a decision matrix. Profit as well as benefit-to-cost ratio can be compared and showed a similar result (see Table 16.30).

ALTERNATIVE 2: BUILDING A 575 MW NATURAL GAS-FUELED POWER PLANT

Tables 16.11–16.15 present the spreadsheet for alternative 2 of building a natural gas-fueled power plant that can generate 500 MW of power.

ALTERNATIVE 3: BUILDING A 543 MW GEOTHERMAL POWER PLANT

Tables 16.16–16.20 present the spreadsheet for alternative 3 of building a geothermal power plant that can generate 500 MW of power.

ALTERNATIVE 4: BUILDING A 2000 MW CONCENTRATING SOLAR POWER PLANT

Tables 16.21–16.25 present the spreadsheet for alternative 4 of building a concentrating solar power plant that can generate 500 MW of power.

ALTERNATIVE 5: NOT BUILDING A NEW POWER PLANT

Tables 16.26–16.29 present the spreadsheet for alternative 5 of not building a new power plant and reselling power purchased from other power companies.

TABLE 16.11

Alternative 2 Natural Gas Plant Costs

A2A	A2B	A2C	A2D	A2E	A2F	A2G
	Elect Selling Price ($/kwh)	Capital Cost Related Interest or Principal + Interest	Operating & Maintenance Cost	Property Taxes	Insurance Cost	Safety Costs
Years						
1		$(7,187,500)	$0	$(17,968,750)	$(21,562,500)	$(3,593,750)
2		$(14,375,000)	$0	$(17,968,750)	$(21,562,500)	$(3,593,750)
3		$(21,562,500)	$0	$(17,968,750)	$(21,562,500)	$(3,593,750)
4	$0.130	$(36,670,093)	$(154,264,073)	$(17,968,750)	$(21,562,500)	$(3,593,750)
5	$0.131	$(36,670,093)	$(154,264,073)	$(17,968,750)	$(21,562,500)	$(3,593,750)
6	$0.133	$(36,670,093)	$(154,264,073)	$(17,968,750)	$(21,562,500)	$(3,593,750)
7	$0.134	$(36,670,093)	$(154,264,073)	$(17,968,750)	$(21,562,500)	$(3,593,750)
8	$0.135	$(36,670,093)	$(154,264,073)	$(17,968,750)	$(21,562,500)	$(3,593,750)
9	$0.137	$(36,670,093)	$(154,264,073)	$(17,968,750)	$(21,562,500)	$(3,593,750)
10	$0.138	$(36,670,093)	$(154,264,073)	$(17,968,750)	$(21,562,500)	$(3,593,750)
11	$0.139	$(36,670,093)	$(154,264,073)	$(17,968,750)	$(21,562,500)	$(3,593,750)
12	$0.141	$(36,670,093)	$(154,264,073)	$(17,968,750)	$(21,562,500)	$(3,593,750)
13	$0.142	$(36,670,093)	$(154,264,073)	$(17,968,750)	$(21,562,500)	$(3,593,750)
14	$0.144	$(36,670,093)	$(154,264,073)	$(17,968,750)	$(21,562,500)	$(3,593,750)
15	$0.145	$(36,670,093)	$(154,264,073)	$(17,968,750)	$(21,562,500)	$(3,593,750)
16	$0.146	$(36,670,093)	$(154,264,073)	$(17,968,750)	$(21,562,500)	$(3,593,750)
17	$0.148	$(36,670,093)	$(154,264,073)	$(17,968,750)	$(21,562,500)	$(3,593,750)
18	$0.149	$(36,670,093)	$(154,264,073)	$(17,968,750)	$(21,562,500)	$(3,593,750)
19	$0.151	$(36,670,093)	$(154,264,073)	$(17,968,750)	$(21,562,500)	$(3,593,750)
20	$0.152	$(36,670,093)	$(154,264,073)	$(17,968,750)	$(21,562,500)	$(3,593,750)
21	$0.154	$(36,670,093)	$(154,264,073)	$(17,968,750)	$(21,562,500)	$(3,593,750)

(Continued)

TABLE 16.11 (Continued)
Alternative 2 Natural Gas Plant Costs

A2A Years	A2B Elect Selling Price ($/kwh)	A2C Capital Cost Related Interest or Principal + Interest	A2D Operating & Maintenance Cost	A2E Property Taxes	A2F Insurance Cost	A2G Safety Costs
22	$0.155	$(36,670,093)	$(154,264,073)	$(17,968,750)	$(21,562,500)	$(3,593,750)
23	$0.157	$(36,670,093)	$(154,264,073)	$(17,968,750)	$(21,562,500)	$(3,593,750)
24	$0.159	$(36,670,093)	$(154,264,073)	$(17,968,750)	$(21,562,500)	$(3,593,750)
25	$0.160	$(36,670,093)	$(154,264,073)	$(17,968,750)	$(21,562,500)	$(3,593,750)
26	$0.162	$(36,670,093)	$(154,264,073)	$(17,968,750)	$(21,562,500)	$(3,593,750)
27	$0.163	$(36,670,093)	$(154,264,073)	$(17,968,750)	$(21,562,500)	$(3,593,750)
28	$0.165	$(36,670,093)	$(154,264,073)	$(17,968,750)	$(21,562,500)	$(3,593,750)
29	$0.167	$(36,670,093)	$(154,264,073)	$(17,968,750)	$(21,562,500)	$(3,593,750)
30	$0.168	$(36,670,093)	$(154,264,073)	$(17,968,750)	$(21,562,500)	$(3,593,750)
31	$0.170	$(36,670,093)	$(154,264,073)	$(17,968,750)	$(21,562,500)	$(3,593,750)
32	$0.172	$(36,670,093)	$(154,264,073)	$(17,968,750)	$(21,562,500)	$(3,593,750)
33	$0.173	$(36,670,093)	$(154,264,073)	$(17,968,750)	$(21,562,500)	$(3,593,750)

TABLE 16.12
Alternative 2 Natural Gas Plant under Outcome 1

Outcome 1 : 1% Increase in Demand/Year

A2O1A O1 Social Cost of Carbon	A2O1B O1 Total Cost	A2O1C O1 Present Value of Total Cost	A2O1D kW sold	A2O1E kWh Sold (24/7 Operation)	A2O1F kW Produced	A2O1G kWh Produced (24/7 Operation)	A2O1H Btu	A2O1I Pounds of CO2	A2O1J kWh Bought	A2O1K Income from Sale of Electricity	A2O1L Economic Impact Benefit	A2O1M Total Revenue	A2O1N Present Value of Total Revenue
$0	-$50,312,500	-$50,312,500	0	0	0	0	0	0	0	$0	$11,717,783	$11,717,783	$11,717,783
$0	-$57,500,000	-$55,825,243	0	0	0	0	0	0	0	$0	$11,717,783	$11,717,783	$11,376,489
$0	-$64,687,500	-$60,974,173	0	0	0	0	0	0	0	$0	$11,717,783	$11,717,783	$11,045,135
-$54,671,712	-$288,730,878	-$264,229,655	400,000	3,504,000,000	400,000	3,504,000,000	24,528,000,000,000	2,869,776,000	0	$455,520,000	$479,402	$455,999,402	$417,304,049
-$55,218,429	-$289,277,595	-$257,019,396	404,000	3,539,040,000	404,000	3,539,040,000	24,773,280,000,000	2,898,473,760	0	$464,675,952	$479,402	$465,155,354	$413,284,507
-$55,770,614	-$289,829,779	-$250,009,713	408,040	3,574,430,400	408,040	3,574,430,400	25,021,012,800,000	2,927,458,498	0	$474,015,939	$479,402	$474,495,341	$409,303,849
-$56,328,320	-$290,387,485	-$243,194,947	412,120	3,610,174,704	412,120	3,610,174,704	25,271,222,928,000	2,956,733,083	0	$483,543,659	$479,402	$484,023,061	$405,361,693
-$56,891,603	-$290,950,768	-$236,569,600	416,242	3,646,276,451	416,242	3,646,276,451	25,523,935,157,280	2,986,300,413	0	$493,262,887	$479,402	$493,742,288	$401,457,664
-$57,460,519	-$291,519,685	-$230,128,331	420,404	3,682,739,216	420,404	3,682,739,216	25,779,174,508,853	3,016,163,418	0	$503,177,471	$479,402	$503,656,872	$397,591,386
-$58,035,124	-$292,094,290	-$223,865,951	424,608	3,719,566,608	424,608	3,719,566,608	26,036,966,253,941	3,046,325,052	0	$513,291,338	$479,402	$513,770,740	$393,762,491
-$58,615,476	-$292,674,641	-$217,777,419	428,854	3,756,762,274	428,854	3,756,762,274	26,297,335,916,481	3,076,788,302	0	$523,608,494	$479,402	$524,087,896	$389,970,614
-$59,201,630	-$293,260,796	-$211,857,838	433,143	3,794,329,897	433,143	3,794,329,897	26,560,309,275,646	3,107,556,185	0	$534,133,024	$479,402	$534,612,426	$386,215,391
-$59,793,647	-$293,852,812	-$206,102,450	437,474	3,832,273,195	437,474	3,832,273,195	26,825,912,368,402	3,138,631,747	0	$544,869,098	$479,402	$545,348,500	$382,496,466
-$60,391,583	-$294,450,748	-$200,506,632	441,849	3,870,595,927	441,849	3,870,595,927	27,094,171,492,086	3,170,018,065	0	$555,820,967	$479,402	$556,300,369	$378,813,482
-$60,995,499	-$295,054,664	-$195,065,892	446,267	3,909,301,887	446,267	3,909,301,887	27,365,113,207,007	3,201,718,245	0	$566,992,968	$479,402	$567,472,370	$375,166,088
-$61,605,454	-$295,664,619	-$189,775,868	450,730	3,948,394,906	450,730	3,948,394,906	27,638,764,339,077	3,233,735,428	0	$578,389,527	$479,402	$578,868,929	$371,553,938
-$62,221,508	-$296,280,674	-$184,632,321	455,237	3,987,878,855	455,237	3,987,878,855	27,915,151,982,468	3,266,072,782	0	$590,015,157	$479,402	$590,494,558	$367,976,687
-$62,843,723	-$296,902,889	-$179,631,131	459,790	4,027,757,643	459,790	4,027,757,643	28,194,303,502,292	3,298,733,510	0	$601,874,461	$479,402	$602,353,863	$364,433,993
-$63,472,161	-$297,531,326	-$174,768,297	464,388	4,068,035,220	464,388	4,068,035,220	28,476,246,537,315	3,331,720,845	0	$613,972,138	$479,402	$614,451,540	$360,925,521
-$64,106,882	-$298,166,048	-$170,039,931	469,031	4,108,715,572	469,031	4,108,715,572	28,761,009,002,689	3,365,038,053	0	$626,312,978	$479,402	$626,792,380	$357,450,936
-$64,747,951	-$298,807,117	-$165,442,256	473,722	4,149,802,728	473,722	4,149,802,728	29,048,619,092,715	3,398,688,434	0	$638,901,869	$479,402	$639,381,271	$354,009,907
-$65,395,431	-$299,454,596	-$160,971,601	478,459	4,191,300,755	478,459	4,191,300,755	29,339,105,283,643	3,432,675,318	0	$651,743,796	$479,402	$652,223,198	$350,602,108

(Continued)

TABLE 16.12 (Continued)
Alternative 2 Natural Gas Plant under Outcome 1

Outcome 1 : 1% Increase in Demand/Year

	A2O1A	A2O1B	A2O1C	A2O1D	A2O1E	A2O1F	A2O1G	A2O1H	A2O1I	A2O1J	A2O1K	A2O1L	A2O1M	A2O1N
	O1 Social Cost of Carbon	O1 Total Cost	O1 Present Value of Total Cost	kW sold	kWh Sold (24/7 Operation)	kW Produced	kWh Produced (24/7 Operation)	Btu	Pounds of CO_2	kWh Bought	Income from Sale of Electricity	Economic Impact Benefit	Total Revenue	Present Value of Total Revenue
	−$66,049,385	−$300,108,550	−$156,624,402	483,244	4,233,213,762	483,244	4,233,213,762	29,632,496,336,479	3,467,002,071	0	$664,843,847	$479,402	$665,323,249	$347,227,214
	−$66,709,879	−$300,769,044	−$152,397,193	488,076	4,275,545,900	488,076	4,275,545,900	29,928,821,299,844	3,501,672,092	0	$678,207,208	$479,402	$678,686,610	$343,884,905
	−$67,376,978	−$301,436,143	−$148,286,608	492,957	4,318,301,359	492,957	4,318,301,359	30,228,109,512,842	3,536,688,813	0	$691,839,173	$479,402	$692,318,575	$340,574,863
	−$68,050,747	−$302,109,913	−$144,289,377	497,886	4,361,484,373	497,886	4,361,484,373	30,530,390,607,971	3,572,055,701	0	$705,745,140	$479,402	$706,224,542	$337,296,774
	−$68,339,640	−$302,398,806	−$140,220,732	502,865	4,405,099,216	500,000	4,380,000,000	30,660,000,000,000	3,587,220,000	25,099,216	$716,918,712	$479,402	$717,398,113	$332,653,723
	−$68,339,640	−$302,398,806	−$136,136,633	507,894	4,449,150,208	500,000	4,380,000,000	30,660,000,000,000	3,587,220,000	69,150,208	$726,103,198	$479,402	$726,582,600	$327,099,535
	−$68,339,640	−$302,398,806	−$132,171,488	512,973	4,493,641,711	500,000	4,380,000,000	30,660,000,000,000	3,587,220,000	113,641,711	$735,525,682	$479,402	$736,005,084	$321,690,712
	−$68,339,640	−$302,398,806	−$128,321,833	518,103	4,538,578,128	500,000	4,380,000,000	30,660,000,000,000	3,587,220,000	158,578,128	$745,191,482	$479,402	$745,670,884	$316,422,727
	−$68,339,640	−$302,398,806	−$124,584,304	523,284	4,583,963,909	500,000	4,380,000,000	30,660,000,000,000	3,587,220,000	203,963,909	$755,106,028	$479,402	$755,585,430	$311,291,193
	−$68,339,640	−$302,398,806	−$120,955,635	528,516	4,629,803,548	500,000	4,380,000,000	30,660,000,000,000	3,587,220,000	249,803,548	$765,274,863	$479,402	$765,754,265	$306,291,862
	−$68,339,640	−$302,398,806	−$117,432,655	533,802	4,676,101,583	500,000	4,380,000,000	30,660,000,000,000	3,587,220,000	296,101,583	$775,703,650	$479,402	$776,183,052	$301,420,624
			−$5,630,122,005											$10,897,674,310
													Benefit − Costs	$5,267,552,305
													Benefits/Costs	1.94

TABLE 16.13

Alternative 2 Natural Gas Plant under Outcome 2

A2O2A	A2O2B	A2O2C	A2O2D	A2O2E	A2O2F	A2O2G	A2O2H	A2O2I	A2O2J	A2O2K	A2O2L	A2O2M	A2O2N
							Outcome 2 : 2% Increase in Demand/Year						
O2 Social Cost of Carbon	O2 Total Cost	O2 Present Value of Total Cost	kW Sold	kWh Sold (24/7 Operation)	kW Produced	kWh Produced (24/7 Operation)	Btu	Pounds of CO_2	kWh Bought	Income from Sale of Electricity	Economic Impact Benefit	Total Revenue	Present Value of Total Revenue
$0	−$46,718,750	−$46,718,750	0	0	0	0		0	0	0	11,717,783	11,717,783	11,717,783
$0	−$53,906,250	−$52,336,165	0	0	0	0		0	0	0	11,717,783	11,717,783	11,376,489
$0	−$61,093,750	−$57,586,719	0	0	0	0		0	0	0	11,717,783	11,717,783	11,045,135
−$54,671,712	−$285,137,128	−$260,940,864	400,000	3,504,000,000	400,000	3,504,000,000	24,528,000,000,000	2,869,776,000	0	455,520,000	479,402	455,999,402	417,304,049
−$55,765,147	−$286,230,562	−$254,312,147	408,000	3,574,080,000	408,000	3,574,080,000	25,018,560,000,000	2,927,171,520	0	469,276,704	479,402	469,756,106	417,372,216
−$56,880,449	−$287,345,865	−$247,867,067	416,160	3,645,561,600	416,160	3,645,561,600	25,518,931,200,000	2,985,714,950	0	483,448,860	479,402	483,928,262	417,440,770
−$58,018,058	−$288,483,474	−$241,600,368	424,483	3,718,472,832	424,483	3,718,472,832	26,029,309,824,000	3,045,429,249	0	498,049,016	479,402	498,528,418	417,509,702
−$59,178,420	−$289,643,835	−$235,506,944	432,973	3,792,842,289	432,973	3,792,842,289	26,549,896,020,480	3,106,337,834	0	513,090,096	479,402	513,569,498	417,578,999
−$60,361,988	−$290,827,403	−$229,581,838	441,632	3,868,699,134	441,632	3,868,699,134	27,080,893,940,890	3,168,464,591	0	528,585,417	479,402	529,064,819	417,648,654
−$61,569,228	−$292,034,643	−$223,820,237	450,465	3,946,073,117	450,465	3,946,073,117	27,622,511,819,707	3,231,833,883	0	544,548,697	479,402	545,028,099	417,718,654
−$62,800,612	−$293,266,028	−$218,217,467	459,474	4,024,994,579	459,474	4,024,994,579	28,174,962,056,102	3,296,470,561	0	560,994,067	479,402	561,473,469	417,788,992
−$64,056,625	−$294,522,040	−$212,768,988	468,664	4,105,494,471	468,664	4,105,494,471	28,738,461,297,224	3,362,399,972	0	577,936,088	479,402	578,415,490	417,859,657
−$65,337,757	−$295,803,173	−$207,470,394	478,037	4,187,604,360	478,037	4,187,604,360	29,313,230,523,168	3,429,647,971	0	595,389,758	479,402	595,869,160	417,930,640
−$66,644,512	−$297,109,928	−$202,317,403	487,598	4,271,356,448	487,598	4,271,356,448	29,899,495,133,631	3,498,240,931	0	613,370,529	479,402	613,849,931	418,001,933
−$67,977,402	−$298,442,818	−$197,305,861	497,350	4,356,783,577	497,350	4,356,783,577	30,497,485,036,304	3,568,205,749	0	631,894,319	479,402	632,373,721	418,073,527
−$69,336,951	−$299,802,366	−$192,431,730	507,297	4,443,919,248	507,297	4,443,919,248	31,107,434,737,030	3,639,569,864	0	650,977,527	479,402	651,456,929	418,145,413
−$70,723,690	−$301,189,105	−$187,691,093	517,443	4,532,797,633	517,443	4,532,797,633	31,729,583,431,771	3,712,361,262	0	670,637,049	479,402	671,116,451	418,217,584
−$72,138,163	−$302,603,579	−$183,080,142	527,792	4,623,453,586	527,792	4,623,453,586	32,364,175,100,406	3,786,608,487	0	690,890,287	479,402	691,369,689	418,290,032
−$73,580,927	−$304,046,342	−$178,595,182	538,347	4,715,922,657	538,347	4,715,922,657	33,011,458,602,414	3,862,340,656	0	711,755,174	479,402	712,234,576	418,362,749
−$75,052,545	−$305,517,961	−$174,232,624	549,114	4,810,241,111	549,114	4,810,241,111	33,671,687,774,463	3,939,587,470	0	733,250,180	479,402	733,729,582	418,435,728
−$76,553,596	−$307,019,011	−$169,988,983	560,097	4,906,445,933	560,097	4,906,445,933	34,345,121,529,952	4,018,379,219	0	755,394,336	479,402	755,873,738	418,508,962
−$68,339,640	−$298,805,056	−$160,622,441	571,298	5,004,574,852	500,000	4,380,000,000	30,660,000,000,000	3,587,220,000	624,574,852	703,258,263	479,402	703,737,665	378,293,672

(*Continued*)

TABLE 16.13 (Continued)
Alternative 2 Natural Gas Plant under Outcome 2

Outcome 2 : 2% Increase in Demand/Year

	A2O2A	A2O2B	A2O2C	A2O2D	A2O2E	A2O2F	A2O2G	A2O2H	A2O2I	A2O2J	A2O2K	A2O2L	A2O2M	A2O2N
	O2 Social Cost of Carbon	O2 Total Cost	O2 Present Value of Total Cost	kW Sold	kWh Sold (24/7 Operation)	kW Produced	kWh Produced (24/7 Operation)	Btu	Pounds of CO_2	kWh Bought	Income from Sale of Electricity	Economic Impact Benefit	Total Revenue	Present Value of Total Revenue
	-$68,339,640	-$298,805,056	-$155,944,118	582,724	5,104,666,349	500,000	4,380,000,000	30,660,000,000,000	3,587,220,000	724,666,349	714,749,142	479,402	715,228,544	373,272,413
	-$68,339,640	-$298,805,056	-$151,402,056	594,379	5,206,759,676	500,000	4,380,000,000	30,660,000,000,000	3,587,220,000	826,759,676	726,709,557	479,402	727,188,959	368,460,645
	-$68,339,640	-$298,805,056	-$146,992,288	606,267	5,310,894,869	500,000	4,380,000,000	30,660,000,000,000	3,587,220,000	930,894,869	739,156,140	479,402	739,635,542	363,851,676
	-$68,339,640	-$298,805,056	-$142,710,959	618,392	5,417,112,766	500,000	4,380,000,000	30,660,000,000,000	3,587,220,000	1,037,112,766	752,106,071	479,402	752,585,473	359,439,013
	-$68,339,640	-$298,805,056	-$138,554,329	630,760	5,525,455,022	500,000	4,380,000,000	30,660,000,000,000	3,587,220,000	1,145,455,022	765,577,100	479,402	766,056,502	355,216,361
	-$68,339,640	-$298,805,056	-$134,518,766	643,375	5,635,964,122	500,000	4,380,000,000	30,660,000,000,000	3,587,220,000	1,255,964,122	779,587,565	479,402	780,066,967	351,177,611
	-$68,339,640	-$298,805,056	-$130,600,744	656,242	5,748,683,405	500,000	4,380,000,000	30,660,000,000,000	3,587,220,000	1,368,683,405	794,156,410	479,402	794,635,812	347,316,841
	-$68,339,640	-$298,805,056	-$126,796,838	669,367	5,863,657,073	500,000	4,380,000,000	30,660,000,000,000	3,587,220,000	1,483,657,073	809,303,202	479,402	809,782,604	343,628,302
	-$68,339,640	-$298,805,056	-$123,103,727	682,755	5,980,930,214	500,000	4,380,000,000	30,660,000,000,000	3,587,220,000	1,600,930,214	825,048,155	479,402	825,527,557	340,106,423
	-$68,339,640	-$298,805,056	-$119,518,181	696,410	6,100,548,818	500,000	4,380,000,000	30,660,000,000,000	3,587,220,000	1,720,548,818	841,412,148	479,402	841,891,550	336,745,798
	-$68,339,640	-$298,805,056	-$116,037,069	710,338	6,222,559,795	500,000	4,380,000,000	30,660,000,000,000	3,587,220,000	1,842,559,795	858,416,747	479,402	858,896,148	333,541,183
			-$5,621,172,480											11,807,377,607

Benefit – Costs: 6,186,205,127

Benefits/Costs: 2.10

TABLE 16.14

Alternative 2 Natural Gas Plant under Outcome 3

A2O3A	A2O3B	A2O3C	A2O3D	A2O3E	A2O3F	A2O3G	A2O3H	A2O3I	A2O3J	A2O3K	A2O3L	A2O3M	A2O3N
						Outcome 3 : No Change in Demand/Year							
O3 Social Cost of Carbon	O3 Total Cost	O3 Present Value of Total Cost	kW Sold	kWh Sold (24/7 Operation)	kW Produced	kWh Produced (24/7 Operation)	Btu	Pounds of CO2	kWh Bought	Income from Sale of Electricity	Economic Impact Benefit	Total Revenue	Present Value of Total Revenue
$0	−$46,718,750	−$46,718,750	0	0	0	0	0	0	0	$0	$11,717,783	$11,717,783	$11,717,783
$0	−$53,906,250	−$52,336,165	0	0	0	0	0	0	0	$0	$11,717,783	$11,717,783	$11,376,489
$0	−$61,093,750	−$57,586,719	0	0	0	0	0	0	0	$0	$11,717,783	$11,717,783	$11,045,135
−$54,671,712	−$285,137,128	−$260,940,864	400,000	3,504,000,000	400,000	3,504,000,000	24,528,000,000,000	2,869,776,000	0	$455,520,000	$479,402	$455,999,402	$417,304,049
−$54,671,712	−$285,137,128	−$253,340,645	400,000	3,504,000,000	400,000	3,504,000,000	24,528,000,000,000	2,869,776,000	0	$460,075,200	$479,402	$460,554,602	$409,196,799
−$54,671,712	−$285,137,128	−$245,961,791	400,000	3,504,000,000	400,000	3,504,000,000	24,528,000,000,000	2,869,776,000	0	$464,675,952	$479,402	$465,155,354	$401,247,094
−$54,671,712	−$285,137,128	−$238,797,855	400,000	3,504,000,000	400,000	3,504,000,000	24,528,000,000,000	2,869,776,000	0	$469,322,712	$479,402	$469,802,113	$393,451,874
−$54,671,712	−$285,137,128	−$231,842,578	400,000	3,504,000,000	400,000	3,504,000,000	24,528,000,000,000	2,869,776,000	0	$474,015,939	$479,402	$474,495,341	$385,808,134
−$54,671,712	−$285,137,128	−$225,089,882	400,000	3,504,000,000	400,000	3,504,000,000	24,528,000,000,000	2,869,776,000	0	$478,756,098	$479,402	$479,235,500	$378,312,929
−$54,671,712	−$285,137,128	−$218,533,866	400,000	3,504,000,000	400,000	3,504,000,000	24,528,000,000,000	2,869,776,000	0	$483,543,659	$479,402	$484,023,061	$370,963,373
−$54,671,712	−$285,137,128	−$212,168,802	400,000	3,504,000,000	400,000	3,504,000,000	24,528,000,000,000	2,869,776,000	0	$488,379,096	$479,402	$488,858,497	$363,756,633
−$54,671,712	−$285,137,128	−$205,989,128	400,000	3,504,000,000	400,000	3,504,000,000	24,528,000,000,000	2,869,776,000	0	$493,262,887	$479,402	$493,742,288	$356,689,934
−$54,671,712	−$285,137,128	−$199,989,444	400,000	3,504,000,000	400,000	3,504,000,000	24,528,000,000,000	2,869,776,000	0	$498,195,515	$479,402	$498,674,917	$349,760,554
−$54,671,712	−$285,137,128	−$194,164,509	400,000	3,504,000,000	400,000	3,504,000,000	24,528,000,000,000	2,869,776,000	0	$503,177,471	$479,402	$503,656,872	$342,965,822
−$54,671,712	−$285,137,128	−$188,509,232	400,000	3,504,000,000	400,000	3,504,000,000	24,528,000,000,000	2,869,776,000	0	$508,209,245	$479,402	$508,688,647	$336,303,122
−$54,671,712	−$285,137,128	−$183,018,672	400,000	3,504,000,000	400,000	3,504,000,000	24,528,000,000,000	2,869,776,000	0	$513,291,338	$479,402	$513,770,740	$329,769,887
−$54,671,712	−$285,137,128	−$177,688,031	400,000	3,504,000,000	400,000	3,504,000,000	24,528,000,000,000	2,869,776,000	0	$518,424,251	$479,402	$518,903,653	$323,363,601
−$54,671,712	−$285,137,128	−$172,512,652	400,000	3,504,000,000	400,000	3,504,000,000	24,528,000,000,000	2,869,776,000	0	$523,608,494	$479,402	$524,087,896	$317,081,796
−$54,671,712	−$285,137,128	−$167,488,011	400,000	3,504,000,000	400,000	3,504,000,000	24,528,000,000,000	2,869,776,000	0	$528,844,579	$479,402	$529,323,980	$310,922,052
−$54,671,712	−$285,137,128	−$162,609,720	400,000	3,504,000,000	400,000	3,504,000,000	24,528,000,000,000	2,869,776,000	0	$534,133,024	$479,402	$534,612,426	$304,881,996
−$54,671,712	−$285,137,128	−$157,873,514	400,000	3,504,000,000	400,000	3,504,000,000	24,528,000,000,000	2,869,776,000	0	$539,474,355	$479,402	$539,953,756	$298,959,303
−$54,671,712	−$285,137,128	−$153,275,257	400,000	3,504,000,000	400,000	3,504,000,000	24,528,000,000,000	2,869,776,000	0	$544,869,098	$479,402	$545,348,500	$293,151,691
−$54,671,712	−$285,137,128	−$148,810,929	400,000	3,504,000,000	400,000	3,504,000,000	24,528,000,000,000	2,869,776,000	0	$550,317,789	$479,402	$550,797,191	$287,456,923

(Continued)

TABLE 16.14 (Continued)
Alternative 2 Natural Gas Plant under Outcome 3

Outcome 3 : No Change in Demand/Year

	A2O3A	A2O3B	A2O3C	A2O3D	A2O3E	A2O3F	A2O3G	A2O3H	A2O3I	A2O3J	A2O3K	A2O3L	A2O3M	A2O3N
	O3 Social Cost of Carbon	O3 Total Cost	O3 Present Value of Total Cost	kW Sold	kWh Sold (24/7 Operation)	kW Produced	kWh Produced (24/7 Operation)	Btu	Pounds of CO2	kWh Bought	Income from Sale of Electricity	Economic Impact Benefit	Total Revenue	Present Value of Total Revenue
	-$54,671,712	-$285,137,128	-$144,476,630	400,000	3,504,000,000	400,000	3,504,000,000	24,528,000,000,000	2,869,776,000	0	$555,820,967	$479,402	$556,300,369	$281,872,807
	-$54,671,712	-$285,137,128	-$140,268,573	400,000	3,504,000,000	400,000	3,504,000,000	24,528,000,000,000	2,869,776,000	0	$561,379,177	$479,402	$561,858,579	$276,397,190
	-$54,671,712	-$285,137,128	-$136,183,080	400,000	3,504,000,000	400,000	3,504,000,000	24,528,000,000,000	2,869,776,000	0	$566,992,968	$479,402	$567,472,370	$271,027,964
	-$54,671,712	-$285,137,128	-$132,216,583	400,000	3,504,000,000	400,000	3,504,000,000	24,528,000,000,000	2,869,776,000	0	$572,662,898	$479,402	$573,142,300	$265,763,063
	-$54,671,712	-$285,137,128	-$128,365,614	400,000	3,504,000,000	400,000	3,504,000,000	24,528,000,000,000	2,869,776,000	0	$578,389,527	$479,402	$578,868,929	$260,600,457
	-$54,671,712	-$285,137,128	-$124,626,810	400,000	3,504,000,000	400,000	3,504,000,000	24,528,000,000,000	2,869,776,000	0	$584,173,422	$479,402	$584,652,824	$255,538,158
	-$54,671,712	-$285,137,128	-$120,996,903	400,000	3,504,000,000	400,000	3,504,000,000	24,528,000,000,000	2,869,776,000	0	$590,015,157	$479,402	$590,494,558	$250,574,218
	-$54,671,712	-$285,137,128	-$117,472,721	400,000	3,504,000,000	400,000	3,504,000,000	24,528,000,000,000	2,869,776,000	0	$595,915,308	$479,402	$596,394,710	$245,706,724
	-$54,671,712	-$285,137,128	-$114,051,186	400,000	3,504,000,000	400,000	3,504,000,000	24,528,000,000,000	2,869,776,000	0	$601,874,461	$479,402	$602,353,863	$240,933,802
	-$54,671,712	-$285,137,128	-$110,729,307	400,000	3,504,000,000	400,000	3,504,000,000	24,528,000,000,000	2,869,776,000	0	$607,893,206	$479,402	$608,372,608	$236,253,614
			-$5,424,634,422											$9,590,154,971
													Benefit - Costs	$4,165,520,549
													Benefits/Costs	1.77

TABLE 16.15
Alternative 2 Natural Gas Plant under Outcome 4

Outcome 4 : 2% Decrease in Demand/Year

A2O4A	A2O4B	A2O4C	A2O4D	A2O4E	A2O4F	A2O4G	A2O4H	A2O4I	A2O4J	A2O4K	A2O4L	A2O4M	A2O4N
O4 Social Cost of Carbon	O4 Total Cost	O4 Present Value of Total Cost	kW sold	kWh sold (24/7 operation)	kW produced	kWh produced (24/7 operation)	Btu	Pounds of CO$_2$	kWh Bought	Income from Sale of Electricity	Economic Impact Benefit	Total Revenue	Present Value of Total Revenue
$0	−$46,718,750	−$46,718,750	0	0	0	0	0	0	0	$0	$11,717,783	$11,717,783	$11,717,783
$0	−$53,906,250	−$52,336,165	0	0	0	0	0	0	0	$0	$11,717,783	$11,717,783	$11,376,489
$0	−$61,093,750	−$57,586,719	0	0	0	0	0	0	0	$0	$11,717,783	$11,717,783	$11,045,135
−$54,671,712	−$285,137,128	−$260,940,864	400,000	3,504,000,000	400,000	3,504,000,000	24,528,000,000,000	2,869,776,000	0	$455,520,000	$479,402	$455,999,402	$417,304,049
−$53,578,278	−$284,043,693	−$252,369,143	392,000	3,433,920,000	392,000	3,433,920,000	24,037,440,000,000	2,812,380,480	0	$450,873,696	$479,402	$451,353,098	$401,021,382
−$52,506,712	−$282,972,128	−$244,094,243	384,160	3,365,241,600	384,160	3,365,241,600	23,556,691,200,000	2,756,132,870	0	$446,274,784	$479,402	$446,754,186	$385,374,085
−$51,456,578	−$281,921,994	−$236,105,231	376,477	3,297,936,768	376,477	3,297,936,768	23,085,557,376,000	2,701,010,213	0	$441,722,782	$479,402	$442,202,183	$370,337,367
−$50,427,447	−$280,892,862	−$228,391,602	368,947	3,231,978,033	368,947	3,231,978,033	22,623,846,228,480	2,646,990,009	0	$437,217,209	$479,402	$437,696,611	$355,887,399
−$49,418,898	−$279,884,313	−$220,943,261	361,568	3,167,338,472	361,568	3,167,338,472	22,171,369,303,910	2,594,050,209	0	$432,757,594	$479,402	$433,236,995	$342,001,285
−$48,430,520	−$278,895,935	−$213,750,511	354,337	3,103,991,703	354,337	3,103,991,703	21,727,941,917,832	2,542,169,204	0	$428,343,466	$479,402	$428,822,868	$328,657,021
−$47,461,909	−$277,927,325	−$206,804,031	347,250	3,041,911,868	347,250	3,041,911,868	21,293,383,079,476	2,491,325,820	0	$423,974,363	$479,402	$424,453,765	$315,833,463
−$46,512,671	−$276,978,087	−$200,094,863	340,305	2,981,073,631	340,305	2,981,073,631	20,867,515,417,886	2,441,499,304	0	$419,649,824	$479,402	$420,129,226	$303,510,292
−$45,582,418	−$276,047,833	−$193,614,396	333,499	2,921,452,159	333,499	2,921,452,159	20,450,165,109,528	2,392,669,318	0	$415,369,396	$479,402	$415,848,798	$291,667,980
−$44,670,769	−$275,136,185	−$187,354,354	326,829	2,863,023,115	326,829	2,863,023,115	20,041,161,807,338	2,344,815,931	0	$411,132,628	$479,402	$411,612,030	$280,287,763
−$43,777,354	−$274,242,769	−$181,306,778	320,293	2,805,762,653	320,293	2,805,762,653	19,640,338,571,191	2,297,919,613	0	$406,939,075	$479,402	$407,418,477	$269,351,610
−$42,901,807	−$273,367,222	−$175,464,018	313,887	2,749,647,400	313,887	2,749,647,400	19,247,531,799,767	2,251,961,221	0	$402,788,297	$479,402	$403,267,699	$258,842,190
−$42,043,771	−$272,509,186	−$169,818,716	307,609	2,694,654,452	307,609	2,694,654,452	18,862,581,163,772	2,206,921,996	0	$398,679,856	$479,402	$399,159,258	$248,742,853
−$41,202,895	−$271,668,311	−$164,363,796	301,457	2,640,761,363	301,457	2,640,761,363	18,485,329,540,496	2,162,783,556	0	$394,613,322	$479,402	$395,092,724	$239,037,595
−$40,378,837	−$270,844,253	−$159,092,454	295,428	2,587,946,136	295,428	2,587,946,136	18,115,622,949,687	2,119,527,885	0	$390,588,266	$479,402	$391,067,668	$229,711,039
−$39,571,261	−$270,036,676	−$153,998,143	289,519	2,536,187,213	289,519	2,536,187,213	17,753,310,490,693	2,077,137,327	0	$386,604,266	$479,402	$387,083,667	$220,748,407
−$38,779,836	−$269,245,251	−$149,074,567	283,729	2,485,463,469	283,729	2,485,463,469	17,398,244,280,879	2,035,594,581	0	$382,660,902	$479,402	$383,140,304	$212,135,497
−$38,004,239	−$268,469,654	−$144,315,668	278,054	2,435,754,199	278,054	2,435,754,199	17,050,279,395,261	1,994,882,689	0	$378,757,761	$479,402	$379,237,163	$203,858,662
−$37,244,154	−$267,709,569	−$139,715,617	272,493	2,387,039,115	272,493	2,387,039,115	16,709,273,807,356	1,954,985,035	0	$374,894,432	$479,402	$375,373,834	$195,904,789

(Continued)

TABLE 16.15 (Continued)
Alternative 2 Natural Gas Plant under Outcome 4

Outcome 4 : 2% Decrease in Demand/Year

	A2O4A	A2O4B	A2O4C	A2O4D	A2O4E	A2O4F	A2O4G	A2O4H	A2O4I	A2O4J	A2O4K	A2O4L	A2O4M	A2O4N
	O4 Social Cost of Carbon	O4 Total Cost	O4 Present Value of Total Cost	kW sold	kWh sold (24/7 operation)	kW produced	kWh produced (24/7 operation)	Btu	Pounds of CO_2	kWh Bought	Income from Sale of Electricity	Economic Impact Benefit	Total Revenue	Present Value of Total Revenue
	−$36,499,271	−$266,964,686	−$135,268,804	267,043	2,339,298,333	267,043	2,339,298,333	16,375,088,331,209	1,915,885,335	0	$371,070,508	$479,402	$371,549,910	$188,261,274
	−$35,769,286	−$266,234,701	−$130,969,831	261,702	2,292,512,366	261,702	2,292,512,366	16,047,586,564,585	1,877,567,628	0	$367,285,589	$479,402	$367,764,991	$180,916,006
	−$35,053,900	−$265,519,315	−$126,813,504	256,468	2,246,662,119	256,468	2,246,662,119	15,726,634,833,293	1,840,016,275	0	$363,539,276	$479,402	$364,018,678	$173,857,348
	−$34,352,822	−$264,818,237	−$122,794,820	251,339	2,201,728,877	251,339	2,201,728,877	15,412,102,136,627	1,803,215,950	0	$359,831,176	$479,402	$360,310,578	$167,074,115
	−$33,665,765	−$264,131,181	−$118,908,967	246,312	2,157,694,299	246,312	2,157,694,299	15,103,860,093,895	1,767,151,631	0	$356,160,898	$479,402	$356,640,300	$160,555,560
	−$32,992,450	−$263,457,866	−$115,151,308	241,386	2,114,540,413	241,386	2,114,540,413	14,801,782,892,017	1,731,808,598	0	$352,528,056	$479,402	$353,007,458	$154,291,354
	−$32,332,601	−$262,798,017	−$111,517,382	236,558	2,072,249,605	236,558	2,072,249,605	14,505,747,234,176	1,697,172,426	0	$348,932,270	$479,402	$349,411,672	$148,271,572
	−$31,685,949	−$262,151,364	−$108,002,891	231,827	2,030,804,613	231,827	2,030,804,613	14,215,632,289,493	1,663,228,978	0	$345,373,161	$479,402	$345,852,563	$142,486,677
	−$31,052,230	−$261,517,646	−$104,603,696	227,190	1,990,188,521	227,190	1,990,188,521	13,931,319,643,703	1,629,964,398	0	$341,850,355	$479,402	$342,329,757	$136,927,502
	−$30,431,185	−$260,896,601	−$101,315,812	222,647	1,950,384,750	222,647	1,950,384,750	13,652,693,250,829	1,597,365,110	0	$338,363,481	$479,402	$338,842,883	$131,585,240
			−$5,213,600,907											$7,488,580,784

Benefit – Costs $2,274,979,877

Benefits/Costs 1.44

TABLE 16.16
Alternative 3 Geothermal Power Plant

A3A	A3B	A3C	A3D	A3E	A3F	A3G	A3H	A3I	A3J
Years	Elect Selling Price ($/kwh)	Capital Cost Related Interest or Principal + Interest	Operating & Maintenance Cost	Property Taxes	Land Lease	Insurance Cost	Safety Costs	Total Cost	Present Value of Total Cost
1		−$16,467,727	$0	−$8,233,863	−$12,100	−$49,403,180	−$28,818,522	−$102,935,392	−$102,935,392
2		−$32,935,454	$0	−$8,233,863	−$12,100	−$49,403,180	−$28,818,522	−$119,403,119	−$115,925,358
3		−$49,403,180	$0	−$8,233,863	−$12,100	−$49,403,180	−$28,818,522	−$135,870,846	−$128,071,304
4	$0.130	−$84,017,122	−$488,648,427	−$8,233,863	−$12,100	−$49,403,180	−$28,818,522	−$659,133,215	−$603,200,264
5	$0.131	−$84,017,122	−$488,648,427	−$8,233,863	−$12,100	−$49,403,180	−$28,818,522	−$659,133,215	−$585,631,324
6	$0.133	−$84,017,122	−$488,648,427	−$8,233,863	−$12,100	−$49,403,180	−$28,818,522	−$659,133,215	−$568,574,101
7	$0.134	−$84,017,122	−$488,648,427	−$8,233,863	−$12,100	−$49,403,180	−$28,818,522	−$659,133,215	−$552,013,691
8	$0.135	−$84,017,122	−$488,648,427	−$8,233,863	−$12,100	−$49,403,180	−$28,818,522	−$659,133,215	−$535,935,622
9	$0.137	−$84,017,122	−$488,648,427	−$8,233,863	−$12,100	−$49,403,180	−$28,818,522	−$659,133,215	−$520,325,846
10	$0.138	−$84,017,122	−$488,648,427	−$8,233,863	−$12,100	−$49,403,180	−$28,818,522	−$659,133,215	−$505,170,725
11	$0.139	−$84,017,122	−$488,648,427	−$8,233,863	−$12,100	−$49,403,180	−$28,818,522	−$659,133,215	−$490,457,014
12	$0.141	−$84,017,122	−$488,648,427	−$8,233,863	−$12,100	−$49,403,180	−$28,818,522	−$659,133,215	−$476,171,859
13	$0.142	−$84,017,122	−$488,648,427	−$8,233,863	−$12,100	−$49,403,180	−$28,818,522	−$659,133,215	−$462,302,775
14	$0.144	−$84,017,122	−$488,648,427	−$8,233,863	−$12,100	−$49,403,180	−$28,818,522	−$659,133,215	−$448,837,646
15	$0.145	−$84,017,122	−$488,648,427	−$8,233,863	−$12,100	−$49,403,180	−$28,818,522	−$659,133,215	−$435,764,705
16	$0.146	−$84,017,122	−$488,648,427	−$8,233,863	−$12,100	−$49,403,180	−$28,818,522	−$659,133,215	−$423,072,529
17	$0.148	−$84,017,122	−$488,648,427	−$8,233,863	−$12,100	−$49,403,180	−$28,818,522	−$659,133,215	−$410,750,028
18	$0.149	−$84,017,122	−$488,648,427	−$8,233,863	−$12,100	−$49,403,180	−$28,818,522	−$659,133,215	−$398,786,435
19	$0.151	−$84,017,122	−$488,648,427	−$8,233,863	−$12,100	−$49,403,180	−$28,818,522	−$659,133,215	−$387,171,296

(Continued)

TABLE 16.16 (Continued)
Alternative 3 Geothermal Power Plant

A3A	A3B	A3C	A3D	A3E	A3F	A3G	A3H	A3I	A3J
Years	Elect Selling Price ($/kwh)	Capital Cost Related Interest or Principal+Interest	Operating & Maintenance Cost	Property Taxes	Land Lease	Insurance Cost	Safety Costs	Total Cost	Present Value of Total Cost
20	$0.152	−$84,017,122	−$488,648,427	−$8,233,863	−$12,100	−$49,403,180	−$28,818,522	−$659,133,215	−$375,894,462
21	$0.154	−$84,017,122	−$488,648,427	−$8,233,863	−$12,100	−$49,403,180	−$28,818,522	−$659,133,215	−$364,946,080
22	$0.155	−$84,017,122	−$488,648,427	−$8,233,863	−$12,100	−$49,403,180	−$28,818,522	−$659,133,215	−$354,316,582
23	$0.157	−$84,017,122	−$488,648,427	−$8,233,863	−$12,100	−$49,403,180	−$28,818,522	−$659,133,215	−$343,996,682
24	$0.159	−$84,017,122	−$488,648,427	−$8,233,863	−$12,100	−$49,403,180	−$28,818,522	−$659,133,215	−$333,977,361
25	$0.160	−$84,017,122	−$488,648,427	−$8,233,863	−$12,100	−$49,403,180	−$28,818,522	−$659,133,215	−$324,249,865
26	$0.162	−$84,017,122	−$488,648,427	−$8,233,863	−$12,100	−$49,403,180	−$28,818,522	−$659,133,215	−$314,805,694
27	$0.163	−$84,017,122	−$488,648,427	−$8,233,863	−$12,100	−$49,403,180	−$28,818,522	−$659,133,215	−$305,636,596
28	$0.165	−$84,017,122	−$488,648,427	−$8,233,863	−$12,100	−$49,403,180	−$28,818,522	−$659,133,215	−$296,734,560
29	$0.167	−$84,017,122	−$488,648,427	−$8,233,863	−$12,100	−$49,403,180	−$28,818,522	−$659,133,215	−$288,091,805
30	$0.168	−$84,017,122	−$488,648,427	−$8,233,863	−$12,100	−$49,403,180	−$28,818,522	−$659,133,215	−$279,700,782
31	$0.170	−$84,017,122	−$488,648,427	−$8,233,863	−$12,100	−$49,403,180	−$28,818,522	−$659,133,215	−$271,554,157
32	$0.172	−$84,017,122	−$488,648,427	−$8,233,863	−$12,100	−$49,403,180	−$28,818,522	−$659,133,215	−$263,644,813
33	$0.173	−$84,017,122	−$488,648,427	−$8,233,863	−$12,100	−$49,403,180	−$28,818,522	−$659,133,215	−$255,965,838
									-$12,524,613,193

TABLE 16.17

Alternative 3 Geothermal Power Plant under Outcome 1

A3O1A	A3O1B	A3O1C	A3O1D	A3O1E	A3O1F	A3O1G	A3O1H	A3O1I
				Outcome 1 : 1% Increase in Demand/Year				
kW Sold	kWh Sold (24/7 Operation)	kW Produced	kWh Produced (24/7 Operation)	kWh Bought	Income from Sale of Electricity	Economic Impact Benefits	Total Revenue	Present Value of Total Revenue
0	0	0	0	0	0	$16,854,520	$16,854,520	$16,854,520
0	0	0	0	0	0	$16,854,520	$16,854,520	$16,363,612
0	0	0	0	0	0	$16,854,520	$16,854,520	$15,887,002
400,000	3,504,000,000	400,000	3,504,000,000	0	455,520,000	$922,845	$456,442,845	$417,709,863
404,000	3,539,040,000	404,000	3,539,040,000	0	464,675,952	$922,845	$465,598,797	$413,678,501
408,040	3,574,430,400	408,040	3,574,430,400	0	474,015,939	$922,845	$474,938,784	$409,686,367
412,120	3,610,174,704	412,120	3,610,174,704	0	483,543,659	$922,845	$484,466,504	$405,733,070
416,242	3,646,276,451	416,242	3,646,276,451	0	493,262,887	$922,845	$494,185,732	$401,818,224
420,404	3,682,739,216	420,404	3,682,739,216	0	503,177,471	$922,845	$504,100,316	$397,941,444
424,608	3,719,566,608	424,608	3,719,566,608	0	513,291,338	$922,845	$514,214,183	$394,102,354
428,854	3,756,762,274	428,854	3,756,762,274	0	523,608,494	$922,845	$524,531,339	$390,300,577
433,143	3,794,329,897	433,143	3,794,329,897	0	534,133,024	$922,845	$535,055,870	$386,535,744
437,474	3,832,273,195	437,474	3,832,273,195	0	544,869,098	$922,845	$545,791,943	$382,807,488
441,849	3,870,595,927	441,849	3,870,595,927	0	555,820,967	$922,845	$556,743,812	$379,115,445
446,267	3,909,301,887	446,267	3,909,301,887	0	566,992,968	$922,845	$567,915,814	$375,459,257
450,730	3,948,394,906	450,730	3,948,394,906	0	578,389,527	$922,845	$579,312,372	$371,838,567
455,237	3,987,878,855	455,237	3,987,878,855	0	590,015,157	$922,845	$590,938,002	$368,253,026
459,790	4,027,757,643	459,790	4,027,757,643	0	601,874,461	$922,845	$602,797,306	$364,702,284
464,388	4,068,035,220	464,388	4,068,035,220	0	613,972,138	$922,845	$614,894,983	$361,185,997
469,031	4,108,715,572	469,031	4,108,715,572	0	626,312,978	$922,845	$627,235,823	$357,703,825

(Continued)

TABLE 16.17 (Continued)
Alternative 3 Geothermal Power Plant under Outcome 1

A3O1A	A3O1B	A3O1C	A3O1D	A3O1E	A3O1F	A3O1G	A3O1H	A3O1I
				Outcome 1 : 1% Increase in Demand/Year				
kW Sold	kWh Sold (24/7 Operation)	kW Produced	kWh Produced (24/7 Operation)	kWh Bought	Income from Sale of Electricity	Economic Impact Benefits	Total Revenue	Present Value of Total Revenue
473,722	4,149,802,728	473,722	4,149,802,728	0	638,901,869	$922,845	$639,824,714	$354,255,431
478,459	4,191,300,755	478,459	4,191,300,755	0	651,743,796	$922,845	$652,666,642	$350,840,481
483,244	4,233,213,762	483,244	4,233,213,762	0	664,843,847	$922,845	$665,766,692	$347,458,644
488,076	4,275,545,900	488,076	4,275,545,900	0	678,207,208	$922,845	$679,130,053	$344,109,594
492,957	4,318,301,359	492,957	4,318,301,359	0	691,839,173	$922,845	$692,762,018	$340,793,008
497,886	4,361,484,373	497,886	4,361,484,373	0	705,745,140	$922,845	$706,667,985	$337,508,565
502,865	4,405,099,216	500,000	4,380,000,000	25,099,216	716,918,712	$922,845	$717,841,557	$332,859,345
507,894	4,449,150,208	500,000	4,380,000,000	69,150,208	726,103,198	$922,845	$727,026,043	$327,299,168
512,973	4,493,641,711	500,000	4,380,000,000	113,641,711	735,525,682	$922,845	$736,448,527	$321,884,531
518,103	4,538,578,128	500,000	4,380,000,000	158,578,128	745,191,482	$922,845	$746,114,327	$316,610,901
523,284	4,583,963,909	500,000	4,380,000,000	203,963,909	755,106,028	$922,845	$756,028,873	$311,473,885
528,516	4,629,803,548	500,000	4,380,000,000	249,803,548	765,274,863	$922,845	$766,197,708	$306,469,234
533,802	4,676,101,583	500,000	4,380,000,000	296,101,583	775,703,650	$922,845	$776,626,495	$301,592,830
								$10,920,832,784

Benefits – Costs $-1,603,780,409

Benefits/Costs 0.87

TABLE 16.18

Alternative 3 Geothermal Power Plant under Outcome 2

Outcome 2 : 2% Increase in Demand/Year

A3O2A	A3O2B	A3O2C	A3O2D	A3O2E	A3O2F	A3O2G	A3O2H	A3O2I
kW Sold	kWh Sold (24/7 Operation)	kW Produced	kWh Produced (24/7 Operation)	kWh Bought	Income from Sale of Electricity	Economic Impact Benefits	Total Revenue	Present Value of Total Revenue
0	0	0	0	0	$0	$16,854,520	$16,854,520	$16,854,520
0	0	0	0	0	$0	$16,854,520	$16,854,520	$16,363,612
0	0	0	0	0	$0	$16,854,520	$16,854,520	$15,887,002
400,000	3,504,000,000	400,000	3,504,000,000	0	$455,520,000	$922,845	$456,442,845	$417,709,863
408,000	3,574,080,000	408,000	3,574,080,000	0	$469,276,704	$922,845	$470,199,549	$417,766,209
416,160	3,645,561,600	416,160	3,645,561,600	0	$483,448,860	$922,845	$484,371,706	$417,823,288
424,483	3,718,472,832	424,483	3,718,472,832	0	$498,049,016	$922,845	$498,971,861	$417,881,078
432,973	3,792,842,289	432,973	3,792,842,289	0	$513,090,096	$922,845	$514,012,942	$417,939,559
441,632	3,868,699,134	441,632	3,868,699,134	0	$528,585,417	$922,845	$529,508,262	$417,998,712
450,465	3,946,073,117	450,465	3,946,073,117	0	$544,548,697	$922,845	$545,471,542	$418,058,517
459,474	4,024,994,579	459,474	4,024,994,579	0	$560,994,067	$922,845	$561,916,913	$418,118,955
468,664	4,105,494,471	468,664	4,105,494,471	0	$577,936,088	$922,845	$578,858,934	$418,180,010
478,037	4,187,604,360	478,037	4,187,604,360	0	$595,389,758	$922,845	$596,312,603	$418,241,662
487,598	4,271,356,448	487,598	4,271,356,448	0	$613,370,529	$922,845	$614,293,374	$418,303,896
497,350	4,356,783,577	497,350	4,356,783,577	0	$631,894,319	$922,845	$632,817,164	$418,366,695
507,297	4,443,919,248	507,297	4,443,919,248	0	$650,977,527	$922,845	$651,900,372	$418,430,043
517,443	4,532,797,633	517,443	4,532,797,633	0	$670,637,049	$922,845	$671,559,894	$418,493,924
527,792	4,623,453,586	527,792	4,623,453,586	0	$690,890,287	$922,845	$691,813,133	$418,558,323
538,347	4,715,922,657	538,347	4,715,922,657	0	$711,755,174	$922,845	$712,678,019	$418,623,226
549,114	4,810,241,111	549,114	4,810,241,111	0	$733,250,180	$922,845	$734,173,026	$418,688,618

(Continued)

TABLE 16.18 (Continued)
Alternative 3 Geothermal Power Plant under Outcome 2

Outcome 2 : 2% Increase in Demand/Year

A3O2A	A3O2B	A3O2C	A3O2D	A3O2E	A3O2F	A3O2G	A3O2H	A3O2I
kW Sold	kWh Sold (24/7 Operation)	kW Produced	kWh Produced (24/7 Operation)	kWh Bought	Income from Sale of Electricity	Economic Impact Benefits	Total Revenue	Present Value of Total Revenue
560,097	4,906,445,933	560,097	4,906,445,933	0	$755,394,336	$922,845	$756,317,181	$418,754,486
571,298	5,004,574,852	500,000	4,380,000,000	624,574,852	$703,258,263	$922,845	$704,181,108	$378,532,045
582,724	5,104,666,349	500,000	4,380,000,000	724,666,349	$714,749,142	$922,845	$715,671,987	$373,503,843
594,379	5,206,759,676	500,000	4,380,000,000	826,759,676	$726,709,557	$922,845	$727,632,403	$368,685,334
606,267	5,310,894,869	500,000	4,380,000,000	930,894,869	$739,156,140	$922,845	$740,078,985	$364,069,820
618,392	5,417,112,766	500,000	4,380,000,000	1,037,112,766	$752,106,071	$922,845	$753,028,916	$359,650,804
630,760	5,525,455,022	500,000	4,380,000,000	1,145,455,022	$765,577,100	$922,845	$766,499,945	$355,421,983
643,375	5,635,964,122	500,000	4,380,000,000	1,255,964,122	$779,587,565	$922,845	$780,510,411	$351,377,245
656,242	5,748,683,405	500,000	4,380,000,000	1,368,683,405	$794,156,410	$922,845	$795,079,255	$347,510,659
669,367	5,863,657,073	500,000	4,380,000,000	1,483,657,073	$809,303,202	$922,845	$810,226,047	$343,816,476
682,755	5,980,930,214	500,000	4,380,000,000	1,600,930,214	$825,048,155	$922,845	$825,971,000	$340,289,116
696,410	6,100,548,818	500,000	4,380,000,000	1,720,548,818	$841,412,148	$922,845	$842,334,993	$336,923,169
710,338	6,222,559,795	500,000	4,380,000,000	1,842,559,795	$858,416,747	$922,845	$859,339,592	$333,713,388
								$11,830,536,080

Benefits − Costs   −$694,077,113
Benefits/Costs   0.94

TABLE 16.19

Alternative 3 Geothermal Power Plant under Outcome 3

A3O3A	A3O3B	A3O3C	A3O3D	A3O3E	A3O3F	A3O3G	A3O3H	A3O3I
			Outcome 3 : No Change in Demand/Year					
kW Sold	kWh Sold (24/7 Operation)	kW Produced	kWh Produced (24/7 Operation)	kWh Bought	Income from Sale of Electricity	Economic Impact Benefits	Total Revenue	Present Value of Total Revenue
0	0	0	0	0	$0	$16,854,520	$16,854,520	$16,854,520
0	0	0	0	0	$0	$16,854,520	$16,854,520	$16,363,612
0	0	0	0	0	$0	$16,854,520	$16,854,520	$15,887,002
400,000	3,504,000,000	400,000	3,504,000,000	0	$455,520,000	$922,845	$456,442,845	$417,709,863
400,000	3,504,000,000	400,000	3,504,000,000	0	$460,075,200	$922,845	$460,998,045	$409,590,792
400,000	3,504,000,000	400,000	3,504,000,000	0	$464,675,952	$922,845	$465,598,797	$401,629,612
400,000	3,504,000,000	400,000	3,504,000,000	0	$469,322,712	$922,845	$470,245,557	$393,823,251
400,000	3,504,000,000	400,000	3,504,000,000	0	$474,015,939	$922,845	$474,938,784	$386,168,694
400,000	3,504,000,000	400,000	3,504,000,000	0	$478,756,098	$922,845	$479,678,943	$378,662,987
400,000	3,504,000,000	400,000	3,504,000,000	0	$483,543,659	$922,845	$484,466,504	$371,303,235
400,000	3,504,000,000	400,000	3,504,000,000	0	$488,379,096	$922,845	$489,301,941	$364,086,597
400,000	3,504,000,000	400,000	3,504,000,000	0	$493,262,887	$922,845	$494,185,732	$357,010,287
400,000	3,504,000,000	400,000	3,504,000,000	0	$498,195,515	$922,845	$499,118,361	$350,071,576
400,000	3,504,000,000	400,000	3,504,000,000	0	$503,177,471	$922,845	$504,100,316	$343,267,786
400,000	3,504,000,000	400,000	3,504,000,000	0	$508,209,245	$922,845	$509,132,090	$336,596,291
400,000	3,504,000,000	400,000	3,504,000,000	0	$513,291,338	$922,845	$514,214,183	$330,054,517
400,000	3,504,000,000	400,000	3,504,000,000	0	$518,424,251	$922,845	$519,347,096	$323,639,940
400,000	3,504,000,000	400,000	3,504,000,000	0	$523,608,494	$922,845	$524,531,339	$317,350,086
400,000	3,504,000,000	400,000	3,504,000,000	0	$528,844,579	$922,845	$529,767,424	$311,182,528
400,000	3,504,000,000	400,000	3,504,000,000	0	$534,133,024	$922,845	$535,055,870	$305,134,886

(Continued)

TABLE 16.19 (Continued)
Alternative 3 Geothermal Power Plant under Outcome 3

Outcome 3 : No Change in Demand/Year

A3O3A kW Sold	A3O3B kWh Sold (24/7 Operation)	A3O3C kW Produced	A3O3D kWh Produced (24/7 Operation)	A3O3E kWh Bought	A3O3F Income from Sale of Electricity	A3O3G Economic Impact Benefits	A3O3H Total Revenue	A3O3I Present Value of Total Revenue
400,000	3,504,000,000	400,000	3,504,000,000	0	$539,474,355	$922,845	$540,397,200	$299,204,827
400,000	3,504,000,000	400,000	3,504,000,000	0	$544,869,098	$922,845	$545,791,943	$293,390,064
400,000	3,504,000,000	400,000	3,504,000,000	0	$550,317,789	$922,845	$551,240,634	$287,688,353
400,000	3,504,000,000	400,000	3,504,000,000	0	$555,820,967	$922,845	$556,743,812	$282,097,496
400,000	3,504,000,000	400,000	3,504,000,000	0	$561,379,177	$922,845	$562,302,022	$276,615,335
400,000	3,504,000,000	400,000	3,504,000,000	0	$566,992,968	$922,845	$567,915,814	$271,239,755
400,000	3,504,000,000	400,000	3,504,000,000	0	$572,662,898	$922,845	$573,585,743	$265,968,685
400,000	3,504,000,000	400,000	3,504,000,000	0	$578,389,527	$922,845	$579,312,372	$260,800,090
400,000	3,504,000,000	400,000	3,504,000,000	0	$584,173,422	$922,845	$585,096,268	$255,731,977
400,000	3,504,000,000	400,000	3,504,000,000	0	$590,015,157	$922,845	$590,938,002	$250,762,391
400,000	3,504,000,000	400,000	3,504,000,000	0	$595,915,308	$922,845	$596,838,153	$245,889,417
400,000	3,504,000,000	400,000	3,504,000,000	0	$601,874,461	$922,845	$602,797,306	$241,111,174
400,000	3,504,000,000	400,000	3,504,000,000	0	$607,893,206	$922,845	$608,816,051	$236,425,820
								$9,613,313,444
							Benefits – Costs	−2,911,299,749
							Benefits/Costs	0.77

TABLE 16.20

Alternative 3 Geothermal Power Plant under Outcome 4

A3O4A	A3O4B	A3O4C	A3O4D	A3O4E	A3O4F	A3O4G	A3O4H	A3O4I
					Outcome 4 : 2% Decrease in Demand/Year			
kW Sold	kWh Sold (24/7 Operation)	kW Produced	kWh Produced (24/7 Operation)	kWh Bought	Income from Sale of Electricity	Economic Impact Benefits	Total Revenue	Present Value of Total Revenue
0	0	0	0	0	$0	$16,854,520	$16,854,520	$16,854,520
0	0	0	0	0	$0	$16,854,520	$16,854,520	$16,363,612
0	0	0	0	0	$0	$16,854,520	$16,854,520	$15,887,002
400,000	3,504,000,000	400,000	3,504,000,000	0	$455,520,000	$922,845	$456,442,845	$417,709,863
392,000	3,433,920,000	392,000	3,433,920,000	0	$450,873,696	$922,845	$451,796,541	$401,415,375
384,160	3,365,241,600	384,160	3,365,241,600	0	$446,274,784	$922,845	$447,197,630	$385,756,604
376,477	3,297,936,768	376,477	3,297,936,768	0	$441,722,782	$922,845	$442,645,627	$370,708,744
368,947	3,231,978,033	368,947	3,231,978,033	0	$437,217,209	$922,845	$438,140,054	$356,247,959
361,568	3,167,338,472	361,568	3,167,338,472	0	$432,757,594	$922,845	$433,680,439	$342,351,343
354,337	3,103,991,703	354,337	3,103,991,703	0	$428,343,466	$922,845	$429,266,311	$328,996,884
347,250	3,041,911,868	347,250	3,041,911,868	0	$423,974,363	$922,845	$424,897,208	$316,163,427
340,305	2,981,073,631	340,305	2,981,073,631	0	$419,649,824	$922,845	$420,572,669	$303,830,645
333,499	2,921,452,159	333,499	2,921,452,159	0	$415,369,396	$922,845	$416,292,241	$291,979,002
326,829	2,863,023,115	326,829	2,863,023,115	0	$411,132,628	$922,845	$412,055,473	$280,589,727
320,293	2,805,762,653	320,293	2,805,762,653	0	$406,939,075	$922,845	$407,861,921	$269,644,778
313,887	2,749,647,400	313,887	2,749,647,400	0	$402,788,297	$922,845	$403,711,142	$259,126,820
307,609	2,694,654,452	307,609	2,694,654,452	0	$398,679,856	$922,845	$399,602,701	$249,019,192
301,457	2,640,761,363	301,457	2,640,761,363	0	$394,613,322	$922,845	$395,536,167	$239,305,886
295,428	2,587,946,136	295,428	2,587,946,136	0	$390,588,266	$922,845	$391,511,111	$229,971,515
289,519	2,536,187,213	289,519	2,536,187,213	0	$386,604,266	$922,845	$387,527,111	$221,001,296

(*Continued*)

TABLE 16.20 (Continued)
Alternative 3 Geothermal Power Plant under Outcome 4

A3O4A kW Sold	A3O4B kWh Sold (24/7 Operation)	A3O4C kW Produced	A3O4D kWh Produced (24/7 Operation)	A3O4E kWh Bought	A3O4F Income from Sale of Electricity	A3O4G Economic Impact Benefits	A3O4H Total Revenue	A3O4I Present Value of Total Revenue
			Outcome 4 : 2% Decrease in Demand/Year					
283,729	2,485,463,469	283,729	2,485,463,469	0	$382,660,902	$922,845	$383,583,747	$212,381,021
278,054	2,435,754,199	278,054	2,435,754,199	0	$378,757,761	$922,845	$379,680,606	$204,097,035
272,493	2,387,039,115	272,493	2,387,039,115	0	$374,894,432	$922,845	$375,817,277	$196,136,218
267,043	2,339,298,333	267,043	2,339,298,333	0	$371,070,508	$922,845	$371,993,354	$188,485,963
261,702	2,292,512,366	261,702	2,292,512,366	0	$367,285,589	$922,845	$368,208,434	$181,134,151
256,468	2,246,662,119	256,468	2,246,662,119	0	$363,539,276	$922,845	$364,462,121	$174,069,139
251,339	2,201,728,877	251,339	2,201,728,877	0	$359,831,176	$922,845	$360,754,021	$167,279,737
246,312	2,157,694,299	246,312	2,157,694,299	0	$356,160,898	$922,845	$357,083,743	$160,755,193
241,386	2,114,540,413	241,386	2,114,540,413	0	$352,528,056	$922,845	$353,450,902	$154,485,173
236,558	2,072,249,605	236,558	2,072,249,605	0	$348,932,270	$922,845	$349,855,116	$148,459,746
231,827	2,030,804,613	231,827	2,030,804,613	0	$345,373,161	$922,845	$346,296,006	$142,669,369
227,190	1,990,188,521	227,190	1,990,188,521	0	$341,850,355	$922,845	$342,773,200	$137,104,874
222,647	1,950,384,750	222,647	1,950,384,750	0	$338,363,481	$922,845	$339,286,326	$131,757,446
								$7,511,739,257
							Benefits − Costs	−$5,012,873,936
							Benefits/Costs	$0.60

TABLE 16.21

Alternative 4 Concentrating Solar Power Plant

A4A	A4B	A4C	A4D	A4E	A4F	A4G	A4H	A4I
	Elect Selling Price ($/kwh)	Capital Cost Related Interest or Principal + Interest	Operating & Maintenance Cost	Property Taxes	Insurance Cost	Safety Costs	Total Cost	Present Value of Total Cost
Years								
1		-$114,114,027	$0	-$26,559,394	-$57,057,013	-$57,057,013	-$254,787,447	-$254,787,447
2		-$228,228,053	$0	-$26,559,394	-$57,057,013	-$57,057,013	-$368,901,474	-$358,156,771
3		-$342,342,080	$0	-$26,559,394	-$57,057,013	-$57,057,013	-$483,015,500	-$455,288,435
4	$0.130	-$582,201,311	-$56,505,388	-$26,559,394	-$57,057,013	-$57,057,013	-$779,380,120	-$713,243,216
5	$0.131	-$582,201,311	-$56,505,388	-$26,559,394	-$57,057,013	-$57,057,013	-$779,380,120	-$692,469,142
6	$0.133	-$582,201,311	-$56,505,388	-$26,559,394	-$57,057,013	-$57,057,013	-$779,380,120	-$672,300,138
7	$0.134	-$582,201,311	-$56,505,388	-$26,559,394	-$57,057,013	-$57,057,013	-$779,380,120	-$652,718,580
8	$0.135	-$582,201,311	-$56,505,388	-$26,559,394	-$57,057,013	-$57,057,013	-$779,380,120	-$633,707,360
9	$0.137	-$582,201,311	-$56,505,388	-$26,559,394	-$57,057,013	-$57,057,013	-$779,380,120	-$615,249,864
10	$0.138	-$582,201,311	-$56,505,388	-$26,559,394	-$57,057,013	-$57,057,013	-$779,380,120	-$597,329,965
11	$0.139	-$582,201,311	-$56,505,388	-$26,559,394	-$57,057,013	-$57,057,013	-$779,380,120	-$579,932,005
12	$0.141	-$582,201,311	-$56,505,388	-$26,559,394	-$57,057,013	-$57,057,013	-$779,380,120	-$563,040,781
13	$0.142	-$582,201,311	-$56,505,388	-$26,559,394	-$57,057,013	-$57,057,013	-$779,380,120	-$546,641,535
14	$0.144	-$582,201,311	-$56,505,388	-$26,559,394	-$57,057,013	-$57,057,013	-$779,380,120	-$530,719,937
15	$0.145	-$582,201,311	-$56,505,388	-$26,559,394	-$57,057,013	-$57,057,013	-$779,380,120	-$515,262,075
16	$0.146	-$582,201,311	-$56,505,388	-$26,559,394	-$57,057,013	-$57,057,013	-$779,380,120	-$500,254,442
17	$0.148	-$582,201,311	-$56,505,388	-$26,559,394	-$57,057,013	-$57,057,013	-$779,380,120	-$485,683,924
18	$0.149	-$582,201,311	-$56,505,388	-$26,559,394	-$57,057,013	-$57,057,013	-$779,380,120	-$471,537,790
19	$0.151	-$582,201,311	-$56,505,388	-$26,559,394	-$57,057,013	-$57,057,013	-$779,380,120	-$457,803,680

(Continued)

TABLE 16.21 (Continued)
Alternative 4 Concentrating Solar Power Plant

A4A	A4B	A4C	A4D	A4E	A4F	A4G	A4H	A4I
Years	Elect Selling Price ($/kwh)	Capital Cost Related Interest or Principal + Interest	Operating & Maintenance Cost	Property Taxes	Insurance Cost	Safety Costs	Total Cost	Present Value of Total Cost
20	$0.152	−$582,201,311	−$56,505,388	−$26,559,394	−$57,057,013	−$57,057,013	−$779,380,120	−$444,469,592
21	$0.154	−$582,201,311	−$56,505,388	−$26,559,394	−$57,057,013	−$57,057,013	−$779,380,120	−$431,523,876
22	$0.155	−$582,201,311	−$56,505,388	−$26,559,394	−$57,057,013	−$57,057,013	−$779,380,120	−$418,955,219
23	$0.157	−$582,201,311	−$56,505,388	−$26,559,394	−$57,057,013	−$57,057,013	−$779,380,120	−$406,752,640
24	$0.159	−$582,201,311	−$56,505,388	−$26,559,394	−$57,057,013	−$57,057,013	−$779,380,120	−$394,905,476
25	$0.160	−$582,201,311	−$56,505,388	−$26,559,394	−$57,057,013	−$57,057,013	−$779,380,120	−$383,403,374
26	$0.162	−$582,201,311	−$56,505,388	−$26,559,394	−$57,057,013	−$57,057,013	−$779,380,120	−$372,236,286
27	$0.163	−$582,201,311	−$56,505,388	−$26,559,394	−$57,057,013	−$57,057,013	−$779,380,120	−$361,394,452
28	$0.165	−$582,201,311	−$56,505,388	−$26,559,394	−$57,057,013	−$57,057,013	−$779,380,120	−$350,868,400
29	$0.167	−$582,201,311	−$56,505,388	−$26,559,394	−$57,057,013	−$57,057,013	−$779,380,120	−$340,648,932
30	$0.168	−$582,201,311	−$56,505,388	−$26,559,394	−$57,057,013	−$57,057,013	−$779,380,120	−$330,727,119
31	$0.170	−$582,201,311	−$56,505,388	−$26,559,394	−$57,057,013	−$57,057,013	−$779,380,120	−$321,094,290
32	$0.172	−$582,201,311	−$56,505,388	−$26,559,394	−$57,057,013	−$57,057,013	−$779,380,120	−$311,742,029
33	$0.173	−$582,201,311	−$56,505,388	−$26,559,394	−$57,057,013	−$57,057,013	−$779,380,120	−$302,662,164
								−$15,467,510,935

TABLE 16.22

Alternative 4 Concentrating Solar Power Plant under Outcome 1

A4O1A	A4O1B	A4O1C	A4O1D	A4O1E	A4O1F	A4O1G	A4O1H	A4O1I
					Outcome 1 : 1% Increase in Demand/Year			
kW Sold	kWh Sold (24/7 Operation)	kW Produced	kWh Produced (24/7 Operation)	kWh Bought	Income from Sale of Electricity	Economic Impact Benefits	Total Revenue	Present Value of Total Revenue
0	0	0	0	0	$0	$304,788,870	$304,788,870	$304,788,870
0	0	0	0	0	$0	$304,788,870	$304,788,870	$295,911,525
0	0	0	0	0	$0	$304,788,870	$304,788,870	$287,292,742
400,000	3,504,000,000	400,000	3,504,000,000	0	$455,520,000	$3,612,523	$459,132,523	$420,171,299
404,000	3,539,040,000	404,000	3,539,040,000	0	$464,675,952	$3,612,523	$468,288,475	$416,068,245
408,040	3,574,430,400	408,040	3,574,430,400	0	$474,015,939	$3,612,523	$477,628,461	$412,006,507
412,120	3,610,174,704	412,120	3,610,174,704	0	$483,543,659	$3,612,523	$487,156,182	$407,985,633
416,242	3,646,276,451	416,242	3,646,276,451	0	$493,262,887	$3,612,523	$496,875,409	$404,005,178
420,404	3,682,739,216	420,404	3,682,739,216	0	$503,177,471	$3,612,523	$506,789,993	$400,064,701
424,608	3,719,566,608	424,608	3,719,566,608	0	$513,291,338	$3,612,523	$516,903,861	$396,163,768
428,854	3,756,762,274	428,854	3,756,762,274	0	$523,608,494	$3,612,523	$527,221,016	$392,301,950
433,143	3,794,329,897	433,143	3,794,329,897	0	$534,133,024	$3,612,523	$537,745,547	$388,478,825
437,474	3,832,273,195	437,474	3,832,273,195	0	$544,869,098	$3,612,523	$548,481,621	$384,693,974
441,849	3,870,595,927	441,849	3,870,595,927	0	$555,820,967	$3,612,523	$559,433,490	$380,946,985
446,267	3,909,301,887	446,267	3,909,301,887	0	$566,992,968	$3,612,523	$570,605,491	$377,237,450
450,730	3,948,394,906	450,730	3,948,394,906	0	$578,389,527	$3,612,523	$582,002,050	$373,564,969
455,237	3,987,878,855	455,237	3,987,878,855	0	$590,015,157	$3,612,523	$593,627,679	$369,929,144
459,790	4,027,757,643	459,790	4,027,757,643	0	$601,874,461	$3,612,523	$605,486,984	$366,329,583
464,388	4,068,035,220	464,388	4,068,035,220	0	$613,972,138	$3,612,523	$617,584,661	$362,765,899
469,031	4,108,715,572	469,031	4,108,715,572	0	$626,312,978	$3,612,523	$629,925,501	$359,237,711

(*Continued*)

TABLE 16.22 (Continued)
Alternative 4 Concentrating Solar Power Plant under Outcome 1

Outcome 1 : 1% Increase in Demand/Year

A4O1A kW Sold	A4O1B kWh Sold (24/7 Operation)	A4O1C kW Produced	A4O1D kWh Produced (24/7 Operation)	A4O1E kWh Bought	A4O1F Income from Sale of Electricity	A4O1G Economic Impact Benefits	A4O1H Total Revenue	A4O1I Present Value of Total Revenue
473,722	4,149,802,728	473,722	4,149,802,728	0	$638,901,869	$3,612,523	$642,514,392	$355,744,640
478,459	4,191,300,755	478,459	4,191,300,755	0	$651,743,796	$3,612,523	$655,356,319	$352,286,315
483,244	4,233,213,762	483,244	4,233,213,762	0	$664,843,847	$3,612,523	$668,456,369	$348,862,366
488,076	4,275,545,900	488,076	4,275,545,900	0	$678,207,208	$3,612,523	$681,819,731	$345,472,431
492,957	4,318,301,359	492,957	4,318,301,359	0	$691,839,173	$3,612,523	$695,451,696	$342,116,151
497,886	4,361,484,373	497,886	4,361,484,373	0	$705,745,140	$3,612,523	$709,357,663	$338,793,170
502,865	4,405,099,216	500,000	4,380,000,000	25,099,216	$716,918,712	$3,612,523	$720,531,234	$334,106,534
507,894	4,449,150,208	500,000	4,380,000,000	69,150,208	$726,103,198	$3,612,523	$729,715,721	$328,510,031
512,973	4,493,641,711	500,000	4,380,000,000	113,641,711	$735,525,682	$3,612,523	$739,138,205	$323,060,127
518,103	4,538,578,128	500,000	4,380,000,000	158,578,128	$745,191,482	$3,612,523	$748,804,005	$317,752,256
523,284	4,583,963,909	500,000	4,380,000,000	203,963,909	$755,106,028	$3,612,523	$758,718,551	$312,581,997
528,516	4,629,803,548	500,000	4,380,000,000	249,803,548	$765,274,863	$3,612,523	$768,887,386	$307,545,070
533,802	4,676,101,583	500,000	4,380,000,000	296,101,583	$775,703,650	$3,612,523	$779,316,173	$302,637,331
								$11,809,413,377
							Benefits − Costs	−$3,658,097,557
							Benefits/Costs	0.76

TABLE 16.23

Alternative 4 Concentrating Solar Power Plant under Outcome 2

				Outcome 2 : 2% Increase in Demand/Year				
A4O2A	A4O2B	A4O2C	A4O2D	A4O2E	A4O2F	A4O2G	A4O2H	A4O2I
kW Sold	kWh sold (24/7 Operation)	kW Produced	kWh Produced (24/7 Operation)	kWh Bought	Income from Sale of Electricity	Economic Impact Benefits	Total Revenue	Present Value of Total Revenue
0	0	0	0	0	$0	$304,788,870	$304,788,870	$304,788,870
0	0	0	0	0	$0	$304,788,870	$304,788,870	$295,911,525
0	0	0	0	0	$0	$304,788,870	$304,788,870	$287,292,742
400,000	3,504,000,000	400,000	3,504,000,000	0	$455,520,000	$3,612,523	$459,132,523	$420,171,299
408,000	3,574,080,000	408,000	3,574,080,000	0	$469,276,704	$3,612,523	$472,889,227	$420,155,953
416,160	3,645,561,600	416,160	3,645,561,600	0	$483,448,860	$3,612,523	$487,061,383	$420,143,428
424,483	3,718,472,832	424,483	3,718,472,832	0	$498,049,016	$3,612,523	$501,661,539	$420,133,641
432,973	3,792,842,289	432,973	3,792,842,289	0	$513,090,096	$3,612,523	$516,702,619	$420,126,514
441,632	3,868,699,134	441,632	3,868,699,134	0	$528,585,417	$3,612,523	$532,197,940	$420,121,968
450,465	3,946,073,117	450,465	3,946,073,117	0	$544,548,697	$3,612,523	$548,161,220	$420,119,931
459,474	4,024,994,579	459,474	4,024,994,579	0	$560,994,067	$3,612,523	$564,606,590	$420,120,328
468,664	4,105,494,471	468,664	4,105,494,471	0	$577,936,088	$3,612,523	$581,548,611	$420,123,090
478,037	4,187,604,360	478,037	4,187,604,360	0	$595,389,758	$3,612,523	$599,002,281	$420,128,148
487,598	4,271,356,448	487,598	4,271,356,448	0	$613,370,529	$3,612,523	$616,983,052	$420,135,436
497,350	4,356,783,577	497,350	4,356,783,577	0	$631,894,319	$3,612,523	$635,506,842	$420,144,889
507,297	4,443,919,248	507,297	4,443,919,248	0	$650,977,527	$3,612,523	$654,590,050	$420,156,444
517,443	4,532,797,633	517,443	4,532,797,633	0	$670,637,049	$3,612,523	$674,249,571	$420,170,042
527,792	4,623,453,586	527,792	4,623,453,586	0	$690,890,287	$3,612,523	$694,502,810	$420,185,622
538,347	4,715,922,657	538,347	4,715,922,657	0	$711,755,174	$3,612,523	$715,367,697	$420,203,128
549,114	4,810,241,111	549,114	4,810,241,111	0	$733,250,180	$3,612,523	$736,862,703	$420,222,503

(Continued)

TABLE 16.23 (Continued)
Alternative 4 Concentrating Solar Power Plant under Outcome 2

Outcome 2 : 2% Increase in Demand/Year

A4O2A	A4O2B	A4O2C	A4O2D	A4O2E	A4O2F	A4O2G	A4O2H	A4O2I
kW Sold	kWh sold (24/7 Operation)	kW Produced	kWh Produced (24/7 Operation)	kWh Bought	Income from Sale of Electricity	Economic Impact Benefits	Total Revenue	Present Value of Total Revenue
560,097	4,906,445,933	560,097	4,906,445,933	0	$755,394,336	$3,612,523	$759,006,859	$420,243,695
571,298	5,004,574,852	500,000	4,380,000,000	624,574,852	$703,258,263	$3,612,523	$706,870,785	$379,977,879
582,724	5,104,666,349	500,000	4,380,000,000	724,666,349	$714,749,142	$3,612,523	$718,361,665	$374,907,566
594,379	5,206,759,676	500,000	4,380,000,000	826,759,676	$726,709,557	$3,612,523	$730,322,080	$370,048,172
606,267	5,310,894,869	500,000	4,380,000,000	930,894,869	$739,156,140	$3,612,523	$742,768,663	$365,392,964
618,392	5,417,112,766	500,000	4,380,000,000	1,037,112,766	$752,106,071	$3,612,523	$755,718,594	$360,935,409
630,760	5,525,455,022	500,000	4,380,000,000	1,145,455,022	$765,577,100	$3,612,523	$769,189,623	$356,669,173
643,375	5,635,964,122	500,000	4,380,000,000	1,255,964,122	$779,587,565	$3,612,523	$783,200,088	$352,588,108
656,242	5,748,683,405	500,000	4,380,000,000	1,368,683,405	$794,156,410	$3,612,523	$797,768,933	$348,686,255
669,367	5,863,657,073	500,000	4,380,000,000	1,483,657,073	$809,303,202	$3,612,523	$812,915,725	$344,957,831
682,755	5,980,930,214	500,000	4,380,000,000	1,600,930,214	$825,048,155	$3,612,523	$828,660,678	$341,397,227
696,410	6,100,548,818	500,000	4,380,000,000	1,720,548,818	$841,412,148	$3,612,523	$845,024,671	$337,999,006
710,338	6,222,559,795	500,000	4,380,000,000	1,842,559,795	$858,416,747	$3,612,523	$862,029,269	$334,757,890
								$12,719,116,674
							Benefits − Costs	−$2,748,394,261
							Benefits/Costs	0.82

TABLE 16.24

Alternative 4 Concentrating Solar Power Plant under Outcome 3

A4O3A	A4O3B	A4O3C	A4O3D	A4O3E	A4O3F	A4O3G	A4O3H	A4O3I
					Outcome 3 : No Change in Demand/Year			
kW Sold	kWh sold (24/7 Operation)	kW Produced	kWh Produced (24/7 Operation)	kWh Bought	Income from Sale of Electricity	Economic Impact Benefits	Total Revenue	Present Value of Total Revenue
0	0	0	0	0	$0	$304,788,870	$304,788,870	$304,788,870
0	0	0	0	0	$0	$304,788,870	$304,788,870	$295,911,525
0	0	0	0	0	$0	$304,788,870	$304,788,870	$287,292,742
400,000	3,504,000,000	400,000	3,504,000,000	0	$455,520,000	$3,612,523	$459,132,523	$420,171,299
400,000	3,504,000,000	400,000	3,504,000,000	0	$460,075,200	$3,612,523	$463,687,723	$411,980,536
400,000	3,504,000,000	400,000	3,504,000,000	0	$464,675,952	$3,612,523	$468,288,475	$403,949,752
400,000	3,504,000,000	400,000	3,504,000,000	0	$469,322,712	$3,612,523	$472,935,234	$396,075,813
400,000	3,504,000,000	400,000	3,504,000,000	0	$474,015,939	$3,612,523	$477,628,461	$388,355,648
400,000	3,504,000,000	400,000	3,504,000,000	0	$478,756,098	$3,612,523	$482,368,621	$380,786,244
400,000	3,504,000,000	400,000	3,504,000,000	0	$483,543,659	$3,612,523	$487,156,182	$373,364,649
400,000	3,504,000,000	400,000	3,504,000,000	0	$488,379,096	$3,612,523	$491,991,618	$366,087,969
400,000	3,504,000,000	400,000	3,504,000,000	0	$493,262,887	$3,612,523	$496,875,409	$358,953,368
400,000	3,504,000,000	400,000	3,504,000,000	0	$498,195,515	$3,612,523	$501,808,038	$351,958,062
400,000	3,504,000,000	400,000	3,504,000,000	0	$503,177,471	$3,612,523	$506,789,993	$345,099,325
400,000	3,504,000,000	400,000	3,504,000,000	0	$508,209,245	$3,612,523	$511,821,768	$338,374,484
400,000	3,504,000,000	400,000	3,504,000,000	0	$513,291,338	$3,612,523	$516,903,861	$331,780,919
400,000	3,504,000,000	400,000	3,504,000,000	0	$518,424,251	$3,612,523	$522,036,774	$325,316,059
400,000	3,504,000,000	400,000	3,504,000,000	0	$523,608,494	$3,612,523	$527,221,016	$318,977,386
400,000	3,504,000,000	400,000	3,504,000,000	0	$528,844,579	$3,612,523	$532,457,101	$312,762,430
400,000	3,504,000,000	400,000	3,504,000,000	0	$534,133,024	$3,612,523	$537,745,547	$306,668,772

(Continued)

TABLE 16.24 (Continued)
Alternative 4 Concentrating Solar Power Plant under Outcome 3

Outcome 3 : No Change in Demand/Year

	A4O3A	A4O3B	A4O3C	A4O3D	A4O3E	A4O3F	A4O3G	A4O3H	A4O3I
	kW Sold	kWh sold (24/7 Operation)	kW Produced	kWh Produced (24/7 Operation)	kWh Bought	Income from Sale of Electricity	Economic Impact Benefits	Total Revenue	Present Value of Total Revenue
	400,000	3,504,000,000	400,000	3,504,000,000	0	$539,474,355	$3,612,523	$543,086,877	$300,694,036
	400,000	3,504,000,000	400,000	3,504,000,000	0	$544,869,098	$3,612,523	$548,481,621	$294,835,898
	400,000	3,504,000,000	400,000	3,504,000,000	0	$550,317,789	$3,612,523	$553,930,312	$289,092,076
	400,000	3,504,000,000	400,000	3,504,000,000	0	$555,820,967	$3,612,523	$559,433,490	$283,460,333
	400,000	3,504,000,000	400,000	3,504,000,000	0	$561,379,177	$3,612,523	$564,991,700	$277,938,478
	400,000	3,504,000,000	400,000	3,504,000,000	0	$566,992,968	$3,612,523	$570,605,491	$272,524,360
	400,000	3,504,000,000	400,000	3,504,000,000	0	$572,662,898	$3,612,523	$576,275,421	$267,215,874
	400,000	3,504,000,000	400,000	3,504,000,000	0	$578,389,527	$3,612,523	$582,002,050	$262,010,953
	400,000	3,504,000,000	400,000	3,504,000,000	0	$584,173,422	$3,612,523	$587,785,945	$256,907,572
	400,000	3,504,000,000	400,000	3,504,000,000	0	$590,015,157	$3,612,523	$593,627,679	$251,903,746
	400,000	3,504,000,000	400,000	3,504,000,000	0	$595,915,308	$3,612,523	$599,527,831	$246,997,528
	400,000	3,504,000,000	400,000	3,504,000,000	0	$601,874,461	$3,612,523	$605,486,984	$242,187,010
	400,000	3,504,000,000	400,000	3,504,000,000	0	$607,893,206	$3,612,523	$611,505,729	$237,470,321
									$10,501,894,038
								Benefits − Costs	−$4,965,616,897
								Benefits/Costs	$0.68

318 Decision-Making in Energy Systems

TABLE 16.25

Alternative 4 Concentrating Solar Power Plant under Outcome 4

Outcome 4 : 2% Decrease in Demand/Year

A4O4A kW Sold	A4O4B kWh Sold (24/7 Operation)	A4O4C kW Produced	A4O4D kWh Produced (24/7 Operation)	A4O4E kWh Bought	A4O4F Income from Sale of Electricity	A4O4G Economic Impact Benefits	A4O4H Total Revenue	A4O4I Present Value of Total Revenue
0	0	0	0	0	$0	$304,788,870	$304,788,870	$304,788,870
0	0	0	0	0	$0	$304,788,870	$304,788,870	$295,911,525
0	0	0	0	0	$0	$304,788,870	$304,788,870	$287,292,742
400,000	3,504,000,000	400,000	3,504,000,000	0	$455,520,000	$3,612,523	$459,132,523	$420,171,299
392,000	3,433,920,000	392,000	3,433,920,000	0	$450,873,696	$3,612,523	$454,486,219	$403,805,119
384,160	3,365,241,600	384,160	3,365,241,600	0	$446,274,784	$3,612,523	$449,887,307	$388,076,743
376,477	3,297,936,768	376,477	3,297,936,768	0	$441,722,782	$3,612,523	$445,335,304	$372,961,306
368,947	3,231,978,033	368,947	3,231,978,033	0	$437,217,209	$3,612,523	$440,829,732	$358,434,913
361,568	3,167,338,472	361,568	3,167,338,472	0	$432,757,594	$3,612,523	$436,370,116	$344,474,600
354,337	3,103,991,703	354,337	3,103,991,703	0	$428,343,466	$3,612,523	$431,955,989	$331,058,298
347,250	3,041,911,868	347,250	3,041,911,868	0	$423,974,363	$3,612,523	$427,586,886	$318,164,800
340,305	2,981,073,631	340,305	2,981,073,631	0	$419,649,824	$3,612,523	$423,262,347	$305,773,725
333,499	2,921,452,159	333,499	2,921,452,159	0	$415,369,396	$3,612,523	$418,981,919	$293,865,488
326,829	2,863,023,115	326,829	2,863,023,115	0	$411,132,628	$3,612,523	$414,745,151	$282,421,266
320,293	2,805,762,653	320,293	2,805,762,653	0	$406,939,075	$3,612,523	$410,551,598	$271,422,972
313,887	2,749,647,400	313,887	2,749,647,400	0	$402,788,297	$3,612,523	$406,400,820	$260,853,222
307,609	2,694,654,452	307,609	2,694,654,452	0	$398,679,856	$3,612,523	$402,292,379	$250,695,311
301,457	2,640,761,363	301,457	2,640,761,363	0	$394,613,322	$3,612,523	$398,225,845	$240,933,185
295,428	2,587,946,136	295,428	2,587,946,136	0	$390,588,266	$3,612,523	$394,200,789	$231,551,418
289,519	2,536,187,213	289,519	2,536,187,213	0	$386,604,266	$3,612,523	$390,216,788	$222,535,182

(Continued)

TABLE 16.25 (Continued)
Alternative 4 Concentrating Solar Power Plant under Outcome 4

A4O4A	A4O4B	A4O4C	A4O4D	A4O4E	A4O4F	A4O4G	A4O4H	A4O4I
			Outcome 4 : 2% Decrease in Demand/Year					
kW Sold	kWh Sold (24/7 Operation)	kW Produced	kWh Produced (24/7 Operation)	kWh Bought	Income from Sale of Electricity	Economic Impact Benefits	Total Revenue	Present Value of Total Revenue
283,729	2,485,463,469	283,729	2,485,463,469	0	$382,660,902	$3,612,523	$386,273,425	$213,870,230
278,054	2,435,754,199	278,054	2,435,754,199	0	$378,757,761	$3,612,523	$382,370,284	$205,542,869
272,493	2,387,039,115	272,493	2,387,039,115	0	$374,894,432	$3,612,523	$378,506,954	$197,539,941
267,043	2,339,298,333	267,043	2,339,298,333	0	$371,070,508	$3,612,523	$374,683,031	$189,848,800
261,702	2,292,512,366	261,702	2,292,512,366	0	$367,285,589	$3,612,523	$370,898,112	$182,457,294
256,468	2,246,662,119	256,468	2,246,662,119	0	$363,539,276	$3,612,523	$367,151,799	$175,353,744
251,339	2,201,728,877	251,339	2,201,728,877	0	$359,831,176	$3,612,523	$363,443,698	$168,526,927
246,312	2,157,694,299	246,312	2,157,694,299	0	$356,160,898	$3,612,523	$359,773,420	$161,966,056
241,386	2,114,540,413	241,386	2,114,540,413	0	$352,528,056	$3,612,523	$356,140,579	$155,660,768
236,558	2,072,249,605	236,558	2,072,249,605	0	$348,932,270	$3,612,523	$352,544,793	$149,601,101
231,827	2,030,804,613	231,827	2,030,804,613	0	$345,373,161	$3,612,523	$348,985,684	$143,777,481
227,190	1,990,188,521	227,190	1,990,188,521	0	$341,850,355	$3,612,523	$345,462,878	$138,180,710
222,647	1,950,384,750	222,647	1,950,384,750	0	$338,363,481	$3,612,523	$341,976,004	$132,801,947
								$8,400,319,851

Benefits − Costs   −$7,067,191,084

Benefits/Costs   0.54

TABLE 16.26

Alternative 5 Not Building a New Power Plant and Reselling Power Purchased from Other Power Companies under Outcome 1

Outcome 1 : 1% Increase in Demand/Year

A5O1A	A5O1B	A5O1C	A5O1D	A5O1E	A5O1F	A5O1G	A5O1H	A5O1I	A5O1J	A5O1K
Years	Elect Selling Price ($/kwh)	kWh Bought	Electricity Purchase Cost	Total Cost	Present Value of Total Cost	kW Sold	kWh Sold (24/7 operation)	Revenue from Selling Electricity	Total Revenue	Present Value of Total Revenue
1		0	$0	$0	$0	0	0	$0	$0	$0
2		0	$0	$0	$0	0	0	$0	$0	$0
3		0	$0	$0	$0	0	0	$0	$0	$0
4	$0.130	3,504,000,000	−$420,480,000	−$420,480,000	−$384,798,765	400,000	3,504,000,000	$455,520,000	$455,520,000	$416,865,329
5	$0.131	3,539,040,000	−$424,684,800	−$424,684,800	−$377,326,944	404,000	3,539,040,000	$464,675,952	$464,675,952	$412,858,565
6	$0.133	3,574,430,400	−$428,931,648	−$428,931,648	−$370,000,207	408,040	3,574,430,400	$474,015,939	$474,015,939	$408,890,313
7	$0.134	3,610,174,704	−$433,220,964	−$433,220,964	−$362,815,737	412,120	3,610,174,704	$483,543,659	$483,543,659	$404,960,202
8	$0.135	3,646,276,451	−$437,553,174	−$437,553,174	−$355,770,772	416,242	3,646,276,451	$493,262,887	$493,262,887	$401,067,866
9	$0.137	3,682,739,216	−$441,928,706	−$441,928,706	−$348,862,601	420,404	3,682,739,216	$503,177,471	$503,177,471	$397,212,942
10	$0.138	3,719,566,608	−$446,347,993	−$446,347,993	−$342,088,570	424,608	3,719,566,608	$513,291,338	$513,291,338	$393,395,070
11	$0.139	3,756,762,274	−$450,811,473	−$450,811,473	−$335,446,074	428,854	3,756,762,274	$523,608,494	$523,608,494	$389,613,894
12	$0.141	3,794,329,897	−$455,319,588	−$455,319,588	−$328,932,558	433,143	3,794,329,897	$534,133,024	$534,133,024	$385,869,061
13	$0.142	3,832,273,195	−$459,872,783	−$459,872,783	−$322,545,518	437,474	3,832,273,195	$544,869,098	$544,869,098	$382,160,223
14	$0.144	3,870,595,927	−$464,471,511	−$464,471,511	−$316,282,498	441,849	3,870,595,927	$555,820,967	$555,820,967	$378,487,032
15	$0.145	3,909,301,887	−$469,116,226	−$469,116,226	−$310,141,090	446,267	3,909,301,887	$566,992,968	$566,992,968	$374,849,147
16	$0.146	3,948,394,906	−$473,807,389	−$473,807,389	−$304,118,933	450,730	3,948,394,906	$578,389,527	$578,389,527	$371,246,228
17	$0.148	3,987,878,855	−$478,545,463	−$478,545,463	−$298,213,711	455,237	3,987,878,855	$590,015,157	$590,015,157	$367,677,939
18	$0.149	4,027,757,643	−$483,330,917	−$483,330,917	−$292,423,154	459,790	4,027,757,643	$601,874,461	$601,874,461	$364,143,947
19	$0.151	4,068,035,220	−$488,164,226	−$488,164,226	−$286,745,034	464,388	4,068,035,220	$613,972,138	$613,972,138	$360,643,923
20	$0.152	4,108,715,572	−$493,045,869	−$493,045,869	−$281,177,169	469,031	4,108,715,572	$626,312,978	$626,312,978	$357,177,540

(Continued)

TABLE 16.26 (Continued)
Alternative 5 Not Building a New Power Plant and Reselling Power Purchased from Other Power Companies under Outcome 1

Outcome 1 : 1% Increase in Demand/Year

Years	Elect Selling Price ($/kwh)	kWh Bought	Electricity Purchase Cost	Total Cost	Present Value of Total Cost	kW Sold	kWh Sold (24/7 operation)	Revenue from Selling Electricity	Total Revenue	Present Value of Total Revenue
21	$0.154	4,149,802,728	−$497,976,327	−$497,976,327	−$275,717,419	473,722	4,149,802,728	$638,901,869	$638,901,869	$353,744,474
22	$0.155	4,191,300,755	−$502,956,091	−$502,956,091	−$270,363,682	478,459	4,191,300,755	$651,743,796	$651,743,796	$350,344,406
23	$0.157	4,233,213,762	−$507,985,651	−$507,985,651	−$265,113,902	483,244	4,233,213,762	$664,843,847	$664,843,847	$346,977,018
24	$0.159	4,275,545,900	−$513,065,508	−$513,065,508	−$259,966,059	488,076	4,275,545,900	$678,207,208	$678,207,208	$343,641,996
25	$0.160	4,318,301,359	−$518,196,163	−$518,196,163	−$254,918,175	492,957	4,318,301,359	$691,839,173	$691,839,173	$340,339,029
26	$0.162	4,361,484,373	−$523,378,125	−$523,378,125	−$249,968,307	497,886	4,361,484,373	$705,745,140	$705,745,140	$337,067,809
27	$0.163	4,405,099,216	−$528,611,906	−$528,611,906	−$245,114,554	502,865	4,405,099,216	$719,930,617	$719,930,617	$333,828,031
28	$0.165	4,449,150,208	−$533,898,025	−$533,898,025	−$240,355,048	507,894	4,449,150,208	$734,401,223	$734,401,223	$330,619,393
29	$0.167	4,493,641,711	−$539,237,005	−$539,237,005	−$235,687,959	512,973	4,493,641,711	$749,162,687	$749,162,687	$327,441,595
30	$0.168	4,538,578,128	−$544,629,375	−$544,629,375	−$231,111,494	518,103	4,538,578,128	$764,220,858	$764,220,858	$324,294,341
31	$0.170	4,583,963,909	−$550,075,669	−$550,075,669	−$226,623,892	523,284	4,583,963,909	$779,581,697	$779,581,697	$321,177,337
32	$0.172	4,629,803,548	−$555,576,426	−$555,576,426	−$222,223,428	528,516	4,629,803,548	$795,251,289	$795,251,289	$318,090,293
33	$0.173	4,676,101,583	−$561,132,190	−$561,132,190	−$217,908,410	533,802	4,676,101,583	$811,235,840	$811,235,840	$315,032,920
					−$8,812,761,667				$10,909,717,863	$2,096,956,196

Benefits − Costs
Benefits/Costs 1.24

TABLE 16.27

Alternative 5 Not Building a New Power Plant and Reselling Power Purchased from Other Power Companies under Outcome 2

Outcome 2 : 2% Increase in Demand/Year

A5O2A	A5O2B	A5O2C	A5O2D	A5O2E	A5O2F	A5O2G	A5O2H	A5O2I
kWh Bought	Purchase Cost	Total Cost	Present Value of Total Cost	kW Sold	kWh Sold (24/7 Operation)	Revenue from Selling Electricity	Total Revenue	Present Value of Total Revenue
0	$0	$0	$0	0	0	$0	$0	$0
0	$0	$0	$0	0	0	$0	$0	$0
0	$0	$0	$0	0	0	$0	$0	$0
3,504,000,000	−$420,480,000	−$420,480,000	−$384,798,765	400,000	3,504,000,000	$455,520,000	$455,520,000	$416,865,329
3,574,080,000	−$428,889,600	−$428,889,600	−$381,062,855	408,000	3,574,080,000	$469,276,704	$469,276,704	$416,946,273
3,645,561,600	−$437,467,392	−$437,467,392	−$377,363,215	416,160	3,645,561,600	$483,448,860	$483,448,860	$417,027,234
3,718,472,832	−$446,216,740	−$446,216,740	−$373,699,495	424,483	3,718,472,832	$498,049,016	$498,049,016	$417,108,210
3,792,842,289	−$455,141,075	−$455,141,075	−$370,071,344	432,973	3,792,842,289	$513,090,096	$513,090,096	$417,189,202
3,868,699,134	−$464,243,896	−$464,243,896	−$366,478,419	441,632	3,868,699,134	$528,585,417	$528,585,417	$417,270,209
3,946,073,117	−$473,528,774	−$473,528,774	−$362,920,376	450,465	3,946,073,117	$544,548,697	$544,548,697	$417,351,233
4,024,994,579	−$482,999,350	−$482,999,350	−$359,396,877	459,474	4,024,994,579	$560,994,067	$560,994,067	$417,432,272
4,105,494,471	−$492,659,337	−$492,659,337	−$355,907,587	468,664	4,105,494,471	$577,936,088	$577,936,088	$417,513,327
4,187,604,360	−$502,512,523	−$502,512,523	−$352,452,173	478,037	4,187,604,360	$595,389,758	$595,389,758	$417,594,397
4,271,356,448	−$512,562,774	−$512,562,774	−$349,030,308	487,598	4,271,356,448	$613,370,529	$613,370,529	$417,675,484
4,356,783,577	−$522,814,029	−$522,814,029	−$345,641,664	497,350	4,356,783,577	$631,894,319	$631,894,319	$417,756,586
4,443,919,248	−$533,270,310	−$533,270,310	−$342,285,920	507,297	4,443,919,248	$650,977,527	$650,977,527	$417,837,703
4,532,797,633	−$543,935,716	−$543,935,716	−$338,962,755	517,443	4,532,797,633	$670,637,049	$670,637,049	$417,918,837
4,623,453,586	−$554,814,430	−$554,814,430	−$335,671,855	527,792	4,623,453,586	$690,890,287	$690,890,287	$417,999,986
4,715,922,657	−$565,910,719	−$565,910,719	−$332,412,905	538,347	4,715,922,657	$711,755,174	$711,755,174	$418,081,151

(Continued)

TABLE 16.27 (Continued)
Alternative 5 Not Building a New Power Plant and Reselling Power Purchased from Other Power Companies under Outcome 2

Outcome 2 : 2% Increase in Demand/Year

A5O2A kWh Bought	A5O2B Purchase Cost	A5O2C Total Cost	A5O2D Present Value of Total Cost	A5O2E kW Sold	A5O2F kWh Sold (24/7 Operation)	A5O2G Revenue from Selling Electricity	A5O2H Total Revenue	A5O2I Present Value of Total Revenue
4,810,241,111	−$577,228,933	−$577,228,933	−$329,185,595	549,114	4,810,241,111	$733,250,180	$733,250,180	$418,162,332
4,906,445,933	−$588,773,512	−$588,773,512	−$325,989,618	560,097	4,906,445,933	$755,394,336	$755,394,336	$418,243,529
5,004,574,852	−$600,548,982	−$600,548,982	−$322,824,671	571,298	5,004,574,852	$778,207,245	$778,207,245	$418,324,741
5,104,666,349	−$612,559,962	−$612,559,962	−$319,690,450	582,724	5,104,666,349	$801,709,104	$801,709,104	$418,405,969
5,206,759,676	−$624,811,161	−$624,811,161	−$316,586,660	594,379	5,206,759,676	$825,920,719	$825,920,719	$418,487,213
5,310,894,869	−$637,307,384	−$637,307,384	−$313,513,003	606,267	5,310,894,869	$850,863,524	$850,863,524	$418,568,473
5,417,112,766	−$650,053,532	−$650,053,532	−$310,469,187	618,392	5,417,112,766	$876,559,603	$876,559,603	$418,649,748
5,525,455,022	−$663,054,603	−$663,054,603	−$307,454,923	630,760	5,525,455,022	$903,031,703	$903,031,703	$418,731,039
5,635,964,122	−$676,315,695	−$676,315,695	−$304,469,924	643,375	5,635,964,122	$930,303,260	$930,303,260	$418,812,346
5,748,683,405	−$689,842,009	−$689,842,009	−$301,513,905	656,242	5,748,683,405	$958,398,419	$958,398,419	$418,893,669
5,863,657,073	−$703,638,849	−$703,638,849	−$298,586,586	669,367	5,863,657,073	$987,342,051	$987,342,051	$418,975,008
5,980,930,214	−$717,711,626	−$717,711,626	−$295,687,687	682,755	5,980,930,214	$1,017,159,781	$1,017,159,781	$419,056,362
6,100,548,818	−$732,065,858	−$732,065,858	−$292,816,933	696,410	6,100,548,818	$1,047,878,006	$1,047,878,006	$419,137,732
6,222,559,795	−$746,707,175	−$746,707,175	−$289,974,050	710,338	6,222,559,795	$1,079,523,922	$1,079,523,922	$419,219,118
			−$10,056,919,702					$12,541,234,711
							Benefits − Costs	$2,484,315,009
							Benefits/Costs	1.25

TABLE 16.28
Alternative 5 Not Building a New Power Plant and Reselling Power Purchased from Other Power Companies under Outcome 3

Outcome 3 : 0% Increase in Demand/Year

A5O3A	A5O3B	A5O3C	A5O3D	A5O3E	A5O3F	A5O3G	A5O3H	A5O3I
kWh Bought	Purchase Cost	Total Cost	Present Value of Total Cost	kW Sold	kWh sold (24/7 operation)	Revenue from Selling Electricity	Total Revenue	Present Value of Total Revenue
0	$0	$0	$0	0	0	$0	$0	$0
0	$0	$0	$0	0	0	$0	$0	$0
0	$0	$0	$0	0	0	$0	$0	$0
3,504,000,000	−$420,480,000	−$420,480,000	−$384,798,765	400,000	3,504,000,000	$455,520,000	$455,520,000	$416,865,329
3,504,000,000	−$420,480,000	−$420,480,000	−$373,591,034	400,000	3,504,000,000	$460,075,200	$460,075,200	$408,770,856
3,504,000,000	−$420,480,000	−$420,480,000	−$362,709,742	400,000	3,504,000,000	$464,675,952	$464,675,952	$400,833,558
3,504,000,000	−$420,480,000	−$420,480,000	−$352,145,380	400,000	3,504,000,000	$469,322,712	$469,322,712	$393,050,382
3,504,000,000	−$420,480,000	−$420,480,000	−$341,888,719	400,000	3,504,000,000	$474,015,939	$474,015,939	$385,418,336
3,504,000,000	−$420,480,000	−$420,480,000	−$331,930,795	400,000	3,504,000,000	$478,756,098	$478,756,098	$377,934,485
3,504,000,000	−$420,480,000	−$420,480,000	−$322,262,908	400,000	3,504,000,000	$483,543,659	$483,543,659	$370,595,951
3,504,000,000	−$420,480,000	−$420,480,000	−$312,876,609	400,000	3,504,000,000	$488,379,096	$488,379,096	$363,399,913
3,504,000,000	−$420,480,000	−$420,480,000	−$303,763,698	400,000	3,504,000,000	$493,262,887	$493,262,887	$356,343,604
3,504,000,000	−$420,480,000	−$420,480,000	−$294,916,212	400,000	3,504,000,000	$498,195,515	$498,195,515	$349,424,311
3,504,000,000	−$420,480,000	−$420,480,000	−$286,326,419	400,000	3,504,000,000	$503,177,471	$503,177,471	$342,639,373
3,504,000,000	−$420,480,000	−$420,480,000	−$277,986,815	400,000	3,504,000,000	$508,209,245	$508,209,245	$335,986,181
3,504,000,000	−$420,480,000	−$420,480,000	−$269,890,112	400,000	3,504,000,000	$513,291,338	$513,291,338	$329,462,178
3,504,000,000	−$420,480,000	−$420,480,000	−$262,029,235	400,000	3,504,000,000	$518,424,251	$518,424,251	$323,064,854
3,504,000,000	−$420,480,000	−$420,480,000	−$254,397,315	400,000	3,504,000,000	$523,608,494	$523,608,494	$316,791,750
3,504,000,000	−$420,480,000	−$420,480,000	−$246,987,685	400,000	3,504,000,000	$528,844,579	$528,844,579	$310,640,454

(Continued)

TABLE 16.28 (Continued)

Alternative 5 Not Building a New Power Plant and Reselling Power Purchased from Other Power Companies under Outcome 3

Outcome 3 : 0% Increase in Demand/Year

	A5O3A	A5O3B	A5O3C	A5O3D	A5O3E	A5O3F	A5O3G	A5O3H	A5O3I
	kWh Bought	Purchase Cost	Total Cost	Present Value of Total Cost	kW Sold	kWh sold (24/7 operation)	Revenue from Selling Electricity	Total Revenue	Present Value of Total Revenue
	3,504,000,000	−$420,480,000	−$420,480,000	−$239,793,869	400,000	3,504,000,000	$534,133,024	$534,133,024	$304,608,600
	3,504,000,000	−$420,480,000	−$420,480,000	−$232,809,581	400,000	3,504,000,000	$539,474,355	$539,474,355	$298,693,870
	3,504,000,000	−$420,480,000	−$420,480,000	−$226,028,720	400,000	3,504,000,000	$544,869,098	$544,869,098	$292,893,989
	3,504,000,000	−$420,480,000	−$420,480,000	−$219,445,359	400,000	3,504,000,000	$550,317,789	$550,317,789	$287,206,727
	3,504,000,000	−$420,480,000	−$420,480,000	−$213,053,746	400,000	3,504,000,000	$555,820,967	$555,820,967	$281,629,898
	3,504,000,000	−$420,480,000	−$420,480,000	−$206,848,297	400,000	3,504,000,000	$561,379,177	$561,379,177	$276,161,356
	3,504,000,000	−$420,480,000	−$420,480,000	−$200,823,590	400,000	3,504,000,000	$566,992,968	$566,992,968	$270,798,999
	3,504,000,000	−$420,480,000	−$420,480,000	−$194,974,359	400,000	3,504,000,000	$572,662,898	$572,662,898	$265,540,766
	3,504,000,000	−$420,480,000	−$420,480,000	−$189,295,494	400,000	3,504,000,000	$578,389,527	$578,389,527	$260,384,635
	3,504,000,000	−$420,480,000	−$420,480,000	−$183,782,033	400,000	3,504,000,000	$584,173,422	$584,173,422	$255,328,623
	3,504,000,000	−$420,480,000	−$420,480,000	−$178,429,158	400,000	3,504,000,000	$590,015,157	$590,015,157	$250,370,785
	3,504,000,000	−$420,480,000	−$420,480,000	−$173,232,193	400,000	3,504,000,000	$595,915,308	$595,915,308	$245,509,217
	3,504,000,000	−$420,480,000	−$420,480,000	−$168,186,595	400,000	3,504,000,000	$601,874,461	$601,874,461	$240,742,047
	3,504,000,000	−$420,480,000	−$420,480,000	−$163,287,956	400,000	3,504,000,000	$607,893,206	$607,893,206	$236,067,445
				−$7,768,492,392					$9,547,158,472
								Benefits − Costs	$1,778,666,080
								Benefits/Costs	1.23

TABLE 16.29

Alternative 5 Not Building a New Power Plant and Reselling Power Purchased from Other Power Companies under Outcome 4

A5O4A	A5O4B	A5O4C	A5O4D	A5O4E	A5O4F	A5O4G	A5O4H	A5O4I
			Outcome 4 : 2% Decrease in Demand/Year					
kWh Bought	Purchase Cost	Total Cost	Present Value of Total Cost	kW Sold	kWh Sold (24/7 Operation)	Revenue from Selling Electricity	Total Revenue	Present Value of Total Revenue
0	$0	$0	$0	0	0	$0	$0	$0
0	$0	$0	$0	0	0	$0	$0	$0
0	$0	$0	$0	0	0	$0	$0	$0
3,504,000,000	−$420,480,000	−$420,480,000	−$384,798,765	400,000	3,504,000,000	$455,520,000	$455,520,000	$416,865,329
3,433,920,000	−$412,070,400	−$412,070,400	−$366,119,213	392,000	3,433,920,000	$450,873,696	$450,873,696	$400,595,439
3,365,241,600	−$403,828,992	−$403,828,992	−$348,346,436	384,160	3,365,241,600	$446,274,784	$446,274,784	$384,960,549
3,297,936,768	−$395,752,412	−$395,752,412	−$331,436,415	376,477	3,297,936,768	$441,722,782	$441,722,782	$369,935,875
3,231,978,033	−$387,837,364	−$387,837,364	−$315,347,268	368,947	3,231,978,033	$437,217,209	$437,217,209	$355,497,601
3,167,338,472	−$380,080,617	−$380,080,617	−$300,039,149	361,568	3,167,338,472	$432,757,594	$432,757,594	$341,622,841
3,103,991,703	−$372,479,004	−$372,479,004	−$285,474,141	354,337	3,103,991,703	$428,343,466	$428,343,466	$328,289,600
3,041,911,868	−$365,029,424	−$365,029,424	−$271,616,173	347,250	3,041,911,868	$423,974,363	$423,974,363	$315,476,743
2,981,073,631	−$357,728,836	−$357,728,836	−$258,430,922	340,305	2,981,073,631	$419,649,824	$419,649,824	$303,163,962
2,921,452,159	−$350,574,259	−$350,574,259	−$245,885,732	333,499	2,921,452,159	$415,369,396	$415,369,396	$291,331,737
2,863,023,115	−$343,562,774	−$343,562,774	−$233,949,531	326,829	2,863,023,115	$411,132,628	$411,132,628	$279,961,314
2,805,762,653	−$336,691,518	−$336,691,518	−$222,592,758	320,293	2,805,762,653	$406,939,075	$406,939,075	$269,034,669
2,749,647,400	−$329,957,688	−$329,957,688	−$211,787,284	313,887	2,749,647,400	$402,788,297	$402,788,297	$258,534,481
2,694,654,452	−$323,358,534	−$323,358,534	−$201,506,348	307,609	2,694,654,452	$398,679,856	$398,679,856	$248,444,106
2,640,761,363	−$316,891,364	−$316,891,364	−$191,724,486	301,457	2,640,761,363	$394,613,322	$394,613,322	$238,747,549
2,587,946,136	−$310,553,536	−$310,553,536	−$182,417,473	295,428	2,587,946,136	$390,588,266	$390,588,266	$229,429,441

(Continued)

TABLE 16.29 (Continued)
Alternative 5 Not Building a New Power Plant and Reselling Power Purchased from Other Power Companies under Outcome 4

Outcome 4 : 2% Decrease in Demand/Year

A5O4A	A5O4B	A5O4C	A5O4D	A5O4E	A5O4F	A5O4G	A5O4H	A5O4I
kWh Bought	Purchase Cost	Total Cost	Present Value of Total Cost	kW Sold	kWh Sold (24/7 Operation)	Revenue from Selling Electricity	Total Revenue	Present Value of Total Revenue
2,536,187,213	-$304,342,466	-$304,342,466	-$173,562,255	289,519	2,536,187,213	$386,604,266	$386,604,266	$220,475,011
2,485,463,469	-$298,255,616	-$298,255,616	-$165,136,903	283,729	2,485,463,469	$382,660,902	$382,660,902	$211,870,064
2,435,754,199	-$292,290,504	-$292,290,504	-$157,120,549	278,054	2,435,754,199	$378,757,761	$378,757,761	$203,600,960
2,387,039,115	-$286,444,694	-$286,444,694	-$149,493,338	272,493	2,387,039,115	$374,894,432	$374,894,432	$195,654,592
2,339,298,333	-$280,715,800	-$280,715,800	-$142,236,379	267,043	2,339,298,333	$371,070,508	$371,070,508	$188,018,365
2,292,512,366	-$275,101,484	-$275,101,484	-$135,331,701	261,702	2,292,512,366	$367,285,589	$367,285,589	$180,680,172
2,246,662,119	-$269,599,454	-$269,599,454	-$128,762,201	256,468	2,246,662,119	$363,539,276	$363,539,276	$173,628,383
2,201,728,877	-$264,207,465	-$264,207,465	-$122,511,609	251,339	2,201,728,877	$359,831,176	$359,831,176	$166,851,819
2,157,694,299	-$258,923,316	-$258,923,316	-$116,564,443	246,312	2,157,694,299	$356,160,898	$356,160,898	$160,339,738
2,114,540,413	-$253,744,850	-$253,744,850	-$110,905,975	241,386	2,114,540,413	$352,528,056	$352,528,056	$154,081,818
2,072,249,605	-$248,669,953	-$248,669,953	-$105,522,190	236,558	2,072,249,605	$348,932,270	$348,932,270	$148,068,140
2,030,804,613	-$243,696,554	-$243,696,554	-$100,399,753	231,827	2,030,804,613	$345,373,161	$345,373,161	$142,289,169
1,990,188,521	-$238,822,622	-$238,822,622	-$95,525,979	227,190	1,990,188,521	$341,850,355	$341,850,355	$136,735,748
1,950,384,750	-$234,046,170	-$234,046,170	-$90,888,796	222,647	1,950,384,750	$338,363,481	$338,363,481	$131,399,071
			-$6,145,434,165					$7,445,584,285
							Benefits – Costs	$1,300,150,120
							Benefits/Costs	1.21

SUMMARY OF RESULTS

Table 16.30 presents the decision matrix obtained by compiling the present values of the benefits minus costs obtained from the above five spreadsheets shown in Tables 16.6–16.29. With the occurrence probabilities of the outcomes, an expected value was calculated for each alternative. The expected value is determined by multiplying each value in a column (represented by an outcome) by the associated outcome probability and then taking the sum of all the multiplication values over all columns (represented by an alternative). The alternatives can be compared with one another using their expected values.

The last three columns of Table 16.30 show calculated values using three other principles described in Chapter 6, namely maximin (most pessimistic), maxmax (most optimistic), and most probable outcome principle. The data shows that alternative 2 has the highest values of benefits minus the costs using all the four principles.

Figure 16.1 shows a plot of the expected values of the benefits minus the costs for the five alternatives. The alternative 2 has the highest value of $ 5.357 billion.

Table 16.31 presents the decision matrix obtained by compiling the values of the benefit-to-cost ratios obtained from the five spreadsheets shown in Tables 16.6–16.29. The expected values of the ratios for the five alternatives are presented in the last column of this table. The data show that alternative 2 has the highest value of 1.96. The solar and geothermal plants had benefit-to-cost ratios lower than 1.0, indicating that they will not be profitable. The expected values of the benefits minus costs for these two plants were also negative (see Figure 16.1).

Figure 16.2 presents a plot of the expected values of the benefit-to-the cost ratios for the five alternatives. The alternative 2 has the highest value of 1.96.

CONCLUSIONS

The cost–benefit analysis presented above showed that the natural-gas-fueled power plant is the most desirable alternative for adding 500 MW of supply in the mid-west portion of the United States. This alternative is closely followed by a wind turbine power plant.

Alternative 5 of not building a new plant and instead negotiating a purchase agreement from an external electric energy provider is the third desirable alternative with the ratio of benefit-to-cost value of 1.24 (over 1.0). This alternative is only better than concentrated solar and geothermal, as the benefit-to-cost ratios of the solar and geothermal are 0.76 and 0.87 (both below 1.0), respectively.

The wind turbine plant alternative is technically feasible overall, but if energy demand is decreasing year over year, it will not be able to generate a profit. The concentrating solar alternative is not feasible due to its enormous cost to construct the solar plant to generate 500 MW (2,000 MW nameplate capacity) (see Figure 16.3). If the concentrating solar plant were implemented in a sunny state closer to the equator, there would be less initial capital cost needed, as the reflector area would be decreased.

TABLE 16.30
Decision Matrix Using Net Cumulative Benefits Minus Costs

Occurrence Probability →	0.25	0.5	0.2	0.05				
	Outcomes & Occurrence Probabilities							
Alternatives ↓	O1	O2	O3	O4	Expected Value	Maximin	Maxmax	Most Probable Outcome = O2
A1: Wind	$4,776,732,815	$5,686,436,111	$3,469,213,475	$1,367,639,288	$4,799,625,919	$1,367,639,288	$5,686,436,111	$5,686,436,111
A2: Natural gas	$5,267,552,305	$6,186,205,127	$4,165,520,549	$2,274,979,877	$5,356,843,743	$2,274,979,877	$6,186,205,127	$6,186,205,127
A3: Geothermal	−$1,603,780,409	−$694,077,113	−$2,911,299,749	−$5,012,873,936	−$1,580,887,305	−$5,012,873,936	−$694,077,113	−$694,077,113
A4: Concentrating solar	−$3,658,097,557	−$2,748,394,261	−$4,965,616,897	−$7,067,191,084	−$3,635,204,453	−$7,067,191,084	−$2,748,394,261	−$2,748,394,261
A5: Purchasing agreement	$2,096,956,196	$2,484,315,009	$1,778,666,080	$1,300,150,120	$2,187,137,276	$1,300,150,120	$2,484,315,009	$2,484,315,009

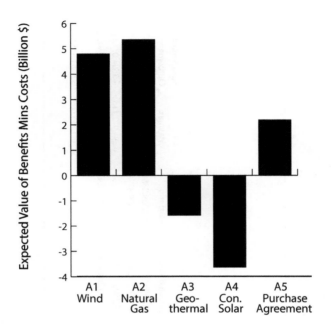

FIGURE 16.1   Expected values of benefits minus costs of the five alternatives.

TABLE 16.31

Benefit-to-Cost Ratios for the Five Alternatives

	Outcomes & Occurrence Probabilities				
Occurrence Probability →	0.25	0.5	0.2	0.05	
Alternatives ↓	O1	O2	O3	O4	Expected Value
A1: Wind	1.77	1.92	1.56	1.22	1.77
A2: Natural Gas	1.94	2.10	1.77	1.44	1.96
A3: Geothermal	0.87	0.94	0.77	0.60	0.87
A4: Concentrating Solar	0.76	0.82	0.68	0.54	0.76
A5: Purchasing Agreement	1.24	1.25	1.23	1.21	1.24

OBSERVATIONS FROM THIS ANALYSIS AND IMPROVEMENTS

1. *Usefulness of the analysis*: The cost–benefit analysis was found to be a useful tool to compare technology alternatives for a large-scale project such as
building a new power plant. As always, the quality of the data will impact
the results of the analysis. Additional research and analyses to obtain more
reliable estimates of benefits and costs of each alternative, as well as the
possibility of including other factors such as geographic, weather, and geological data (to identify suitable locations for wind/geothermal/solar as well

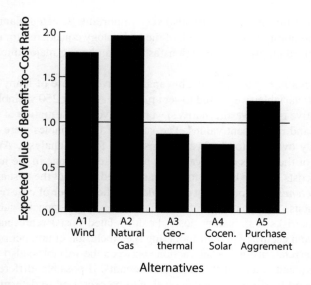

FIGURE 16.2 Expected value of benefit to-cost ratios for the five alternatives.

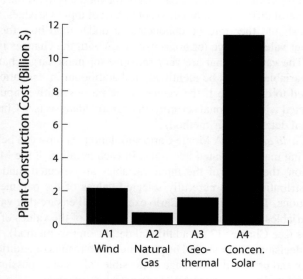

FIGURE 16.3 Comparison of construction costs of the four plants.

as availability of rebates) are also very important. Selected plant locations and their ability to support each of the technology considered for each alternative need careful considerations as they can have a major impact on the results.

2. *Large number of variables*: This analysis involved use of many variables. Over 60 variables were used to set up inputs. Another 250 variables (50 per alternative times five alternatives) were involved in estimation of costs, revenues, and net present values. The values of the variables were computed annually over the total 33 years considered for the analysis. Any change in any of the values of the variables can make a change in the net benefits minus costs and the benefit-to-cost ratios used to select the best alternative.

3. *Completeness and accuracy of the model*: The number of cost-related considerations, representation of technical issues (e.g., capacity factors), and types of risks (e.g., safety costs, health-related costs) represented in the model and accuracy in their modeling (i.e., inclusion of number of variables and interrelationships or interactions between the variables) also will affect the predicted values of the costs and benefits. If possible, different models (developed by different modelers) should be exercised to determine differences and agreements in the predicted alternatives and values of the output variables.

4. *Sensitivity analysis*: One method to understand the effect of changes in the values of input variables on the outputs of the model is to exercise the model under several different combinations of values of input variables. Sensitivity analysis should therefore be conducted to understand how the values of the output values change (or are sensitive to) with the changes in the input values. The variables that are very sensitive (or make large change) to the output variables should be identified, and additional investigations may be warranted to determine if the values of the more sensitive input variables are selected with additional scrutiny and careful analysis (see Chapter 17 for additional details on this method).

5. *Monte Carlo analysis*: A Monte Carlo simulation exercise can be conducted by creating many simulated iterations. In each iteration of the Monte Carlo simulation, the values of the input variables are selected randomly from their distributions with carefully selected values of the parameters of the distributions. Thus, the Monte Carlo exercise will ensure that variability in input variables is accounted for in the prediction of the values of the output variables (see Chapter 17 for additional details on this method).

6. Thus, a decision-maker should be presented with results of additional model exercises and provided with range of possible values (e.g., pessimistic, most likely, and optimistic estimates) of the decision variables.

CONCLUDING REMARKS

The cost–benefit model presented in this chapter was essentially a mathematical model to compute values of benefits and costs. The models are useful tools as they can be exercised under a variety of "what if" situations. The model itself relies of the modeler's abilities to understand alternatives and outcomes, determine relevant parameters (independent) variables, dependent (response) variables, and model the relationships between them. The decision-makers such as the company management who review the outputs of the model also need to understand the details of the relationships modeled and values of the input variables selected for the analysis. They also need to understand limitations of such models. Generally, a good decision-maker gathers results from a number of exercises of different models, experiments, and subjective methods by consulting several subject matter experts before a final decision is made. The final decision also can be changed if later data shows substantial differences in the assumptions made during earlier exercises.

REFERENCES

Emmen, G. 2020. *Cost-Benefit Analysis of Energy Systems.* Project P-4 Report, ESE 504 course held in winter 2020 term, Dearborn, MI: University of Michigan-Dearborn.

National Renewable Energy Laboratory (NREL). 2020a. JEDI: Jobs & Economic Development Impact Models. Website: https://www.nrel.gov/analysis/jedi/models.html (Accessed: May 22, 2020).

National Renewable Energy Laboratory (NREL). 2020b. PVWatts Calculator. Website: NREL: https://pvwatts.nrel.gov/ (Accessed: May 22, 2020).

17 Sensitivity and Monte Carlo Analyses

INTRODUCTION

The cost–benefit analysis presented in the last chapter was based on fixed estimates of costs and benefits. However, in the real world, the values of the estimates vary depending upon how other factors, such as assumptions and methods used in determining estimates, changes in economic and political conditions, government regulations and other uncontrollable variables (e.g., weather-related situations), affect the energy systems. Therefore, this chapter covers two important methods, namely sensitivity analysis and Monte Carlo simulation. These methods allow introduction of changes and variability in the values of input (or independent) variables included in the cost–benefit model, and the resultant variations in the outputs (e.g., benefit minus cost or benefit-to-cost ratio) can be used to get additional information for decision-making.

SENSITIVITY ANALYSIS

Sensitivity analysis involves exercising the evaluation model (decision model) in several iterations by changing values of input variables in each iteration to determine if the output values and final decision will change. In each new iteration of the model, values of one or more variables are typically changed (increased or decreased) by a small amount, e.g., 10%–20%. The sensitivity analysis thus provides an understanding the sensitivity of the output to small changes in one or more input variables. The small changes in input values are usually made to account for uncertainties or inaccuracies in estimating the selected values of input variables.

The sensitivity analysis is a useful and an excellent method to understand effects of different input variables on the output of the whole system. It helps in identifying the combinations of input variables and their values that can affect outputs of the whole system. The following example will illustrate the above points.

AN EXAMPLE OF PROFIT FROM SALE OF ELECTRICITY

Let us assume that a utility company sells its output, i.e., electricity (A kWh), and it charges E \$/kWh to its customers. The electricity generating plant operates at the capacity factor of C %. The utility's operating and maintenance costs are B \$/kWh and electricity distribution costs are D \$/kWh. The variables are described below.

A = Electricity (sold) demand (kWh)

B = Plant operating and maintenance cost (e.g., fuel cost) in \$/kWh

C = Plant efficiency (capacity factor) in percent

DOI: 10.1201/9781003107514-22

D = Electricity distribution costs in $/kWh

E = Electricity purchase price paid by the customer in $/kWh

The revenue generated, costs, and profits of the utility company can be computed as follows:

$$\text{Revenue} = \text{Amount received by selling demanded electricity} = A \times E$$

$$\text{Costs} = \text{Cost of producing and distributing the demanded electricity}$$
$$= A \times B / (C/100) + (A \times D)$$

$$\text{Profit} = \text{Revenue} - \text{Costs}$$

Tables 17.1 and 17.2 present values of input variables and outputs (profits) for different combinations of 10% changes in some of the five input variables (namely, A through E). Under nominal values of input variables (shown in the third column) of the tables show that the output (profit) value was $867.86. Table 17.1 shows that with 10% increase in value of variables A, B, and E, the profit changes to $954.64, $782.14, and $1,047.86, respectively. Table 17.2 shows that under the 10% changes in input variables, the worse and best revenues are $440.89 and $1,332.32, respectively. Thus, by conducting iterations of the model, a decision-maker can get a good understanding into the magnitude of changes in outputs possible with small changes in input variables (e.g., 10%).

AN ILLUSTRATIVE EXAMPLE OF SENSITIVITY ANALYSIS WITH FOUR ALTERNATIVES AND FOUR OUTCOMES

The following example illustrates how the sensitivity analysis was used in a problem that was formulated by using a decision matrix (see Chapter 6). The problem involved developing answers for five questions. The last question (i.e., Q5) involved use of the sensitivity analysis. In the sensitivity analysis, the values of three important input variables were varied by increasing and decreasing their values by 15%. And the resulting expected values of the decision variable (benefits minus costs) were computed for each alternative to determine the best alternative.

Problem

Assume that you have been hired as an energy systems engineering director to study possible development of an electric energy generating plant that can supply up to 350 MW power to a new community that will grow to about 250,000 homes at one of the following three locations: (a) on the west coast of lake Michigan, (b) in the state of Nevada, or (c) west coast of Florida about 200 miles north of Tampa.

To complete your assignment, prepare a report answering the following five questions:

Q1. Select an area for location of the proposed plant and describe (a) the types of power plants (or sources) that you will consider, (b) reasons for your selection of the power plant types in your selected area, and (c) any assumptions that you made in answering the question.

TABLE 17.1

Illustration of Sensitivity Analysis – Outputs with 10% changes in Demand, Plant Operating and Maintenance Costs, and Electricity Purchase Price

Variable	Description of Variable	Input Value	Illustration #1 Change (%)	Illustration #1 New Value	Illustration #2 Change (%)	Illustration #2 New Value	Illustration #3 Change (%)	Illustration #3 New Value
A	Electricity Demand in kWh	15,000	10	16,500	0	15,000	0	16,500
B	Plant operating and maintenance cost (e.g., fuel) $ per kWh	0.020	0	0.020	10	0.022	0	0.020
C	Plant efficiency (capacity factor)(%)	35.000	0	35.000	0	35.000	0	35.000
D	Distribution costs ($/kWh)	0.005	0	0.005	0	0.005	0	0.005
E	Electricity purchase price ($/kWh)	0.120	0	0.120	0	0.120	10	0.132
Output	Profit = Revenue − Costs	$867.86		$954.64		$782.14		$1,047.86

TABLE 17.2

Illustration of Sensitivity Analysis – Nominal, Worst, and Best Outputs with 10% Changes in Combinations of Input Variables

Variable	Description of Variable	Input Value	Illustration #1 Change (%)	Illustration #1 New Value	Illustration #2 Change (%)	Illustration #2 New Value	Illustration #3 Change (%)	Illustration #3 New Value
A	Electricity demand in kWh	15,000	0	15,000	−10	13,500	10	16,500
B	Plant operating and maintenance cost (e.g., fuel) $ per kWh	0.020	0	0.020	10	0.022	−10	0.018
C	Plant efficiency (capacity factor) (%)	35.000	0	35.000	−10	31.500	10	38.500
D	Distribution costs ($/kWh)	0.005	0	0.005	10	0.006	−10	0.0045
E	Electricity purchase price ($/kWh)	0.120	0	0.120	−10	0.108	10	0.132
Output	Profit = −Revenue−Costs	$867.86		$867.86		$440.89		$1,332.32

Q2. Briefly describe at least five important issues and at least five types of risks (each issue/risk in 2–3 sentences) that you will face in undertaking this assignment.

Q3. Describe at least three alternatives and at least four possible outcomes that you will seriously consider in your proposal to develop the 350 MW plant to provide power initially to the 250,000 homes in the community.

Q4. Conduct a cost–benefit analysis (to select one of the alternatives from the alternatives by considering the outcomes described in Q3 above) to recommend power plant type for the new community.

Your cost–benefit analysis should include estimates of benefits (e.g., revenue generated, economic benefits to the community in terms of new jobs generated) and estimates of costs (e.g., annual payments for repaying the capital costs of the plant at 3% interest rate over 30-year period with no down payment, plant operating and maintenance costs, power distribution costs, estimates of social cost of carbon at 3% discount rate to the cover costs due to the GHG emissions, and costs to cover risks associated with other health and safety-related causes).

Make sure that you describe how you obtained estimates of capacity factors, costs, benefits, probabilities of outcomes, and decision-making principles used in your analysis. Use data available from various data sources (e.g., EPA, NREL, EIA, and other references). Further, make the following assumptions: (a) Generated electricity can be sold at 12 cents/kWh to the homeowners in the community in the first year followed with 1% increase per year thereafter. (b) Excess generated energy can be sold to other utility companies at 10% below your selling price. (c) Any power shortages can be accommodated by purchasing additional power from other utility companies at 5% below your selling price. (d) The life of the electricity generating plant is 30 years.

Q5. Conduct at least three sensitivity analyses on your above formulated cost–benefit model by changing combinations costs and benefits by plus and minus 15% to evaluate robustness of your final recommendation (i.e., to determine if your recommendation on power plant selection will be affected by changes in your estimates of costs and benefits). Present results of the sensitivity analyses and discuss your findings.

Selection of Plant Location (Q1)

Specific Location Considerations for: West Coast of Lake Michigan

a. Michigan currently generates a larger percentage (about 49%) of electricity through coal than Florida (about 20% from coal) or Nevada (about 12% from coal) (EIA, 2020a).
b. EIA's U.S. map databases on locations of geothermal sources, annual average wind speed at 80 m, and sun illumination show that geothermal, land-based wind, and concentrating solar systems would not be favorable in Michigan (NREL, 2020a, b, c).
c. The offshore wind electricity generation system may be feasible on or near the western coast of Lake Michigan (NREL, 2020b).

Specific Location Considerations for: Nevada

a. The average direct sunlight data shows that the southern part of Nevada receives a large amount of direct sunlight (over 7.5 kWh/m²/day), which could potentially make solar electricity generation a feasible option (NREL, 2020c).
b. EIA's geothermal map database data also shows that there are favorable locations for electricity generating geothermal systems in Nevada (NREL, 2020a).
c. About 73% of electric energy generated currently in Nevada is from natural-gas-fired power plants (EIA, 2020b).

Specific Location Considerations for: West coast of Florida 200 miles north of Tampa

a. Using the geothermal resource map of the United States, the western coast of Florida is included in the areas of least favorable locations for a geothermal system (NREL, 2020a).
b. Using the annual average direct nominal solar resource data map, the western coast of Florida only receives 4.5–5 kWh/m²/day, which is less favorable than other locations in the United States (NREL, 2020b).

c. The wind resource map database show that land-based and offshore wind is not favorable relative to other locations around the United States (NREL 2020c).

d. About 68% of the electric power generated in Florida currently comes from natural-gas-fired power plants (EIA, 2020a).

Thus, the possible location selected for the analysis is the southern section of Nevada because most of the electricity generated in Nevada is produced by using natural gas. In particular, the objective was to see if this cost–benefit analysis shows that natural gas is truly the most desirable option in Nevada. However, both the solar electricity generation and geothermal electricity generation are also possible in southern Nevada. The analysis will include coal-fired, natural-gas-fired, geothermal, and solar photovoltaic electricity production plants as alternatives to provide 350 MW to 250,000 homes.

Other important considerations for plant locations are as follows:

a. *Fossil plant*: To reduce fuel transportation costs, the plant should be located close to the fuel source, i.e., near coal mines or natural gas pipelines.

b. *Hydro*: The site for the hydroelectric project should have availability of large quantity of water (reservoir/lake, dam) with a high enough difference in height of top water level in the reservoir and the height at the water turbines.

c. *Wind turbines*: The site should provide wind speeds over 20 mph throughout day and nighttime.

d. *Solar*: The plant should be located closer to equator to receive high levels (over 6,000 Wh/m^2/day) of sun illumination.

e. *Geothermal*: The site should provide large amount of hot (over 320° F) underground water sources.

Energy-Source-Related Issues
a. Availability of one or more choices in selecting energy sources
b. Technical considerations related to energy source choice, e.g., efficiency
c. Availability of inputs (e.g., fuel, water, electricity to start the turbine)
d. Capital costs (funding source, financing details)
e. Operation and maintenance costs (dependent on plant technology)
f. Environmental emissions generated by the plant
g. Capacity factor

Energy Demand issues
a. Variability in demand as a function of time
b. Day vs. night variations in demand
c. Seasonal variations in demand
d. Effects of energy needs of large industrial users

Issues and Risks (Q2)

Issues
1. *Air quality*: In 2004, the Las Vegas Valley (a section of southern Nevada) was identified by the EPA as a region that violated air quality standards.

Power plants, which produce less air pollutants, will play a role in complying with the increasingly stringent air quality standards.

2. *Water*: Power plants use large amounts of water for cleaning and cooling. Western states have been under extreme drought conditions. Water use must be considered when constructing a new power plant in Nevada.

3. *Carbon dioxide and greenhouse gas emissions*: Burning fossil fuels releases carbon dioxide into the atmosphere. This release increases the greenhouse effect of the atmosphere causing environmental changes.

4. *Fossil fuel availability*: Nevada imports much of its coal and natural gas from other states because it does not have coal mines or any dedicated natural gas wells. It only produces a small amount of petroleum. This makes it more difficult to obtain the necessary fuel to operate a fossil fuel burning power plant.

5. Nevada's energy portfolio requires 25% of the state's electricity to come from renewable sources by 2025. This requirement will force companies to construct renewable electricity generation plants even if it is not the most beneficial option.

Risks

1. *Economic risks*: Construction costs for building a new plant will increase every year, and the customers will be at risk for paying more for electricity. This will affect lower-income homeowners more because a larger portion of their income will be used to pay electric bills.

2. *Health risks*: Health risks are created by harmful emissions released by power plants. Coal and natural gas plants produce more harmful emissions than renewable energy sources.

3. *Energy import*: Nevada imports 90% of its energy from other states. If something were to happen to energy prices in other states, Nevada customers would be directly affected by the price change.

4. *Plant safety*: Each type of electricity generation has a different safety record (accident risk). These safety-related risks must be considered in selection of alternatives.

5. Drop in demand: Demand dropping is always a risk if many residents were to begin moving away from the region. This would increase financial risk associated with the new power plant.

Alternatives and Outcomes (Q3)

Alternatives

1. *A1*: To build a coal-fueled power plant to generate the 350 MW
2. *A2*: To build a natural-gas-fueled power plant to generate the 350 MW
3. *A3*: To build a geothermal plant to generate the 350 MW
4. *A4*: To build a solar photovoltaic plant to generate the 350 MW

The alternatives were selected based on the data from Q1. These are also the four most feasible electricity generation options for the southern section of Nevada.

Outcomes

1. *O1*: The economy will continue with the current 1% annual rate of increase in electricity demand. Occurrence probability $= 0.6$
2. *O2*: The economy will accelerate to 2% annual rate of increase in electricity demand. Occurrence probability $= 0.1$
3. *O3*: The economy will be stagnant with no increase in electricity demand. Occurrence probability $= 0.1$
4. *O4*: The economy will get worse and decelerate at 2% annual rate of electricity demand. Occurrence probability $= 0.2$

Cost–Benefit Analysis (Q4)

Assumptions

The electricity generating plant will produce up to 350 MW and will be financed using a 3% interest rate over a 30-year period and no down payment. Capacity factors used for each alternative are shown in Table 17.3.

These capacity factors were used to determine the nameplate size of each plant to produce 350 MW. It was assumed that the plant would always produce 350 MW and any excess or shortage of power would be sold or purchased, respectively. Since the power plant will always be producing 350 MW of electricity, the social cost of carbon will be determined for the 350 MW output. The $39/metric ton of CO_2 was used as the social cost of carbon (*SCC*) figure from the EPA website (EPA, 2017). JEDI models were used to determine annual operations and maintenance costs and economic benefits (NREL, 2020d). It was assumed that economic benefits were 5% of the economic output during construction and operation of the plant. The annual net profit for all four outcomes with each alternative was calculated using Equation 17.1. Net profits are equal to the sum of electricity revenue (*ER*) and economic development benefits (*EDB*) minus the operations and maintenance (*O & M*) and social cost of carbon (*SCC*). The profits or losses due to selling or purchasing electricity (*ESP*) to match demand were also accounted for.

$$\text{Net Profits} = \text{ER} + \text{EDB} - \text{O\&M} - \text{SCC} \pm \text{ESP} \qquad (17.1)$$

Electricity revenue is the revenue earned from selling electricity. The electricity purchased cost is the cost of electricity purchased from other utilities to make up for the demand greater than the generated electricity. The economic development benefit is

TABLE 17.3
Capacity Factors for the Four Alternatives

Alternative	Capacity Factor (%)
Coal-Fueled	85
Natural-Gas-Fueled	87
Geothermal	92
Solar Photovoltaic	25

the monetary benefit from the community for constructing and operating the plant. The operations and maintenance costs include a wide range of expenses including:

a. *Personnel expenses*: Salaries and benefits of field, administrative, and management personnel
b. *Materials and services*: Fuel and material needed to run the plant including vehicles, fees, permits, licenses, insurance, tools, and repair/maintenance costs (e.g., spare parts)
c. *Debt payment*: Interest payments on the financed amount to construct the plant
d. *Equity payment*: Repayment of principal for capital costs related to equipment purchase, construction, installation, and testing
e. Property Taxes

Spreadsheets

Four large spreadsheets were created, one for each alternative. Each spreadsheet had 30 rows for the 30 years (2 construction years plus 28 plant operating years). Each spreadsheet had the following columns:

1. Year (1–30)
2. Annual operations and maintenance cost
3. Amount of electricity sold in kWh in each of the four outcomes (includes annual increase or decrease depending upon the outcome)
4. Selling price of electricity ($/kWh)
5. Revenue generated from selling the electricity in each of the four outcomes
6. Electric energy purchased from other sources (when demand exceeded 350 MW/h) in each of the four outcomes
7. Cost of electricity purchased from other sources in each of the four outcomes ($/kWh)
8. Economic impact benefits from JEDI model for each outcome
9. Social cost of carbon for each outcome
10. Net annual profit for each of the four outcomes (annual revenue minus annual costs)
11. Present value of net annual profit for each of the four outcomes
12. Sums of the present values of net annual profits for each of the four outcomes over 30 years

Results

Using the assumptions above, the cost–benefit analysis yielded the results shown in Table 17.4. The net present value was computed by adding net present values (revenues minus costs) over the 30 years (2 years for construction and 28 years for operation) (item 12 above under Spreadsheets section).

From the data presented in above table, the coal-fueled and natural-gas-fueled power plants have almost identical expected net present values. The geothermal electricity production was not an economically feasible option. The solar photovoltaic plant was not as profitable as the coal or natural gas alternatives.

TABLE 17.4
Results of Initial Cost–Benefit Analysis

Alternatives	Expected Net Present Value (million $)
Coal-fueled plant	4,671.60
Natural-gas-fueled plant	4,575.27
Geothermal plant	−159.78
Solar photovoltaic plant	944.01

TABLE 17.5
Design of Experiment Table to Show the Factors That Were Changed and the Net Expected Present Values of Each Trial for the Sensitivity Analysis ($ values in billions)

Iteration	1	2	3	4	5	6	7	8
Price	−	+	−	+	−	+	−	−
Social cost of carbon (SCC)	−	−	+	+	−	−	+	+
O&M Costs	−	−	−	−	+	+	+	+
Coal-fueled plant	$3.60	$5.64	$3.60	$5.54	$2.80	$4.84	$2.80	$4.84
Natural-gas-fueled plant	$3.52	$5.56	$3.52	$5.55	$2.71	$4.75	$2.71	$4.75
Geothermal plant	−$0.09	$1.95	−$0.09	$1.95	−$2.21	−$0.16	−$2.21	−$0.16
Solar PV plant	$0.81	$2.86	$0.8:	$2.86	−S1.2	$0.92	−$1.12	$0.92

Sensitivity Analysis (Q5)

For the sensitivity analysis, values of three factors, namely price of electric energy sold, SCC, and annual operations and maintenance costs (O&M) were increased by 15% (+) or decreased by 15% (−). The experiment with eight combinations of the two values (+ and −) of the three factors was conducted by running the model in eight separate iterations. The output of each iteration was measured by computing the net present value over the 30 years for the four alternatives. The results are shown in Table 17.5.

The data in Table 17.5 show that when the three different factors were increased or decreased by 15%, the coal and natural gas plants were always the most profitable options. The geothermal option was always the least desirable. The largest driver of change in the net expected present value was the price of electricity, while O&M costs have less of an impact, and SCC has almost no impact at all.

MONTE CARLO SIMULATION TECHNIQUE

This method is used in three steps. In the first step, a computer model of a given decision-making situation (or a system) is created by modeling relationships between input (or independent) variables and the output (or response) variable of the system. In the second step, the model is exercised by conducting a series of iterations by using randomly selected values of input variables from predetermined distributions of the input variables. And in the third step, the output values obtained in each of the iterations are compared with a predetermined acceptance criterion value. The decision is generally made based on the percentage of iterations in which the output variable meets a predetermined acceptance criterion value.

AN EXAMPLE: PROFIT FROM SALE OF ELECTRICITY

Table 17.6 presents an example of a Monte Carlo simulation for a profit computation problem from sale of electricity shown earlier in Tables 17.1 and 17.2. Here the values of the input variables A through E are selected randomly from uniform distributions defined by the range of values provided in the third column of Table 17.6. Ten iterations were run and the randomly drawn values of the input variables in the ten iterations along with the profits generated by using the randomly generated input values are shown in the last ten columns of this table. The acceptance criterion value was selected as $700. The acceptance value was exceeded in five out of the 10 iterations. The Monte Carlo simulation thus allows a decision-maker to introduce variability in the selected values of the input variables and determine the chances (or probability) of meeting a given acceptance value to make the decision. The maximum and minimum values of profit obtained from the 10 iterations were $1,191.00 and $609.70, respectively. The variability in profit also helps the decision-maker understand that the profits from sale of electricity can also vary by factor of about 2 (i.e., 1,191.00/609.70 = 1.95).

Depending upon the assumption of distribution of values of input variables, different functions can be selected for generation of random values of input variables. For example, for random selection of values of input variables, the following functions in Excel can be used: RAND(), RANDBETWEEN(x1,x2), and NORM.INV(p, μ, σ).

CONCLUDING REMARKS

The sensitivity analysis and Mont Carlo analysis are useful techniques to understand the effects of changes in values of independent variables on the response variable values. Based on the results of these analyses, the decision-maker gets additional information on the magnitude of variability in the response variables as the values of independent variables are varied. If the results do not change over the range of changes in the independent variables, the decision-maker will be more confident that the results are stable, i.e., not very sensitive to the changes in the independent variables. The decision-maker needs to decide on the magnitude of changes in independent variables that should be used to generate different iterations. The magnitude of changes in independent variables is generally selected based on the decision-maker's experience in selecting acceptable level of accuracies (or possible errors) in determining estimates of values of the independent variables. Some decision-makers will create different combinations of input variables to define the best, nominal, or worse-case scenarios and run iterations to apply other decision-making principles

TABLE 17.6
Example of Monte Carlo Simulation with 10 Iterations

Variable	Description of Variable	Range of Possible Values	Iterations									
			1	2	3	4	5	6	7	8	9	10
A	Electricity demand in kWh	10,000–20,000 kWh	13,181	13,855	14,238	18,839	15,288	11,598	11,366	16,502	18,318	13,899
B	Plant operating and maintenance cost (e.g, fuel) $ per kWh	2–2.5 cents/kWh	2.30	2.2	2.4	2.4	2.2	2.5	2	2.2	2	2.4
C	Plant efficiency (capacity factor)(%)	35%–40%	40.00	38	39	38	35	35	40	35	40	36
D.	Distribution costs ($/kWh)	0.5–0.6 cents/kWh	0.50	0.5	0.6	0.5	0.6	0.6	0.6	0.5	0.5	0.5
E	Electricity purchase price ($/kWh)	11–13 cents/kWh	13.00	11	13	11	11	13	11	13	12	12
Output	Profit($) =Revenue −Costs		889.72	652.6	889.3	788.3	629	609.7	613.8	1025	1191	671.8
	Is Profit>$700? 1=yes. 2=No		1	0	1	1	0	0	0	1	1	0
	Percentage of iterations with Profit>$700		50									

such as maximin (most pessimistic), maxmax (most optimistic), Hurwicz, or other aspiration level principles covered in Chapter 6.

REFERENCES

EIA. 2020a. State Electricity Profiles. Website: http://www.eia.gov/electricity/state/ (Accessed: November 1, 2020).

EIA. 2020b. Nevada State Profiles and Energy Estimates. Website: http://www.eia.gov/state/analysis.cfm?sid=NV (Accessed: November 1, 2020).

Environmental Protection Agency. 2017. The Social Cost of Carbon. Retrieved from epa.gov: https://19january2017snapshot.epa.gov/climatechange/social-cost-carbon_.html.

NREL. 2020a. Geothermal Technologies. Website: http://www.nrel.gov/gis/data_geothermal.html (Accessed: November 1, 2020).

NREL. 2020b. Wind Resource Data, Tools and Maps. Website: http://www.nrel.gov/gis/wind.html (Accessed: November 1, 2020).

NREL. 2020c. Solar Resource Data, Tools and Maps. Website: http://www.nrel.gov/gis/solar.html (Accessed: November 1, 2020).

NREL. 2020d. JEDI: Jobs & Economic Development Impact Models. Website: https://www.nrel.gov/analysis/jedi/models.html (Accessed: May 22, 2020).

Appendix 1
Acronyms, Abbreviation, and Units

ACRONYMS

AHM	Analytical Hierarchical Method
BWR	Boiling water (nuclear) reactor
CAA	Clean Air Act
CAD	Computer Aided Design
CAFE	Corporate Average Fuel Economy
CBA	Cost–Benefit Analysis
CC	Combined Cycle System (uses two compressors – gas and steam for power generation)
CCS	Carbon Capture and Sequestration
CH₄	Methane gas
CIT	Critical Incident Technique
CNG	Compressed Natural Gas
CO₂	Caron dioxide
CT	Combustion Turbine
CWA	Clean Water Act
DAMES	Define, Analyze, Make search for possible solutions, Evaluate solutions, and Specify – Steps in scientific method as it is applied in engineering problem-solving solution
DCFC	Direct Current Fast Charging
DESIRE	Database of State Incentives for Renewables
DOE	Department of Energy
EPA	Environmental Protection Agency
EV	Electric Vehicle
EVSE	Electric Vehicle Supply Equipment
FMEA	Failure Modes and Effects Analysis
FMVSS	Federal Motor Vehicle Safety Standards
FTA	Fault Tree Analysis
GHG	Greenhouse Gases
HRSG	Heat Recovery Steam Generator
HVAC	Heating Ventilating and Air-conditioning System
ICE	Internal Combustion Engine
IEA	International Energy Agency
IEC	International Electrotechnical Commission
IGCC	Integrated Gasification Combined Cycle Power Plant
IR	Incident Rate (for measurement of injury and illness incident rate based on incidences per 200,000 hours of work)

JEDI	Jobs Economic Development Impact Models of NREL
LACE	Levelized Avoided Cost of Electricity
LED	Light-Emitting Diode
LNG	Liquified Natural Gas
LOCE	Levelized Cost of Energy
MATS	Mercury and Air Toxics standards
MSW	Municipal Solid Waste
NATA	National Air Toxics Assessment
NG	Natural Gas
NGCC	Natural Gas Combined Cycle Power Plant
NHTSA	National Highway Traffic Safety Administration
NO_x	Nitrous oxides
NREL	National Renewable Energy Laboratory
NSC	National Safety Council
O&M	Operating and maintenance
OSHA	Occupational Safety and Health Administration
PC	Pulverized Coal
PHS	Pumped Hydro Storage
PM	Particulate matter (made up of small airborne particles such as dust, soot, and drops of liquids)
PP	Project Plan
PPA	Power Purchase Agreement
PV	Photovoltaic
PWR	Pressurized water (nuclear) reactor
RPN	Risk Priority Number
SAE	Society of Automotive Engineers, Inc. (SAE International)
$SC\text{-}CH_4$	Social cost of Methane
$SC\text{-}CO_2$	Social Cost of Carbon
$SC\text{-}N_2O$	Social Cost of Nitrous Oxide
SE	Systems Engineering
SEMP	Systems Engineering Management Plan
SO_x	Sulfur oxides
SREC	Solar Renewable Energy Credits
USDOT	U.S. Department of Transportation
WBS	Work Breakdown Structure
WPT	Wireless Power Transfer

UNITS

Barrel	42 gallons
Billion	10^9
Btu	A "British thermal unit" (Btu) is a measure of the heat content of fuels. It is the quantity of heat required to raise the temperature of 1 pound of liquid water by 1°F at the temperature that water has its greatest density (approximately 39°F).
C	100 or centum (natural gas measurement unit)

Ccf	100 cubic feet (cf) of natural gas
GW	Giga Watts (billion watts)
kW	Kilo Watt (10^3 W)
kWh	Kilo Watt hour (1,000 Wh)
M	1,000 (natural gas measurement unit)
Mcf	1,000 cubic feet (cf) of natural gas
Million	10^6
MM	1 million (natural gas measurement unit)
MW	Mega Watts (10^6 watts)
ppb	Parts per billion
Quadrillion	10^{15}
Short Ton	1 short ton = 2,000 lbs
Ton	In the United States, 1 ton = 2,000 lbs
toe	One tonne of oil equivalent (toe) is a unit of energy defined as the amount of energy released by burning 1 tonne of crude oil. It is approximately 42 gigajoules or 11,630 kWh, although as different crude oils have different calorific values, the exact value is defined by convention; several slightly different definitions exist. The *toe* is sometimes used for large amounts of energy.
Tonne	Metric ton = 1,000 kg = 2204 lbs
Trillion	10^{12}
TW	Terra watts(10^{12} W)
TWh	Terra watt hours (10^{12} Wh)

Appendix 2
Cost–Benefit Analysis of PV Solar Electric Energy Systems for Residential Use

Conduct cost–benefit analyses for installing 10 kW residential PV solar energy systems to generate electric energy in two average size homes, one in Detroit, MI, and other in Phoenix, AZ. Use the analysis procedure similar to one presented in the class and described in file: "TSF_SolarAcct_Final" (see https://www.thesolarfoundation.org/solar-accounting-measuring-the-costs-benefits-of-going-solar/).

Prepare two spread sheets (one for each city) by using assumptions and data from the websites given below and compute benefit-to-cost ratio for homes in both cities and discuss your findings.

Assumptions
1. Life of Equipment: 25 years
2. Financing Period: 25 years
3. Loan Interest (or discount rate): 5% per year

Websites
1. Solar Photovoltaic System Costs:
   See file called, "Solar Photovoltaic System Costs" (Ref: Fu, R., et al. "U.S. Solar Photovoltaic System Cost Benchmark: Q1:2017". NREL Report no. TP/-6A20-68925. August 2017.)
   2. Energy Consumption:
   http://www.eia.gov/consumption/residential/reports/2009/state_briefs/
   http://www.eia.gov/consumption/residential/reports/2009/state_briefs/pdf/mi.pdf
   http://www.eia.gov/consumption/residential/reports/2009/state_briefs/pdf/az.pdf
2. PV Watts
   http://rredc.nrel.gov/solar/calculators/PVWATTS/version1/
3. Financial Incentives for Solar PV
   http://programs.dsireusa.org/system/program?state=MI
   http://programs.dsireusa.org/system/program?state=AZ

Appendix 3
Fault Tree and Reliability Analysis

Q1. Develop a fault tree to study how a home loses its electric power. Your fault tree considerations should include events such as overload in a home circuit, short circuit, the utility company fails to provide power during severe weather conditions (such as high temperatures causing excessive power draw to run air-conditioning units and very cold temperatures requiring excessive power to run heating furnaces, wind storm, or ice storm breaking power lines), transformer failures, power lines breakage due to car accidents, falling trees, and so forth. You fault tree should include at least 15 failures causes.

Q2. Determine reliability of the power grid network presented in Figure A.1. Your reliability analysis should include all possible combinations of power distribution to maintain full power to loads L1, L2, and L4. Assume that the generators G1, G2, G31, and G32 can run up to their full rated capacity provided in the figure (e.g., G1 can produce up to 10 KW). Further, assume that each generator has only two states (i.e., operates normally [provides up to its full rated capacity] or fails to generate any output). B1, B2, B3, and B4 are buses to distribute power to the loads in the four grids. The

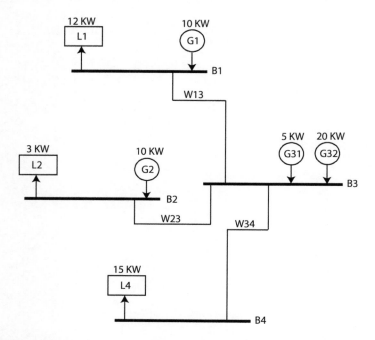

FIGURE A.1 Power distribution system with four buses, four generators, and three loads.

grids can also supply power to the buses through transmission lines (wires) shown in the figure as W12, W13, W23, W24, and W34. Assume that reliabilities of the generators, buses, and wires are 0.999, 0.9999, and 0.97, respectively. (Hint: First, determine possible combinations of working generators needed to meet total power demand from all the loads. Then, make sure that buses and wires needed for the combinations of generators are operational.)

Q3. Develop a reliability block diagram for the above problem (Q2).

Q4. Develop an event tree to illustrate all combinations of states of the four generators in Figure A.1 above. Assume that each generator has only two states (i.e., operates normally [provides up to its full rated capacity] or fails to generate any output. Identify the paths that can maintain the full power load. (Assume that buses, wires, and loads do not fail.)

Appendix 4
Decision-Making in Energy Systems

Q1. Conduct a Pugh Analysis for design of an electricity generation plant by considering at least four alternate energy sources (except a coal-fired source) for the operation of the plant. In your analysis, use a coal-fired electric energy generation plant as the datum. Your Pugh analysis should be based on consideration at least eight key attributes of electric energy power plants. Include brief descriptions of the attributes, assumptions made in conducting the analysis, and discuss results of the analysis.

Q2. Determine relative preferences (or importance) of the attributes used in Q1 above by applying Thurstone's method of paired comparisons. Use at least six subjects to collect data needed for your analysis.

Q3. Using five of the most important attributes obtained from the results of Q2 above, apply Analytical Hierarchical Method to determine the best alternative among the alternate energy sources considered in Q1 above. Use at least three subjects to obtain data needed for your analysis. Compare the results of the AHP with the results of the Pugh analysis (from Q1 above) and discuss your observations.

Appendix 5
Cost–Benefit Analysis in Energy Systems

Conduct a cost–benefit analysis to determine the best alternative in solving the problem of increasing the electric energy supplying capacity of a utility company supplying electricity in southern and southwest states of the United States by 600 MW. The utility company is considering building the new plant in Texas with the following five alternatives and four outcomes:

A1: Build new land-based wind turbines to generate the additional capacity
A2: Build a natural-gas-fueled power plant to generate the additional capacity
A3: Build a geothermal plant to generate the additional capacity
A4: Build a concentrating solar plant to generate the additional capacity
A5: Do not build a new plant and purchase the needed energy from other utility companies
O1: The economy will grow at 2% annual rate of increase in electricity demand.
O2: The economy will accelerate to 4% annual rate of increase in electricity demand.
O3: The economy will be stagnant with no increase in electricity demand.
O4: The economy will get worse and decelerate at 2% annual rate of electricity demand.

Assume that probabilities of the above O1–O4 outcomes are 0.30, 0.45, 0.20, and 0.05, respectively. Use plant capacity factors of 35% for wind turbine plant, 87% for natural-gas-fueled plant, 92% for geothermal plant, and 25% for the concentrating solar plant.

Apply the JEDI models to predict project costs, operating and maintenance costs (including finance costs), and associated impact of economic development on the local community. Assume that the utility company can finance 100% of the project costs at fixed 3.0% interest rate from the federal and state governments over the next 34 years (4 years for construction and 30 years of plant operation).

Assume that in the first year following the plant completion, the company will begin selling 600 MW of electricity at $0.13/kWh and the rate will increase 1% per year in the subsequence years. Assume that the output of each plant will degrade 1% per year. Additional energy can be purchased from other utility companies at a pre-negotiated fixed rate of $0.12/kWh over the next 34 years. Further assume that the utility company can gain benefits (incentives) equivalent to 5%

of the economic impact (earnings of the added heads) generated during the construction (over the first 4 years) and operation (from the fifth year up to 34 years of plant operation) phases of the new plant. Include any additional costs (e.g., social cost of carbon, accident costs) and benefits (e.g., incentives and/or rebates offered by state, local, and federal governments) that can be apply to selected power plant technologies.

Index

For Product Safety Concerns and Information please contact our EU
representative GPSR@taylorandfrancis.com
Taylor & Francis Verlag GmbH, Kaufingerstraße 24, 80331 München, Germany